T0309294

Decomposition Analysis Method in Linear and Nonlinear Differential Equations

Decomposition Analysis Method in Linear and Nonlinear Differential Equations

Kansari Haldar
Retired Professor
Indian Statistical Institute
Kolkata, India

CRC Press is an imprint of the
Taylor & Francis Group, an **informa** business
A CHAPMAN & HALL BOOK

CRC Press
Taylor & Francis Group
6000 Broken Sound Parkway NW, Suite 300
Boca Raton, FL 33487-2742

Printed on acid-free paper
Version Date: 20150831

International Standard Book Number-13: 978-1-4987-1633-8 (Hardback)

Visit the Taylor & Francis Web site at
http://www.taylorandfrancis.com

and the CRC Press Web site at
http://www.crcpress.com

To Santwana, Swati, and Jerry for their unwavering support

Contents

Preface

Decomposition Analysis Method in Linear and Nonlinear Differential Equations is written mainly for use by research students and scientists in all branches of science and technology. It deals with the powerful *decomposition methodology of Adomian*, which is capable of solving all types of differential equations for the series solutions of fundamental problems of physics, astrophysics, chemistry, biology, medicine, and other disciplines of science. It has been discovered that a few terms of the series solution are sufficient for an exact solution to the real problem. The method discussed in this book represents a remarkable achievement in the modern research field and deserves to be compared to any contemporary superfast computer in terms of solving differential equations.

This book serves as a reference that has been written in such a way that each topic of this book is depicted elaborately and elegantly in very lucid language.

The decomposition method is divided into four classes, namely, regular or ordinary decomposition, double decomposition, modified decomposition, and asymptotic decomposition. These classes have been applied to Laplace and Navier–Stokes equations in Cartesian and polar coordinates for obtaining partial solutions of the equations. The applicability of these classes in solving several physical problems has been demonstrated.

The advantage of Adomian's decomposition method is to avoid restrictions and assumptions, used while simplifying a real problem to reduce it to a mathematically tractable form. Solutions obtained from solving the simplified form are not consistent with the numerical solution of the *mother problem*.

In Chapter 1, a brief description of the decomposition method is provided, where the partial solutions of a differential equation are depicted clearly. In Chapter 2, the asymptotic decomposition method is discussed considering a linear differential equation. Here, the Ramanujan's integral formula, gamma integral, and Laplace transform are derived with the help of the regular decomposition method. Using the decomposition method, the Bessel's functions are solved in Chapter 3. Moreover, some physical problems, such as pulsatile motion in a rigid tube, periodic motion of a visco-elastic fluid in a rigid tube, tidal waves in a channel, and temperature distribution in an infinitely long circular cylinder, are considered for clear illustration of the method. Chapters 4 and 5 discuss the applications of decomposition methods to the Navier–Stokes equations in Cartesian and cylindrical polar coordinates, respectively. In these chapters, several physical problems are solved using the decomposition method. Many physiological problems such as steady flow of blood in a constricted artery, flow of blood in a constricted artery in the presence of a magnetic field, etc., are considered in Chapter 6. These problems are addressed by the decomposition method. In Chapter 7, the problems of subsonic flow past a wavy wall and axisymmetric subsonic flow past a corrugated circular cylinder are discussed in depth, applying the decomposition method to the Navier–Stokes equation. In Chapter 8, the decomposition method is applied to a linearized transonic equation for the flow past a wave-shaped wall. Chapter 9 documents the use of the decomposition method in solving Laplace's equation in polar coordinates and problems on a circular disc and circular annulus. Chapter 10 discusses flow near a rotating disc in a fluid at rest, while the appendix provides equations of motion in the different coordinate systems.

Acknowledgments

I am thankful to all those who helped me through this undertaking.

I express my deep gratitude to my mentor, Professor Pradip Niyogi. In a true sense, Professor Niyogi is the founder of my career and has been instrumental in shaping my entire career life.

I thank Santwana (my wife) and Swati (my daughter) for their constant support and encouragement. I do not know whether to thank Shonuma! She is my pet dachshund. I surely remember her quietly accompanying me during my several all-nighters.

I owe my gratitude to Subir Mukherjee and Aparesh Chatterjee for technical help with the manuscript preparation.

I thank Taylor & Francis/CRC Press for publishing this work. I earnestly appreciate the help extended by its team members.

About the Author

The start of my career was with fluid dynamics and gas dynamics. My thesis work provided expertise in high-speed gas dynamics. Later on there was a shift in my area of interest, and I started working on problems related to fluid dynamics, biofluid dynamics, biomagneto fluid dynamics, and Adomian's decomposition methodology. This book is a consolidation of all the problems I have worked on throughout my research career. The idea is to publish the work as a collection for students, scholars, and researchers in this field.

My 35-year research career has been quite a substantial length of time for me to gain a deep and thorough insight into the fundamental problems of hydrodynamics. My independent research career began with my appointment as a lecturer at the Indian Statistical Institute, Kolkata. Since then, I have worked on several fundamental aspects of hydrodynamics, which were published in reputed international journals. Toward the last phase of my service as a professor, I focused on the decomposition method in fluid and biofluid dynamics.

The decomposition method has been used regularly in solving problems in physics. However, reports on solving problems of fluid dynamics using the decomposition method are very rare. The efficacy and rigorousness of the decomposition method are validated on several problems of fluid dynamics, borrowed from the book *Boundary Layer Theory* by H. Schlichting, where they have been addressed by different methods.

1

Decomposition Method

1.1 Introduction

In nature, all the fundamental problems are nonlinear in character. These problems exist in physics, astrophysics, engineering, biology, medicine, and other disciplines of sciences. The mathematical models of these nonlinear problems are made by the nonlinear ordinary differential equations, partial differential equations, or systems of them subject to certain initial and boundary conditions.

Because of the nonlinear character of the equations involved in these problems, their exact solutions form are not always possible. To solve these problems, we take the help of some simplifications, such as linearization, perturbation, and other restrictions, so that the original problem is reduced to a mathematically tractable form. The closed-form solution of the reduced problem can be easily obtained, but it is not consistent with and deviates much from the solution of the real problem. As a result we use traditional numerical technique, which gives an exact solution of the real problem, although lengthy computation is a major shortcoming here.

Here, we discuss a powerful and effective method known as the *Adomian decomposition method*, which was developed by George Adomian [1,3,5–8]. This method has been drawing attention of the researchers in the fields of applied mathematics, particularly in the areas of infinite series solutions because of its speed, ease and elegance.

The decomposition method can provide approximate solutions to a wide class of linear and nonlinear ordinary differential equations, partial differential equations, and integral equations as well. The solutions obtained by the Adomian decomposition method converge rapidly [6,8]. The method will be successfully used for solving the partial differential equations that appear in the physical models. This method addresses several physical models without using prior simplification linearization or perturbation and permits only restrictive assumptions which otherwise would change the physical behavior of the models.

The advantage of the Adomian decomposition method is it avoids the restrictions and assumptions that are used in the simplifications of the mother problem. The method is divided into four classes, namely, (1) regular (ordinary or single) decomposition, (2) double decomposition, (3) modified decomposition, and (4) asymptotic decomposition. The details of this method are given in the references [3,6,8], and the readers are advised to go through these references and the references therein. In this chapter, we shall discuss the different classes of the decomposition method, except asymptotic decomposition, by considering a linear partial differential equation with constant coefficients. The asymptotic decomposition is discussed in chapter 2 with the applications of engineering problems.

1.2 Partial Solutions of a Partial Differential Equation

Consider the two-dimensional linear nonhomogenous partial differential equation:

$$\frac{\partial^2 u}{\partial t^2} + \frac{\partial^2 u}{\partial x^2} + su = g \tag{1.1}$$

Let L_t and L_x be two linear partial differential operators defined by $L_t = \frac{\partial^2}{\partial t^2}$ and $L_x = \frac{\partial^2}{\partial x^2}$ Then equation (1.1) takes the operator form as

$$L_t u + L_x u + su = g \tag{1.2}$$

The dependent variable u is a function of x and t. The partial differential operators are invertible. The inverse of L_t is defined as a twofold definite integral denoted by L_t^{-1}, whereas that of L_x is defined as a twofold indefinite integral denoted by L_x^{-1}. These integrals are given by

$$L_t^{-1} = \int_0^t \int_0^t (\cdot)dtdt \tag{1.3}$$

and

$$L_x^{-1} = \iint (\cdot)dxdx \tag{1.4}$$

Solving for $L_t u$ and $L_x u$ we have from (1.2)

$$L_t u = g - (L_x u + su) \tag{1.5}$$

and

$$L_x u = g - (L_t u + su) \tag{1.6}$$

The solution of equation (1.5) obtained by operating L_t^{-1} on both sides of it is known as a t-partial solution whereas that of equation (1.6) obtained by using L_x^{-1} is called an x-partial solution. The number of partial solutions of an equation, ordinary or partial, depends on the number of independent variables, and each partial solution is expressed by "X-partial" where X is only an independent variable.

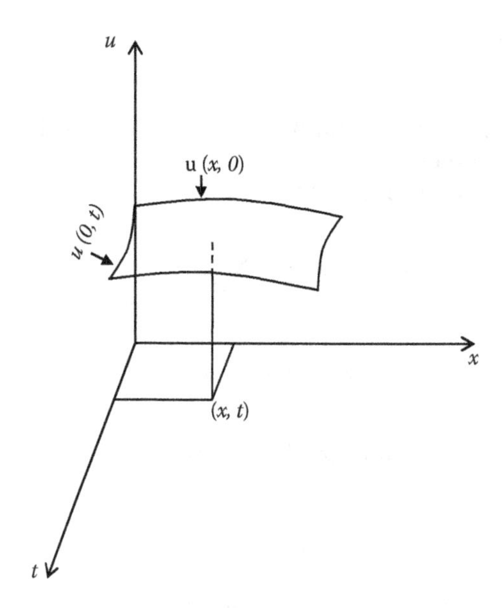

FIGURE 1.1
Equal partial solutions.

Here, it is very important to note that the partial solutions of the partial differential equation are identically equal. To explain it, we return to the operator equation (1.2), and we assume that the initial and boundary conditions are $u(x, 0) = f(x)$ and $u(0, t) = F(t)$, respectively. The physical meaning of the initial condition is the intersection of ux-plane with the surface to be generated by $u(x, t)$ (see Figure 1.1). As t increases from this value, the surface is generated, and that is possible only because of the x-partial solution of equation (1.6). Similarly, the boundary condition $u(0, t) = F(t)$ is the intersection of the ut-plane with the surface to be generated by $u(x, t)$. As x increases from this value, the surface $u(x, t)$, is generated, and that is possible only because of the t-partial solution of equation (1.5). The details of these solutions are given in reference [6].

1.2.1 Solution by Regular Decomposition

The t-Partial Solution

For the t-partial solution, we return to equation (1.5) and then operating on both sides of it with the inverse operator L_t^{-1}, we write

$$u(x, t) = u_0(x, t) + L_t^{-1}[g - L_x u - su] \tag{1.7}$$

where

$$u_0(x, t) = \xi_0(x) + \xi_1(x)t \tag{1.8}$$

is the solution of

$$L_t u = 0 \tag{1.9}$$

$\xi_0(x)$ and $\xi_1(x)$ being the integration constants are to be determined from the condition of the problem.

We now decompose u and write

$$u(x, t) = \sum_{m=0}^{\infty} \lambda^m u_m$$

$$g(x, t) = \sum_{m=0}^{\infty} \lambda_m^{mg} \tag{1.10}$$

where λ is a counting parameter. Then we write equation (1.7) in the parameterized form as

$$u(x, t) = u_0(x, t) + \lambda L^{-1}[g - L_x u - su] \tag{1.11}$$

Using (1.10) in equation (1.11) and comparing the terms for different powers of λ, we get

$$u_1(x, t) = L_t^{-1}[g_0 - L_x u_0 - su_0]$$

$$u_2(x, t) = L_t^{-1}[g_1 - L_x u_1 - su_1]$$

$$u_3(x, t) = L_t^{-1}[g_2 - L_x u_2 - su_2]$$

$$\cdots\cdots\cdots\cdots\cdots\cdots$$

$$\cdots\cdots\cdots\cdots\cdots\cdots$$

$$u_{m+1}(x, t) = L_t^{-1}[g_m - L_x u_m - su_m] \tag{1.12}$$

$$\cdots\cdots\cdots\cdots\cdots\cdots$$

Since $u_0(x, t)$ is known, therefore, all the components of $u(x, t)$ are computable and the final solution is $u(x, t) = \sum_{m=0}^{\infty} u_m$, remembering that $\lambda = 1$.

The x-Partial Solution

Consider equation (1.6) for the x-partial solution. The use of the operator L_x^{-1} on both sides of this equation gives

$$u(x, t) = \varphi_0(x, t) + L_x^{-1}[g - L_t u - su] \tag{1.13}$$

where $\varphi_0(x, t)$, solution of the equation $L_x \varphi = 0$, is given by

$$\varphi_0(x, t) = \eta_0(t) + \eta_1(t)x \tag{1.14}$$

The integration constants $\eta_0(t)$ and $\eta_1(t)$ are to be evaluated by the conditions of the problem. Then we consider decomposition forms of $u(x, t)$ and g given in (1.10) and write the parameterized form of (1.13) as

$$u(x, t) = \varphi_0(x, t) + \lambda L_x^{-1}[g - L_t u - su] \tag{1.15}$$

where λ has its usual meaning. Using (1.10) in (1.15) and comparing like-power terms of λ we get

$$\begin{aligned}
u_0(x, t) &= \varphi_0(x, t) = \eta_0(t) + \eta_1(t)x \\
u_1(x, t) &= L_x^{-1}[g_0 - L_t u_0 - s u_0] \\
u_2(x, t) &= L_x^{-1}[g_1 - L_t u_1 - s u_1] \\
&\cdots\cdots\cdots\cdots\cdots\cdots \\
&\cdots\cdots\cdots\cdots\cdots\cdots \\
u_{m+1}(x, t) &= L_x^{-1}[g_m - L_t u_m - s u_m] \\
&\cdots\cdots\cdots\cdots\cdots\cdots
\end{aligned} \tag{1.16}$$

We see that $u_1 \equiv u_1(u_0),\ u_2 \equiv u_2(u_0, u_1),\ u_3 \equiv u_3(u_0, u_1, u_2)$, etc. Therefore, all the components of $u(x, t)$ are known as u_0. Thus, the final x-partial solution is $u(x, t) = \sum_{m=0}^{\infty} u_m$.

1.2.2 Solution by Double Decomposition

The t-Partial Solution

For the t-partial solution, we consider the operator equation (1.5) and apply the double decomposition technique. Following the regular decomposition technique for the t-partial solution of equation (1.5), we recall (1.11), that is,

$$u(x, t) = u_0(x, t) + \lambda L_t^{-1}[g - L_x u - su] \tag{1.17}$$

where

$$u_0(x, t) = \xi_0(x) + \xi_1(x)t \tag{1.18}$$

For the double decomposition technique we have to decompose $u_0(x, t)$, and its decomposed form is

$$u_0(x, t) = \sum_{m=0}^{\infty} \lambda^m u_{0,m} \tag{1.19}$$

We now use (1.10) and (1.19) in (1.17) and get

$$\sum_{m=0}^{\infty} \lambda^m u_m = \sum_{m=0}^{\infty} \lambda^m u_{0,m} + \lambda L_t^{-1}\left[\sum_{m=0}^{\infty} \lambda^m g_m - L_x \sum_{m=0}^{\infty} \lambda^m u_m - s \sum_{m=0}^{\infty} \lambda^m u_m\right]$$

Then comparing the terms, we obtain

$$\begin{aligned}
u_0(x,t) &= u_{0,0}(x,t) \\
u_1(x,t) &= u_{0,1}(x,t) + L_t^{-1}[g_0 - L_x u_0 - s u_0] \\
u_2(x,t) &= u_{0,2}(x,t) + L_t^{-1}[g_1 - L_x u_1 - s u_1] \\
&\cdots\cdots\cdots \\
&\cdots\cdots\cdots \\
u_{m+1}(x,t) &= u_{0,m+1}(x,t) + L_t^{-1}[g_m - L_x u_m - s u_m] \\
&\cdots\cdots\cdots
\end{aligned} \tag{1.20}$$

Our next task is to find out $u_{0,0}(x,t),\ u_{0,1}(x,t),\ u_{0,2}(x,t),$ etc., and for that we have to equate (1.18) and (1.19). Before equating these relations, we write the parameterized decomposition of the integration constants $\xi_0(x)$ and $\xi_1(x)$ involved in the expression (1.18) for $u_0(x,t)$ as

$$\left.\begin{aligned}
\xi_0(x) &= \sum_{m=0}^{\infty} \lambda^m \xi_{0,m} \\
\xi_1(x) &= \sum_{m=0}^{\infty} \lambda^m \xi_{1,m}
\end{aligned}\right\} \tag{1.21}$$

Now we equate (1.18) and (1.19) after using (1.21) in (1.18) and then compare the terms on both sides of resulting equation to get

$$\begin{aligned}
u_{0,0}(x,t) &= \xi_{0,0}(x) + t\xi_{1,0}(x) \\
u_{0,1}(x,t) &= \xi_{0,1}(x) + t\xi_{1,1}(x) \\
u_{0,2}(x,t) &= \xi_{0,2}(x) + t\xi_{1,2}(x) \\
&\cdots\cdots\cdots \\
&\cdots\cdots\cdots \\
u_{0,m}(x,t) &= \xi_{0,m}(x) + t\xi_{1,m}(x) \\
&\cdots\cdots\cdots
\end{aligned} \tag{1.22}$$

Substituting (1.22) in (1.20) we can have all the components of $u(x,y)$, and each component involves integration constants which are to be determined from the respective boundary conditions. Thus, the components of $u(x,t)$ are computable and hence the final t-partial solution is $u(x,t) = \sum_{m=0}^{\infty} u_m$.

The x-Partial Solution

For the x-partial solution, we return to equation (1.6), and using the operator L_x^{-1}, we consider the solution (1.15); that is, we rewrite it as

$$u(x,t) = \varphi_0(x,t) + \lambda L_x^{-1}[g - L_t u - s u] \tag{1.23}$$

where

$$\varphi_0(x,t) = \eta_0(t) + x\eta_1(t) \tag{1.24}$$

For double decomposition, we decompose $\varphi_0(x, t)$ into

$$\varphi_0(x, t) = \sum_{m=0}^{\infty} \lambda^m \varphi_{0,m} \tag{1.25}$$

Substituting (1.10) and (1.25) in (1.23) and then comparing the terms, we have

$$\begin{aligned}
u_0(x, t) &= \varphi_{0,0}(x, t)\\
u_1(x, t) &= \varphi_{0,1}(x, t) + L_x^{-1}[g_0 - L_t u_0 - s u_0]\\
u_2(x, t) &= \varphi_{0,2}(x, t) + L_x^{-1}[g_1 - L_t u_1 - s u_1]\\
&\cdots\cdots\cdots\cdots\cdots\cdots\cdots\cdots\\
&\cdots\cdots\cdots\cdots\cdots\cdots\cdots\cdots\\
u_{m+1}(x, t) &= \varphi_{0,m+1}(x, t) + L_x^{-1}[g_m - L_t u_m - s u_m]\\
&\cdots\cdots\cdots\cdots\cdots\cdots\cdots\cdots
\end{aligned} \tag{1.26}$$

Again, we decompose the integration constants $\eta_0(t)$ and $\eta_1(t)$ into the following forms:

$$\left.\begin{aligned}
\eta_0(t) &= \sum_{m=0}^{\infty} \lambda^m \eta_{0,m}\\
\eta_1(t) &= \sum_{m=0}^{\infty} \lambda^m \eta_{1,m}
\end{aligned}\right\} \tag{1.27}$$

Using (1.25) and (1.27) in (1.24) and then comparing we obtain

$$\begin{aligned}
\varphi_{0,0}(x, t) &= \eta_{0,0}(t) + x\eta_{1,0}(t)\\
\varphi_{0,1}(x, t) &= \eta_{0,1}(t) + x\eta_{1,1}(t)\\
\varphi_{0,2}(x, t) &= \eta_{0,2}(t) + x\eta_{1,2}(t)\\
&\cdots\cdots\cdots\cdots\cdots\cdots\\
&\cdots\cdots\cdots\cdots\cdots\cdots\\
\varphi_{0,m}(x, t) &= \eta_{0,m}(t) + x\eta_{1,m}(t)
\end{aligned} \tag{1.28}$$

Putting (1.28) in (1.26), we get all the components of $u(x, t)$ that are computable, and hence the final x-partial solution is $u(x, t) = \sum_{m=0}^{\infty} u_m$.

1.2.3 Solution by Modified Decomposition

The t-Partial Solution

For the t-partial solution, we consider equation (1.5) and operate on both sides of it with the inverse operator L_t^{-1} in order to get

$$u(x, t) = \zeta_0(x) + t\zeta_1(x) + L_t^{-1}[g - L_x u - s u] \tag{1.29}$$

where the integration constants ζ_0 and ζ_1 are to be determined from the conditions of the problem

If we assume $u(x, t)$ in the form

$$u(x,t) = \sum_{m=0}^{\infty}\sum_{n=0}^{\infty} a_{m,n} t^m x^n \tag{1.30}$$

then we get the terms

$$a_{0,0},\ a_{0,1}x,\ a_{0,2}x^2, \cdots\cdots\cdots$$

$$a_{1,0}t,\ a_{1,1}tx,\ a_{1,2}tx^2, \cdots\cdots\cdots$$

$$a_{2,0}t^2,\ a_{2,1}t^2x,\ a_{2,2}t^2x^2, \cdots\cdots\cdots$$

The sum of the terms in the first group can be written as

$$a_{0,0} + a_{0,1}x + a_{0,2}x^2 + \cdots\cdots\cdots = t^0 \sum_{n=0}^{\infty} a_{0,n}x^n$$

Similarly, for the second group, third group, etc., we can write

$$a_{1,0}t + a_{1,1}tx + a_{1,2}tx^2 + \cdots\cdots\cdots = t \sum_{n=0}^{\infty} a_{1,n}x^n$$

$$a_{2,0}t^2 + a_{2,1}t^2x + a_{2,2}t^2x^2 + \cdots\cdots\cdots = t^2 \sum_{n=0}^{\infty} a_{2,n}x^n$$

$$\cdots\cdots\cdots\cdots\cdots\cdots\cdots\cdots\cdots\cdots\cdots\cdots$$

Therefore, we write the solution (1.30) in the form

$$u(x,t) = \sum_{m=0}^{\infty}\left[\sum_{n=0}^{\infty} a_{m,n}x^n\right] t^m = \sum_{m=0}^{\infty} a_m(x)t^m \tag{1.31}$$

where

$$a_m(x) = \sum_{n=0}^{\infty} a_{m,n}x^n \tag{1.32}$$

We also write

$$g(x,t) = \sum_{m=0}^{\infty}\sum_{k=0}^{\infty} g_{m,k}t^m x^k = \sum_{m=0}^{\infty}\left[\sum_{k=0}^{\infty} g_{m,k}x^k\right] t^m = \sum_{m=0}^{\infty} g_m(x)t^m \tag{1.33}$$

where

$$g_m(x) = \sum_{k=0}^{\infty} g_{m,k}x^k \tag{1.34}$$

Substituting (1.31) and (1.33) in the t-partial solution (1.29), we have

$$\sum_{m=0}^{\infty} a_m(x)t^m = \zeta_0(x) + t\zeta_1(x) + L_t^{-1}\sum_{m=0}^{\infty} g_m(x)t^m - L_t^{-1}\frac{\partial^2}{\partial x^2}\left[\sum_{m=0}^{\infty} a_m(x)t^m\right] - sL_t^{-1}\sum_{m=0}^{\infty} a_m(x)t^m$$

Then performing integrations on the right-hand side of the previous relation we get

$$\sum_{m=0}^{\infty} a_m(x)t^m = \zeta_0(x) + t\zeta_1(x) + \sum_{m=0}^{\infty} g_m(x)\frac{t^{m+2}}{(m+1)(m+2)} - \sum_{m=0}^{\infty} \frac{\partial^2}{\partial x^2}a_m(x)\frac{t^{m+2}}{(m+1)(m+2)} - s\sum_{m=0}^{\infty} a_m(x)\frac{t^{m+2}}{(m+1)(m+2)} \tag{1.35}$$

Replacing m by $(m' - 2)$ on the right side of (1.35) and then dropping the dash, we write (1.35) as

$$\sum_{m=0}^{\infty} a_m(x)t^m = \zeta_0(x) + t\zeta_1(x) + \sum_{m=2}^{\infty} g_{m-2}(x)\frac{t^m}{m(m-1)} - \sum_{m=2}^{\infty} \frac{\partial^2}{\partial x^2}a_{m-2}(x)\frac{t^m}{m(m-1)} - s\sum_{m=2}^{\infty} a_m(x)\frac{t^m}{m(m-1)} \tag{1.36}$$

Now if we compare the coefficients of like-power terms of t on both sides of (1.36), we have

$$\left.\begin{aligned} a_0(x) &= \zeta_0(x) \\ a_1(x) &= \zeta_1(x) \end{aligned}\right\} \tag{1.37}$$

and for $m \geq 2$ the recurrence relation

$$m(m-1)a_m(x) = g_{m-2}(x) - sa_{m-2}(x) - \frac{\partial^2}{\partial x^2}a_{m-2}(x) \tag{1.38}$$

Therefore, the final t-partial solution is given by (1.31) as the coefficients $a_m^{,s}$ are obtained by (1.37) and (1.38).

The x-Partial Solution

For the x-partial solution, we consider equation (1.6). Then we operate with the inverse operator L_x^{-1} on both sides of this equation and get

$$u(x,t) = \delta_0(t) + x\delta_1(t) + L_x^{-1}[g - L_t u - su] \tag{1.39}$$

where $\delta_0(t)$ and $\delta_1(t)$ have their usual meanings. Just as with the t-partial solution, we assume the x-partial solution as

$$u(x,t) = \sum_{m=0}^{\infty}\sum_{n=0}^{\infty} b_{m,n}x^m t^n = \sum_{m=0}^{\infty}\left[\sum_{n=0}^{\infty} b_{m,n}t^n\right]x^m = \sum_{m=0}^{\infty} b_m(t)x^m \tag{1.40}$$

where

$$b_m(t) = \sum_{n=0}^{\infty} b_{m,n}t^n \tag{1.41}$$

We write the source term $g(x,t)$ as

$$g(x,t) = \sum_{m=0}^{\infty}\sum_{k=0}^{\infty} g_{m,k}x^m t^k = \sum_{m=0}^{\infty}\left[\sum_{k=0}^{\infty} g_{m,k}t^k\right]x^m = \sum_{m=0}^{\infty} g_m(t)x^m \tag{1.42}$$

where

$$g_m(t) = \sum_{k=0}^{\infty} g_{m,k}t^k \tag{1.43}$$

We now substitute (1.40) and (1.42) in (1.39) and get

$$\begin{aligned} u(x,t) &= \delta_0(t) + x\delta_1(t) + L_x^{-1}\sum_{m=0}^{\infty} g_m(t)x^m \\ &\quad - L_x^{-1}\left[L_t\sum_{m=0}^{\infty} b_m(t)x^m\right] - L_x^{-1}\sum_{m=0}^{\infty} sb_m(t)x^m \end{aligned} \tag{1.44}$$

Using (1.4) on the right-hand side of (1.44) and then performing integrations, we obtain

$$\begin{aligned} u(x,t) &= \delta_0(t) + x\delta_1(t) + \sum_{m=0}^{\infty} g_m(t).x^{m+2}\Big/(m+1)(m+2) \\ &\quad - \sum_{m=0}^{\infty}\frac{\partial^2}{\partial t^2}b_m(t).x^{m+2}\Big/(m+1)(m+2) \\ &\quad - \sum_{m=0}^{\infty} sb_m(t).x^{m+2}\Big/(m+1)(m+2) \end{aligned} \tag{1.45}$$

Then we replace m by $(m-2)$ on the right side of (1.45) and get

$$\begin{aligned} \sum_{m=0}^{\infty} b_m(t)x^m &= \delta_0(t) + x\delta_1(t) + \sum_{m=2}^{\infty} g_{m-2}(t).\frac{x^m}{m(m-1)} \\ &\quad - \sum_{m=2}^{\infty}\frac{\partial^2}{\partial t^2}b_{m-2}(t).\frac{x^m}{m(m-1)} - \sum_{m=2}^{\infty} sb_{m-2}(t).\frac{x^m}{m(m-1)} \end{aligned} \tag{1.46}$$

If we compare the like-power terms of (1.46), we have

$$\left.\begin{aligned} b_0(t) &= \delta_0(t) \\ b_1(t) &= \delta_1(t) \end{aligned}\right\} \tag{1.47}$$

and for $m \geq 2$ the recurrence relation

$$m(m-1)b_m(t) = g_{m-2}(t) - sb_{m-2}(t) - \frac{\partial^2}{\partial t^2}b_{m-2}(t) \tag{1.48}$$

Thus, the final x-partial solution is given by (1.40) as the coefficients are obtained by (1.47) and (1.48).

1.3 A Review on the Convergence of the Decomposition Method

Adomian and others have discussed the application of convergence of Adomian's decomposition method (ADM) to algebraic, transcendental, and matrix equations [1]. A broad spectrum of linear and nonlinear ordinary or partial differential equations can be solved easily, effectively, and accurately using the decomposition method [1–3]. Moreover, convergence of Adomian's decomposition can be an elegant procedure for evaluating certain Laplace transforms by their reformulations into easily computable convergent series of differential equations [4]. An approximate solution of the nonharmonic oscillator problem can be arrived at using ADM [5]. Only a few terms of the series solution of the above problem need to be subjected to a subsequent round of decomposition for arriving at the time period of oscillation numerically. Interestingly, this result has been found to be in excellent agreement, with the exact value of the time period pertaining to the rapid convergence of decomposition series [5]. The solution of the flow of fluid through a circular rigid tube can be obtained approximately by means of the decomposition method [19,21,22]. This solution is undoubtedly equal to the exact solution of the problem and shows the rapid convergence for a very small number of terms of the series solution. An adaptation of ADM allows calculation of an integral, which is neither adequately tabulated nor expressible in terms of elementary functions so precisely that solving only a couple of the terms of the series can lead to the solution that is in excellent agreement with the actual one pertaining to the rapid convergence of the series [6]. Integrations by asymptotic decomposition show a rapid convergence and precise corroboration with the known results [7]. Recent work by Cherruault [9] and Gabet [15] on the mathematical framework has provided the rigorous basis for the accuracy and rapid rate of convergence of ADM. Obviously, the rapidity of this convergence means that few terms of the series solution are required, as shown in some problems of mathematical physics [10–12]. Acceleration of convergence can be calculated for some initial and boundary value problems, as shown by Adomian and others [2,8]. Best convergence to the exact solution is obtained with actual forcing function or at least when the majority of the terms of its series are used [13].

We emphasize that a method of approximation does not need infinite precision; a uniform convergence of input and coefficient series is sufficient. Moreover, the series solution can be truncated after a few terms without compromising the precision of the solution. The effect of increased accuracy due to the addition of terms to the series solution is investigated by computation to develop stopping rules. Practically, this means that if we are interested in accuracy to n decimal places, under circumstances where subsequent calculation does not lead to any further change in the n places, the solution is considered to be known and verified by satisfying the equation and given conditions. Accumulated literature on the subject exhibits a consensus on the rapidity of the convergence. This implies that the solution resides in the early terms of the decomposition series. Insight gained from the computation can be used to develop stopping rules and acceleration techniques (Pade Euler, Shanks), investigate convergence rate, substantiate the solution by modified and/or asymptotic decomposition and verify whether it satisfies the given equation or conditions, and determine the region of convergence, the possible use of analytic continuation, and other special techniques. Since the integrations being dealt with are trivial, that is, Green's function of unity is taken into consideration [8] and L and R are chosen, such that R is a lower-order operator than L, convergence is evident as indicated by the evaluation of matrix inverse with $|L^{-1}R| < 1$. Although the series captures the actual result in its entirety, nonlinearity does incorporate complex interactions among the input terms. However, the result can be plotted and a Fourier analysis performed to confirm the components in the output.

The most notable feature that surfaces up from a comparison between perturbation and decomposition is that the results of perturbation converge slowly, while those of decomposition converge rapidly. This ensures that only a few terms are generally sufficient to arrive at the result. (However, when this is not the case, one can use Pade approximations or another acceleration technique or the method of asymptotic decomposition.) As mentioned earlier, the convergence of decomposition series has been investigated by several authors. The theoretical treatment of convergence of the decomposition method was studied by Cherruault [9] and Repaci [23]. The results, thus obtained, were improved by Cherruault and Adomian [11], who put forward a new convergence proof of Adomian's technique based on properties of convergent series. Some results on the speed of convergence were obtained that subsequently facilitated solving linear and nonlinear functional equations. The convergence of this series is also established using fixed-point theorems [8]. The hypotheses for proving convergence are less restrictive and are generally satisfied in physical problems [17].

References

1. Adomian, G.: *Stochastic Systems*, Academic Press (1983).
2. Adomian, G.: On the Convergence Region for Decomposition Solutions, *J. Comput. and Appl. Math.*, *11*(3), 379–380 (1984).
3. Adomian, G.: *Nonlinear Stochastic Operator Equations*, Academic Press (1986).
4. Adomian, G.: A New Approach to the Heat Equation: An Application of the Decomposition Method, *J. Math. Anal. Appl.*, *113*, 202–209 (1986).
5. Adomian, G.: Application of the Decomposition Method to the Navier–Stokes Equations, *J. Math. Anal. Appl.*, *119*, 340–360 (1986).
6. Adomian, G.: *Nonlinear Stochastic Systems Theory and Applications to Physics*, Kluwer Academic Publishers (1989).
7. Adomian, G.: Nonlinear transport in moving fluids, *Appl. Math. Lett.*, *6*(5), 35–38 (1993).
8. Adomian, G.: *Solving Frontier Problems of Physics: The Decomposition Method*, Kluwer Academic Publishers (1994).
9. Cherruault, Y.: Convergence of Adomian's Method, *Kybernetes*, *18*(2), 31–39 (1990).
10. Cherruault, Y.: Some New Results for Convergence of Adomian's Method Applied to Integral Equations, *Math. Comput. Modelling*, *16*(2), 85–93 (1992).
11. Cherruault, Y. and G. Adomian: Decomposition Methods: A New Proof of Convergence, *Math. Comput. Modelling*, *18*, 103–106 (1993).
12. Datta, B. K.: A New Approach to the Wave Equation: An Application of the Decomposition, *J. Math. Anal. Appl.*, *142*(1), 6–12 (1989).
13. Datta, B. K.: A Technique for Approximate Solutions to Schrödinger-Like Equations, *J. Computers and Math. with Appl.*, *20*(1), 61–65 (1990).
14. Datta, B. K.: The Approximate Solution of Certain Integrals, *Computers Math. Appl.*, *25*(7), 47–49 (1993).
15. Gabet, L.: The Theoretical Foundation of the Adomian Method, *Computers Math. Appl.*, *27*(12), 41–52 (1994).
16. Haldar, K., and B. K. Datta: On the Approximate Solutions of the Anharmonic Oscillator Problem, *Int. J. Math. Ed. Sc. Tech.*, *25*(6), 907–911 (1994).
17. Haldar, K., and B. K. Datta: Integrations by Asymptotic Decomposition, *Appl. Math. Lett.*, *9*(2), 81–83 (1996).
18. Haldar, K., and B. K. Datta: Laplace Transform Evaluations by Ramanujan's Integral Formula, *Int. J. Math. Ed. Sc. Tech.*, *28*(1), 123–128 (1997).
19. Haldar, K., and M. K. Bandyopadhyay: Application of Double Decomposition Method to the Non-linear Fluid Flow Problems, *Rev. Bull. Cal. Math. Soc.*, *14*(2), 109–120 (2006).

20. Haldar, K.: Application of Modified Decomposition Method to Laplace Equation in Two Dimensional Polar Coordinates, *Bull. Cal. Math. Soc.*, *98*(5), 571–582 (2006).
21. Haldar, K.: Some Exact Solutions of Linear Fluid Flow Problems by Modified Decomposition Method, *Bull. Cal. Math. Soc.*, *100*(3), 283–300 (2008).
22. Mamaloukas, C., K. Haldar, and H. P. Mazumdar: Application of Double Decomposition to Pulsatile Flow, *Int. J. Appl. Math. Comp.*, *10*(1–2), 293–207 (2002).
23. Repaci, A.: Non-linear Dynamical Systems on the Accuracy of Adomian's Decomposition Method, *Appl. Math. Lett.*, *3*, 35–39 (1990).

2

Asymptotic Decomposition

2.1 Introduction

In the previous chapter, we mentioned that the decomposition method is divided into four classes. Out of these three methods (except the asymptotic decomposition) are used in chapters 4 and 5 for developing the Navier–Stokes equations in Cartesian and cylindrical polar coordinates, respectively. Several physical problems of fluid dynamics have also been considered in these chapters for clear illustration of the methods as far as possible. In this chapter, we shall discuss the asymptotic decomposition technique in order to derive some important relations that are frequently used in solving real problems in nature.

It is known that an interesting variation of the decomposition method is observed when the forcing term and the differential coefficients of different orders except the nonlinear term are on the right-hand side of the equation. If the forcing term is a polynomial, then we expect to get a terminating series solution because of the differential coefficients, which decrease the powers of the terms in the solution. The procedure that leads here to a rapidly terminating series solution is known as the *asymptotic decomposition method*. And in this method, we still use the basic decompositions of the dependent variable and nonlinear term. Besides this, the asymptotic decomposition results in a solution for sufficiently large values of independent variables. It is also interesting to note that the decomposition method requires the initial condition and integrations, whereas asymptotic decomposition, in which the initial condition is of no importance, requires differentiations only. The details of this method are given in the references [1–3,7].

2.2 Application of Asymptotic Decomposition

The asymptotic decomposition of Adomian [1–4] is an effective procedure for the evaluation of certain difficult integrals by reformulation into differential integrals and consequent evaluation in easily computed convergent series.

To demonstrate the idea, we consider a more general problem of reducing integrations to differential equations because the latter problem is solved easily, quickly, and elegantly by the decomposition method.

Consider the simple first-order differential equation

$$\frac{dy}{dx} + P(x)y = Q(x), \quad y(0) = 0 \tag{2.1}$$

The solution of the problem described by (2.1) is

$$y(x)f(x) = \int Qf(x)dx \tag{2.2}$$

where the integrating factor $f(x)$ is

$$f(x) = e^{\int P(x)dx} \tag{2.3}$$

It may happen that the above integral is complicated and not computable in terms of elementary functions or conveniently tabulated for ready reference.

Now following the analysis of the Adomian [4,7] equation (2.1) is written in the operator form as

$$Ly + P(x)y = Q(x), \quad y(0) = 0 \tag{2.4}$$

where $L = \frac{d}{dx}$. Then we consider the asymptotic decomposition technique [1,2] for the solution of the equation (2.4) as

$$P(x)y = Q(x) - Ly \tag{2.5}$$

Then we decompose y into the following form

$$y = \sum_{m=0}^{\infty} \lambda^m y_m \tag{2.6}$$

Using (2.6) in (2.5), we have on comparison

$$\left.\begin{aligned} y_0 &= \frac{Q(x)}{P(x)} \\ y_1 &= -\frac{1}{P(x)}Ly_0 \\ y_2 &= (-1)^2\frac{1}{P(x)}Ly_1 \\ &\cdots\cdots\cdots\cdots\cdots \\ &\cdots\cdots\cdots\cdots\cdots \end{aligned}\right\} \tag{2.7}$$

Hence,

$$y_{k+1} = (-1)^{k+1}\frac{1}{P(x)}Ly_k \tag{2.8}$$

where $k = 0, 1, 2,$ etc. If $P(x)$ and $Q(x)$ are known, then all the components of y are computable.

The expression

$$\varphi_n = \sum_{i=0}^{n-1} y_i \tag{2.9}$$

is the n-term approximation of the solution y. Convergence is well established [1,5,6] and it is also seen that a rapid stabilization to an acceptable accuracy is evident when numerical computation of the analytic approximation is carried out. Therefore, we obtain from (2.2) and (2.6) remembering that $\lambda = 1$,

$$\int Q(x) f(x) dx = \sum_{i=1}^{\infty} y_i f(x) \tag{2.10}$$

2.2.1 Ramanujan's Integral Formula

As a special case of our analysis, we consider the Ramanujan's integral formula [10] which was established by Haldar and Datta [8] using asymptotic decomposition. For the purpose, we here take $P = -n$ and $Q = \varphi(x)$. Then it follows from (2.3) and (2.7)

$$\left.\begin{aligned} f(x) &= e^{-nx} \\ y_0 &= -\frac{1}{n}\varphi \\ y_1 &= -\frac{1}{n^2}L\varphi \\ y_2 &= -\frac{1}{n^3}L^2\varphi \\ &\cdots\cdots\cdots\cdots\cdots \\ &\cdots\cdots\cdots\cdots\cdots \end{aligned}\right\} \tag{2.11}$$

Substituting (2.11) in (2.10) and remembering that L stands for the differentiation, we finally obtain

$$\int \varphi(x) e^{-nx} dx = -e^{-nx} \left[\frac{\varphi(x)}{n} + \frac{\varphi_1(x)}{n^2} + \frac{\varphi_2(x)}{n^3} + \cdots \right] \tag{2.12}$$

which is the expression for Ramanujan's integral formula and suffixes denote the differentiation with respect to x.

2.2.2 Gamma Integral

To derive the gamma integral [8] from (2.10), we consider $P(x) = -1$ and $Q(x) = x^{m-1}$ where m is an integer. It follows immediately from (2.7)

$$\left.\begin{aligned} f(x) &= e^{-x} \\ y_0 &= -x^{m-1} \\ y_1 &= -L(x^{m-1}) \\ y_2 &= -L^2(x^{m-1}) \\ &\cdots\cdots\cdots\cdots\cdots \\ &\cdots\cdots\cdots\cdots\cdots \end{aligned}\right\} \tag{2.13}$$

From (2.8) we have

$$y_{m-1} = -L^{m-1}(x^{m-1}) = -(m-1)! \tag{2.14}$$

and

$$y_m = -L^m(x^{m-1}) = 0 \tag{2.15}$$

Consequently,

$$y = \sum_{i=0}^{\infty} y_i = -\sum_{i=0}^{\infty} L^i(x^{m-1}) \tag{2.16}$$

Hence, from (2.10) we get

$$\int e^{-x}x^{m-1}dx = -e^{-x}\sum_{i=0}^{\infty} L^i(x^{m-1}) \tag{2.17}$$

Evaluating the integral (2.17) between the limits $x = 0$ to $x = \infty$ we find that

$$\begin{aligned}\int_0^\infty e^{-x}x^{m-1}dx &= -\left[e^{-x}\sum_{i=0}^{\infty} L^i(x^{m-1})\right]_0^\infty \\ &= -\big[e^{-x}\{x^{m-1} + (m-1)x^{m-2} + (m-1)(m-2)x^{m-3} \\ &\quad + \cdots + (m-1)!\}\big]_0^\infty \\ &= (m-1)! = \Gamma(m)\end{aligned} \tag{2.18}$$

This completes the proof of the gamma integral by our analysis.

2.2.3 Laplace Transform

For the derivation of the Laplace transform [9], we evaluate the integral (2.12) between the limits $x = 0$ and $x = \infty$ and obtain

$$\int_0^\infty \exp(-nx)\,\varphi(x)dx = \frac{\varphi(x)}{n} + \frac{\varphi_1(x)}{n^2} + \frac{\varphi_2(x)}{n^3} + \cdots \tag{2.19}$$

For a clear illustration of our analysis, we consider the following examples:

Problem 1. Find the Laplace transform of the function $\exp(-ax)\sin bx$ using Ramanujan's formula.

Solution. Let $f(x) = \exp(-ax)\sin bx$. If $f(s)$ is the Laplace transform of $f(x)$ then

$$\begin{aligned}f(s) &= \int_0^\infty \exp(-sx)f(x)dx \\ &= \int_0^\infty \exp[-(s+a)x]\sin bx\,dx\end{aligned} \tag{2.20}$$

Comparing equations (2.19) and (2.20), we get $n = s + a$, $\varphi(x) = \sin bx$ and hence we get

$$f(s) = \frac{\varphi(0)}{(s+a)} + \frac{\varphi_1(0)}{(s+a)^2} + \frac{\varphi_2(0)}{(s+a)^3} + \cdots \tag{2.21}$$

Since $\varphi(x) = \sin bx$, therefore, the value of the nth derivative of the function $\varphi(x)$ for the zero argument is

$$\varphi_n(0) = b^n \sin(n\pi/2) \tag{2.22}$$

Substituting (2.22) in the series (2.21), we get

$$\begin{aligned} f(s) &= \frac{b}{(s+a)^2} - \frac{b^3}{(s+a)^4} + \frac{b^5}{(s+a)^6} - \cdots \\ &= \frac{b}{(s+a)^2}\left[1 - \frac{b^2}{(s+a)^2} + \frac{b^4}{(s+a)^4} - \cdots\right] \\ &= \frac{b}{(s+a)^2 + b^2} \end{aligned} \tag{2.23}$$

which is the required Laplace transform.

Problem 2. Find the Laplace transform of the function $\exp(-sa)\, J_0(bx)$ by using Ramanujan's formula.

Solution. Let $f(x) = \exp(-ax) J_0(bx)$ where $J_0(x)$ denotes the Bessel's function of order zero. The Laplace transform $f(s)$ of the function $f(x)$ is given by

$$\begin{aligned} f(s) &= \int_0^\infty \exp(-sx).\exp(-ax) J_0(bx) dx \\ &= \int_0^\infty \exp[-(s+a)x]\, J_0(bx) dx \end{aligned} \tag{2.24}$$

If we put $n = s + a$ and $\varphi(x) = J_0(bx)$, then the integral (2.24) with the help of (2.19) becomes

$$f(s) = \frac{\varphi(0)}{s+a} + \frac{\varphi_1(0)}{(s+a)^2} + \frac{\varphi_2(0)}{(s+a)^3} + \cdots \tag{2.25}$$

Since $\varphi(x) = J_0(bx)$, therefore, we have

$$\varphi_1(0) = \varphi_3(0) = \cdots = 0 \tag{2.26}$$

and

$$\varphi(0) = 1,\ \varphi_2(0) = -(b^2/2),\ \varphi_4(0) = \left(\frac{3}{8b^4}\right), etc. \tag{2.27}$$

Substituting relations (2.26) and (2.27) in the series (2.25), we get

$$
\begin{aligned}
f(s) &= \frac{1}{s+a} - \frac{1}{2}.\frac{b^2}{(s+a)^3} + \frac{3}{2^3}.\frac{b^4}{(s+a)^5}\cdots \\
&= \frac{1}{s+a}\left[1 - \frac{1}{2}.\frac{b^2}{(s+a)^2} + \frac{3}{2^3}.\frac{b^4}{(s+a)^4}\cdots\right] \\
&= \frac{1}{s+a}\left[1 + \frac{b^2}{(s+a)^2}\right]^{-1/2} = \frac{1}{[(s+a)^2+b^2]^{1/2}} \qquad (2.28)
\end{aligned}
$$

which is the required Laplace transform of the function.

Problem 3. Use Ramanujan's formula to find the Laplace transform of the function $\sin x/x$.

Solution. Again we consider the function $f(x) = \sin x/x$ whose corresponding Laplace transform $f(s)$ is

$$f(s) = \int_0^\infty \exp(-sx)\frac{\sin x}{x}dx \qquad (2.29)$$

If we put $n = s$, then we can write the equation (2.29) with the help of the equation (2.19) as

$$f(s) = \frac{\varphi(0)}{s} + \frac{\varphi_1(0)}{s^2} + \frac{\varphi_2(0)}{s^3} + \cdots \qquad (2.30)$$

where

$$\varphi(x) = \frac{\sin x}{x} = 1 - \frac{x^2}{3!} + \frac{x^4}{5!} - \frac{x^6}{7!} + \cdots \qquad (2.31)$$

Hence, we find that

$$\varphi_1(0) = \varphi_3(0) = \cdots = 0 \qquad (2.32)$$

and

$$\varphi(0) = 1, \quad \varphi_2(0) = -{}^1/_3, \quad \varphi_4(0) = {}^1/_5, \quad etc. \qquad (2.33)$$

Using the relations (2.32) and (2.33), we have from the series (2.30)

$$
\begin{aligned}
f(s) &= 1 - \frac{1}{3}\left(\frac{1}{s}\right)^3 + \frac{1}{5}\left(\frac{1}{s}\right)^5\cdots \\
&= \tan^{-1}(1/s) \qquad (2.34)
\end{aligned}
$$

which is the required Laplace transform.

Problem 4. Use Ramanujan's formula to find the Laplace transform of the function $f(x) = \sum_0^\infty a_r x^r$.

Solution. Last we consider the expansion of $f(x)$ in the form given by

$$f(x) = a_0 + a_1 x + a_2 x^2 + \cdots = \sum_{r=0}^{\infty} a_r x^r \tag{2.35}$$

Then the Laplace transform $f(s)$ is

$$f(s) = \int_0^{\infty} \exp(-sx) f(x) dx \tag{2.36}$$

Comparing the integral (2.36) with the equation (2.19), we have

$$f(s) = \frac{\varphi(0)}{s} + \frac{\varphi_1(0)}{s^2} + \frac{\varphi_2(0)}{s^3} + \cdots \tag{2.37}$$

where $n = s$ and

$$\varphi(x) = f(x) = \sum_{r=0}^{\infty} a_r x^r \tag{2.38}$$

whose rth derivative for the zero argument is

$$\varphi_r(0) = a_r r \tag{2.39}$$

Substituting the relation (2.39) in the series (2.37), we have

$$f(s) = \frac{a_0}{s} + \frac{a_1}{s^2} + a_2 \frac{2!}{s^3} + a^3 \frac{3!}{s^4} + \cdots = \sum_{r=0}^{\infty} \frac{r! a_r}{s^{r+1}} \tag{2.40}$$

It is worth noticing that although Ramanujan's integral formula, the gamma integral and the Laplace transform have been proved as particular cases of our analysis, this illustrates, nevertheless, the methodology by which complicated integrals may sometimes be handled more easily, quickly, and elegantly than the traditional numerical methods. The number of terms required to obtain a computable and accurate solution is generally small.

References

1. Adomian, G.: *Nonlinear Stochastic Systems Theory and Applications to Physics*, Kluwer Academic Publishers (1989).
2. Adomian, G.: A Review of the Decomposition Method and Some Recent Results for Nonlinear Equations, *Computers Math. Appl., 21*(5), 101–127 (1991).

3. Adomian, G. and R. Rach: Analytic Solution of Nonlinear Boundary-Value Problems in Several Dimensions, *J. Math. Anal. Appl.*, *173*(1), 118–137 (1993).
4. Adomian, G.: *Solving Frontier Problems of Physics: The Decomposition Method*, Kluwer Academic Publishers (1994).
5. Cherruault, Y.: Convergence of Adomian's Method, *Kybernetes*, *18*(2), 31–39 (1989).
6. Cherruault, Y.: Some New Results for Convergence of Adomian's Method Applied to Integral Equations, *Math. Comput. Modelling*, *16*(2), 85–93 (1992).
7. Datta, B. K.: The Approximate Evaluation of Certain Integrals, *Computers Math. Appl.*, *25*(7), 47–49 (1993).
8. Haldar, K. and B. K. Datta: Integrations by Asymptotic Decomposition, *Appl. Math. Lett.*, *9*(2), 81–83 (1996).
9. Haldar, K. and B. K. Datta: Laplace Transform Evaluations by Ramanujan's Integral Formula *Int. J. Math. Ed. Sc. Tech.*, *28*(1), 123–128 (1997).
10. Ramanujan, S.: *Note Book of S. Ramanujan*, Vol. 1, T. I. F. R., Bombay (1957) p. 54.

3

Bessel's Equation

3.1 Introduction

The most general ordinary differential equation of the form

$$x^2\frac{d^2y}{dx^2} + x\frac{dy}{dx} + (x^2 - n^2)y = 0 \tag{3.1}$$

is called Bessel's equation of the order n. This equation is used to study the flow of fluid in a circular pipe, temperature distribution, vibration in cylindrical regions, etc. The equation is also used to discuss the symmetrical vibration of a complete circular membrane and to determine the elevation of a tidal wave in a channel open to the sea.

3.2 Solution of Bessel's General Equation by Modified Decomposition

For solving the equation (3.1) by means of the decomposition method, we write the equation as

$$\frac{d^2y}{dx^2} + \frac{1}{x}\frac{dy}{dx} + \left(1 - \frac{n^2}{x^2}\right) = 0 \tag{3.2}$$

Let $L = \frac{d^2}{dx^2}$ be the second-order linear operator whose inverse form is $L^{-1} = \int\int(-)\, dx\, dx$. Then the operator form of (3.2) is

$$Ly + \frac{1}{x}\frac{dy}{dx} + \left(1 - \frac{n}{x^2}\right)y = 0 \tag{3.3}$$

which on solving for Ly becomes

$$Ly = -\left[\frac{1}{x}\frac{dy}{dx} + \left(1 - \frac{n^2}{x^2}\right)y\right] \tag{3.4}$$

Now we apply the inverse operator L^{-1} on both sides of (3.4) and get

$$y = y_o - L^{-1}\left[\frac{1}{x}\frac{dy}{dx} + \left(1 - \frac{n^2}{x^2}\right)y\right] \tag{3.5}$$

where

$$y_o = \xi_0 + \xi_1 x \tag{3.6}$$

is the solution of the equation

$$\frac{d^2 y}{dx^2} = 0 \tag{3.7}$$

with ξ_0 and ξ_1 being the integration constants.

Then we follow the modified decomposition procedure and for the purpose we set

$$y = \sum_{m=0}^{\infty} a_m x^{m+n} \tag{3.8}$$

Substituting (3.8) into (3.5) we get

$$\begin{aligned}\sum_{m=0}^{\infty} a_m x^{m+n} &= \xi_0 + \xi_1 x - L^{-1}\left[\frac{1}{x}\frac{d}{dx}\sum_{m=0}^{\infty} a_m x^{m+n} + \left(1 - \frac{n^2}{x^2}\right)\sum_{m=0}^{\infty} a_m x^{m+n}\right] \\ &= \xi_0 + \xi_1 x - L^{-1}\left[\sum_{m=0}^{\infty}(m+n+1)\, a_{m+1} x^{m+n-1}\right. \\ &\quad \left. + \sum_{m=0}^{\infty} a_m x^{m+n} - n^2 \sum_{m=0}^{\infty} a_m x^{m+n-2}\right]\end{aligned} \tag{3.9}$$

Performing integrations on the right side of (3.9), we obtain

$$\begin{aligned}\sum_{m=0}^{\infty} a_m x^{m+n} = \xi_0 + \xi_1 x - &\left[\sum_{m=1}^{\infty}(m+n+1) a_{m+1}\frac{x^{m+n+1}}{(m+n)(m+n+1)}\right. \\ &+ \sum_{m=2}^{\infty} a_m \frac{x^{m+n+2}}{(m+n+1)(m+n+2)} \\ &\left. - n^2 \sum_{m=0}^{\infty} a^m \frac{x^{m+n}}{(m+n-1)(m+n)}\right]\end{aligned} \tag{3.10}$$

Equating the coefficients of x^{m+n} from both sides of (3.10), we have the recurrence relation as

$$a_m = -\frac{a_{m-2}}{m(m+2n)} \tag{3.11}$$

The relation (3.11) gives the coefficients a_1, a_2, etc., which are determined in terms of a_0, and a_0 shall never be zero as it is the leading term of the series solution. Putting $m = 1$, we have form (3.11) $a_1 = -a_{-1}/(2n+1) = 0$ as $a_{-1} = 0$. The other coefficients can be obtained by putting $m = 2, 3, 4$, etc., and these are given below:

$$a_1 = a_3 = a_5 = \ldots = 0(each) \tag{3.12}$$

and

$$\left.\begin{aligned} a_2 &= -\frac{a_0}{2^2 \cdot 1!(n+1)} \\ a_4 &= \frac{a_0}{2^4 \cdot 2!(n+1)(n+2)} \\ a_6 &= -\frac{a_0}{2^6 \cdot 3!(n+1)(n+2)(n+3)}, \text{ etc.} \end{aligned}\right\} \tag{3.13}$$

Using (3.12) and (3.13) in the expanded form of (3.8), we obtain

$$\begin{aligned} y = a_0 x^n \Bigg[1 &- \frac{x^2}{2^2 \cdot 1!(n+1)} + \frac{x^4}{2^4 \cdot 2!(n+1)(n+2)} \\ &- \frac{x^6}{2^6 \cdot 3!(n+1)(n+2)(n+3)} + \cdots \Bigg] \end{aligned} \tag{3.14}$$

If we put $a_0 = \frac{1}{2^n \Gamma(n+1)}$ where $\Gamma(n+1)$ is the gamma function, the solution (3.14) is denoted by $J_n(x)$, which is called Bessel's function of the first kind of order n, and it is expressed as

$$J_n(x) = \sum_{r=0}^{\infty} (-1)^r \cdot \left(\frac{x}{2}\right)^{n+2r} \cdot \frac{1}{r!\Gamma(n+r+1)} \tag{3.15}$$

When we change n by $-n$ in the equation (3.2) we see that the equation remains unchanged. Therefore, the other solution of the equation (3.2) can be obtained by replacing n by $-n$ in (3.15) and it is given by

$$J_{-n}(x) = \sum_{r=0}^{\infty} (-1)^r \cdot \left(\frac{x}{2}\right)^{-n+2r} \cdot \frac{1}{r!T^1(-n+r+1)} \tag{3.16}$$

Thus, the most general solution of the Bessel's equation (3.2) is

$$y = AJ_n(x) + BJ_{-n}(x) \tag{3.17}$$

where A and B are two arbitrary constants to be evaluated by the boundary conditions.

In particular, if we put $n = 0$ in (3.15) we obtain the Bessel's function of order zero and it is found to be

$$J_0(x) = 1 - \frac{x^2}{2^2} + \frac{x^4}{2^2 \cdot 4^2} - \frac{x^6}{2^2 \cdot 4^2 \cdot 6^2} + \cdots \tag{3.18}$$

3.3 Second Solution of Bessel's Equation by Regular Decomposition

In the previous section, we discussed the solution of Bessel's general ordinary differential equation. We also saw that the two solutions $J_n(x)$ and $J_{-n}(x)$ are equal when n is zero.

The Bessel's equation for $n = 0$ is

$$\frac{d^2y}{dx^2} + \frac{1}{x}\frac{dy}{dx} + y = 0 \tag{3.19}$$

The equation (3.19) is a secondorder ordinary differential equation, and it must have two solutions. Out of these two solutions, one solution is $J_0(x)$, given by (3.18), and the other one should be found out in order to get the general solution of the equation (3.19).

For this purpose we follow the single or regular decomposition procedure. We now write the equation (3.19) as

$$\frac{1}{x}\frac{d}{dx}\left(x\frac{dy}{dx}\right) + y = 0 \tag{3.20}$$

Let

$$L = \frac{1}{x}\frac{d}{dx}\left(x\frac{d}{dx}\right) \tag{3.21}$$

be a differential operator whose inverse is

$$L^{-1} = L_1^{-1}[x^{-1}(L_1^{-1}x)] \tag{3.22}$$

where L_1^{-1} is defined by

$$L_1^{-1} = \int (\cdot)dx \tag{3.23}$$

Then the operator form of (3.20) is

$$Ly + y = 0 \tag{3.24}$$

Solving for Ly and then operating on both sides of (3.24) with the inverse operator L^{-1}, we get

$$y = y_0 - L^{-1}y \tag{3.25}$$

where y_0, the solution of the equation $Ly = 0$, is given by

$$y_0 = a_0 + a_1 \log x \tag{3.26}$$

Here a_0 and a_1 involved in the solution (3.26) are arbitrarily constants.

We now decompose y into the following form:

$$y = \sum_{m=0}^{\infty} \lambda^m y_m \tag{3.27}$$

where λ has its usual meaning. Substituting (3.27) in (3.25) and then comparing like-power terms of λ on both sides of resulting expression we have

$$\left.\begin{aligned} y_1 &= -L^{-1}y_0 \\ y_2 &= -L^{-1}y_1 \\ &\cdots \\ &\cdots \\ y_{m+1} &= -L^{-1}y_m \\ &\cdots\cdots\cdots \end{aligned}\right\} \tag{3.28}$$

Using (3.26) in each of the expressions given in (3.28), we obtain

$$\begin{aligned} y_1 &= -a_0\frac{x^2}{2^2} - a_1\left[\frac{x^2}{2^2}\log x - \frac{x^2}{2^2}(1)\right] \\ y_2 &= a_0\frac{x^4}{2^2 \cdot 4^2} + a_1\left[\frac{x^4}{2^2 \cdot 4^2}\log x - \frac{x^4}{2^2 \cdot 4^2}\left(1+\frac{1}{2}\right)\right] \\ y_3 &= -a_0\frac{x^6}{2^2 \cdot 4^2 \cdot 6^2} - a_1\left[\frac{x^6}{2^2 \cdot 4^2 \cdot 6^2}\log x - \frac{x^6}{2^2 \cdot 4^2 \cdot 6^2}\left(1+\frac{1}{2}+\frac{1}{3}\right)\right] \\ &\cdots\cdots\cdots\cdots\cdots\cdots\cdots\cdots\cdots\cdots \qquad (3.29) \\ &\cdots\cdots\cdots\cdots\cdots\cdots\cdots\cdots\cdots\cdots \\ y_m &= (-1)^m\left[a_0\frac{x^{2m}}{2^2 \cdot 4^2 \ldots (2m)^2} + a_1\left\{\frac{x^{2m}}{2^2 \cdot 4^2 \cdot 6^2 \ldots (2m)^2}\log x \right.\right. \\ &\left.\left. - \frac{x^{2m}}{2^2 \cdot 4^2 \ldots (2m)^2}\left(1+\frac{1}{2}+\frac{1}{3}+\cdots+\frac{1}{m}\right)\right\}\right] \\ &\cdots\cdots\cdots\cdots\cdots\cdots\cdots\cdots\cdots\cdots \end{aligned}$$

We now add the components y_0, y_1, y_2, etc., given in (3.26) and (3.29), and then we use the expanded form of (3.27), remembering that $\lambda = 1$. Finally, we get

$$\begin{aligned} y &= a_0\left[1 - \frac{x^2}{2^2} + \frac{x^4}{2^2 \cdot 4^2} - \frac{x^6}{2^2 \cdot 4^2 \cdot 6^2} + \cdots\right] \\ &+ a_1\left[\left(1 - \frac{x^2}{2^2} + \frac{x^4}{2^2 \cdot 4^2} - \frac{x^6}{2^2 \cdot 4^2 \cdot 6^2} + \cdots\right)\log x\right. \\ &\left. + \left\{\frac{x^2}{2^2}(1) - \frac{x^4}{2^2 \cdot 4^2}\left(1+\frac{1}{2}\right) + \frac{x^6}{2^2 \cdot 4^2 \cdot 6^2}\left(1+\frac{1}{2}+\frac{1}{3}\right)\ldots\right\}\right] \\ &= a_0 J_0(x) + a_1\left[J_0(x)\log x + \sum_{m=1}^{\infty}\frac{(-1)^{m-1}hm}{2^{2m}(m!)^2} \cdot x^{2m}\right] \end{aligned} \tag{3.30}$$

where

$$hm = \sum_{r=1}^{m} \frac{1}{r} \tag{3.31}$$

The solution (3.30) is a linear combination of two solutions each multiplied by a constant. Since the equation (3.19) is a second-order ordinary differential equation, therefore, it must have two solutions of which the first one is $J_0(x)$ given by (3.18), and the second one denoted by $Y_0(x)$ is

$$Y_0(x) = J_0(x) \log x + \sum_{m=1}^{\infty} \frac{(-1)^{m-1} hm}{2^{2m}(m!)^2} \cdot x^{2m} \tag{3.32}$$

which is called Bessel's function of the second kind of order zero. Thus, the general solution of (3.19) is

$$y = a_0 J_0(x) + a_1 Y_0(x) \tag{3.33}$$

In the following sections, we will consider some physical problems for the application of Bessel's function.

3.4 Pulsatile Flow of Fluid in a Rigid Tube

Consider pulsatile flow of fluid in a rigid tube of constant radius. It is assumed that the flow is laminar and symmetrical about the axis of the tube. It is also assumed that the axial velocity is much more than the radial velocity. Under these assumptions, the continuity equation becomes $\frac{\partial u}{\partial x} = 0$, which shows that u is a function of r only, where u is the axial velocity and (r, x) are radial and axial coordinates, respectively. The Navier–Stokes equations, which govern the flow field in the tube, reduce to the form

$$\rho \frac{\partial u}{\partial t} = -\frac{\partial p}{\partial x} + \mu \left(\frac{\partial^2 u}{\partial r^2} + \frac{1}{r} \frac{\partial u}{\partial r} \right) \tag{3.34}$$

where ρ and μ are density and viscosity of the fluid, respectively.

The appropriate boundary conditions imposed on u are

$$u = 0 \quad \text{at} \quad r = R_0 \tag{3.35}$$

for no slip at the wall and

$$\frac{\partial u}{\partial r} = 0 \quad \text{at} \quad r = 0 \tag{3.36}$$

for symmetry about the axis of the tube.

Before solving the equation (3.34) under the boundary conditions (3.35) and (3.36), we introduce a transformation defined by $s = r/R_0$. Then the basic equation (3.34) for studying the flow fluid takes the form

$$\frac{\partial^2 u}{\partial s^2} + \frac{1}{s} \frac{\partial u}{\partial s} - \frac{\rho R_0^2}{\mu} \frac{\partial u}{\partial t} = \frac{R_0^2}{\mu} \frac{\partial p}{\partial x} \tag{3.37}$$

The corresponding boundary conditions (3.35) and (3.36) become

$$u = 0 \quad \text{at} \quad s = 1 \tag{3.38}$$

and

$$\frac{\partial u}{\partial s} = 0 \quad \text{at} \quad s = 0 \tag{3.39}$$

The simple solution of the pulsatile motion of a viscous fluid will be obtained under a pressure gradient that varies with time. Let the solutions for u and p be set in the forms

$$u(s,t) = \bar{u}(s)e^{i\omega t} \tag{3.40}$$

$$-\frac{\partial p}{\partial x} = Pe^{i\omega t} \tag{3.41}$$

Substituting (3.40) and (3.41) in the equation (3.37), we obtain

$$\frac{d^2\bar{u}}{ds^2} + \frac{1}{s}\frac{d\bar{u}}{ds} - \frac{i\rho\omega R_0^2}{\mu}\bar{u} = -\frac{PR_0^2}{\mu} \tag{3.42}$$

which can be written as

$$\frac{d^2 f}{ds^2} + \frac{1}{s}\frac{df}{ds} + l^2 f = 0 \tag{3.43}$$

where

$$f = \bar{u} - \frac{PR_0^2}{-i\alpha^2\mu} \tag{3.44}$$

and

$$\left.\begin{aligned} l^2 &= i^3\alpha^2 \\ \alpha^2 &= \frac{\rho\omega R_0^2}{\mu} \end{aligned}\right\} \tag{3.45}$$

The boundary conditions (3.38) and (3.39), by virtue of (3.40) become

$$\bar{u} = 0 \quad \text{at} \quad s = 1 \tag{3.46}$$

and

$$\frac{\partial \bar{u}}{\partial s} = 0 \quad \text{at} \quad s = 0 \tag{3.47}$$

Let $L = \frac{d^2}{ds^2} + \frac{1}{s}\frac{d}{ds} = \frac{1}{s}\frac{d}{ds}\left(s\frac{d}{ds}\right)$ whose inverse is $L^{-1} = L_1^{-1}\left[s^{-1}(L,s)\right]$. Then the operator form of (3.43) is

$$Lf + l^2 f = 0 \tag{3.48}$$

The use of the inverse operator L^{-1} on both sides of the equation (3.48) leads to

$$f = f_0 - L^{-1}\left(l^2 f\right) \tag{3.49}$$

where f_0 is given by

$$f_0 = \eta_0 + \eta_1 \log s \tag{3.50}$$

Then we follow the decomposition procedure and write for the purpose

$$f = \sum_{m=0}^{\infty} \lambda^m f_m \tag{3.51}$$

The parameterixed form of (3.49) is

$$f = f_0 - \lambda L^{-1}(l^2 f) \tag{3.52}$$

Substituting (3.51) in (3.52) and then comparing the terms, we obtain

$$f_1(s) = -\eta_0 \frac{(ls)^2}{2^2} - \eta_1 \left[\frac{(ls)^2}{2^2} \log s - \frac{(ls)^2}{2^2}(1) \right]$$

$$f_2(s) = \eta_0 \frac{(ls)^4}{2^2-4^2} + \eta_1 \left[\frac{(ls)^4}{2^2-4^2} \log s - \frac{(ls)^4}{2^2-4^2} \left(1 + \frac{1}{2}\right) \right] \tag{3.53}$$

....

....

$$f_m(s) = (-1)^m \left[a_0 \frac{(ls)^{2m}}{2^2.4^2 \cdots (2m)^2} + \eta_1 \left\{ \frac{(ls)^{2m}}{2^2.4^2 \cdots (2m)^2} \log s - \frac{(ls)^{2m}}{2^2.4^2 \cdots (2m)^2} \left(1 + \frac{1}{2} + \frac{1}{3} \cdots + \frac{1}{n}\right) \right\} \right]$$

....

Adding (3.50) and other components given in (3.53), we get

$$f(s) = \eta_0 J_0(ls) + \eta_1 Y_0(ls) \tag{3.54}$$

where

$$J_0(ls) = \sum_{m=0}^{\infty} (-1)^m \frac{(ls)^m}{2^{2m}(m!)^2} \tag{3.55}$$

$$Y_0(ls) = J_0(ls) \log s + \sum_{m=1}^{\infty} (-1)^{m-1} \frac{hm}{2^{2m}(m!)^2} (ls)^{2m} \tag{3.56}$$

where

$$hm = \sum_{r=1}^{m} \frac{1}{r}$$

The use of (3.54) in (3.44) gives

$$\bar{u}(s) = -\frac{PR_0^2}{i\alpha^2 \mu} + \eta_0 J_0(ls) + \eta_1 Y_0(ls) \tag{3.57}$$

The boundary condition (3.39) states that u that is, $\bar{u}$, should be finite on the axis of the tube that is, $\eta_1 = 0$, and the solution (3.57) reduces to

$$\bar{u}(s) = -\frac{PR_0^2}{i\alpha^2\mu} + \eta_0 J_0(ls) \tag{3.58}$$

Satisfying the condition (3.46) by (3.58), we have

$$\eta_0 = -\frac{PR_0^2}{i\alpha^2\mu} \cdot \frac{1}{J_0(l)} \tag{3.59}$$

and the expression for $\bar{u}(s)$ becomes

$$\bar{u}(s) = \frac{PR_0^2}{i\alpha^2\mu}\left[1 - \frac{J_0(ls)}{J_0(l)}\right] \tag{3.60}$$

Finally, the solution of the original equation is

$$u(r, t) = \frac{PR_0^2}{\alpha^2\mu i}\left[1 - \frac{J_0\left(i^{3/2}\alpha r/R_0\right)}{J_0\left(i^{3/2}\alpha\right)}\right] e^{i\omega t} \tag{3.61}$$

whose real part is the exact solution of the problem.

3.5 Periodic Motion of a Visco-Elastic Fluid in a Rigid Tube

Consider an oscillatory but axially symmetric and laminar flow of a fluid in a rigid tube of constant radius. The density and viscosity of the fluid are assumed to be constant. It is also assumed that the fluid flowing in the tube is visco-elastic of the Maxwell type.

The approximate equation of motion describing the flow field in the tube is

$$\rho\frac{\partial u}{\partial t} = -\frac{\partial p}{\partial x} - \frac{1}{r}\frac{\partial}{\partial r}(r\tau_{rx}) \tag{3.62}$$

and the linearized Maxwell model of a visco-elastic fluid is

$$\tau_{rx} + T_0\frac{\partial \tau_{rx}}{\partial t} = -\mu\frac{\partial u}{\partial r} \tag{3.63}$$

where (r, x) are the radical and axial coordinates respectively, τ_{rx} is the component of stress tensor in the x direction, u is the velocity in the direction parallel to the axis of the tube, p is the pressure, ρ is the density, μ is the viscosity, $T_0 = G/\mu$ is the characteristic time for the fluid, and G is the modulus of shear rigidity.

The appropriate boundary conditions are

$$u = 0 \quad \text{at} \quad r = R_0 \tag{3.64}$$

and

$$\frac{\partial u}{\partial r} = 0 \quad \text{at} \quad r = 0 \tag{3.65}$$

where R_0 is the radius of the tube.

The solution to visco-elastic fluid flow through a rigid tube can be obtained by solving the equations (3.62) and (3.63) under the boundary conditions (3.64) and (3.65), respectively. Since the flow is oscillatory, we set u, τ_{rx}, and $-rp/\partial x$ in the following forms:

$$u(r, t) = \bar{u}(r)e^{i\omega t} \tag{3.66}$$

$$\tau_{rx}(r, t) = \tau(r)e^{i\omega t} \tag{3.67}$$

$$-\frac{\partial p}{\partial x} = Pe^{i\omega t} \tag{3.68}$$

where P is real, $(\bar{u}, \tau)$ are complex and ω is the frequency. Substituting (3.66), (3.67) and (3.68) in (3.62) and (3.63) and then eliminating τ, we have the differential equation for $\bar{u}$ as

$$\frac{1}{r}\frac{d}{dr}\left(r\frac{d\bar{u}}{dr}\right) + k^2\left(\bar{u} - \frac{P}{i\omega\rho}\right) = 0 \tag{3.69}$$

which can be written as

$$\frac{1}{r}\frac{d}{dr}\left(r\frac{df}{dr}\right) + k^2 f = 0 \tag{3.70}$$

where

$$f = \bar{u} - \frac{P}{i\omega\rho} \tag{3.71}$$

and

$$k^2 = -\frac{i\omega\rho\,(1 + i\omega T_0)}{\mu} \tag{3.72}$$

The corresponding boundary conditions (3.64) and (3.65) become

$$\bar{u} = 0 \quad \text{at} \quad r = R_0 \tag{3.73}$$

and

$$\frac{\partial \bar{u}}{\partial r} = 0 \quad \text{at} \quad r = 0 \tag{3.74}$$

Following the procedure of the decomposition method for the solution of Bessel's equation in the previous section we write the general solution of the equation (3.70) in the form given by

$$f(r) = A_1 J_0\,(kr) + B_1 Y_0\,(kr) \tag{3.75}$$

or by virtue of (3.71)

$$\bar{u}(r) = \frac{P}{i\omega\rho} + A_1 J_0(kr) + B_1 Y_0(kr) \tag{3.76}$$

where A_1 and B_1 are arbitrary constants to be determined by the boundary conditions (3.73) and (3.74), respectively. Under these conditions, the solution

(3.76) takes the form

$$\bar{u}(r) = \frac{P}{i\omega\rho}\left[1 - \frac{J_0(kr)}{J_0(kR_0)}\right] \tag{3.77}$$

Using (3.66) we have the final solution as

$$u(r, t) = \frac{P}{i\omega\rho}\left[1 - \frac{J_0(kr)}{J_0(kR_0)}\right] e^{i\omega t} \tag{3.78}$$

whose real part is the exact solution of the problem.

3.6 Tidal Waves in a Channel Open to the Sea

Let O be the origin situated at the bottom of the channel of constant depth h whose length is parallel to the x-axis and the y-axis is vertical upward to the plane bottom of the channel (see Figure 3.1). A dam is constructed at a distance $x = a$ from the origin and $x = l_0$, where $0 < a < l_0$ is the mouth of the channel that is open to the sea. It is assumed that a tidal oscillation $\eta = \eta_0 \cos \omega t$ is maintained at the mouth of the channel. If $\eta(x, t)$ is the elevation of the tidal wave at a distance x from the origin at any time t and if the breadth b of the channel varies as the distance from the origin, then the equation of the tidal wave elevation in the channel is

$$gh\frac{\partial}{\partial x}\left(b\frac{\partial \eta}{\partial x}\right) = b\frac{\partial^2 \eta}{\partial t^2} \tag{3.79}$$

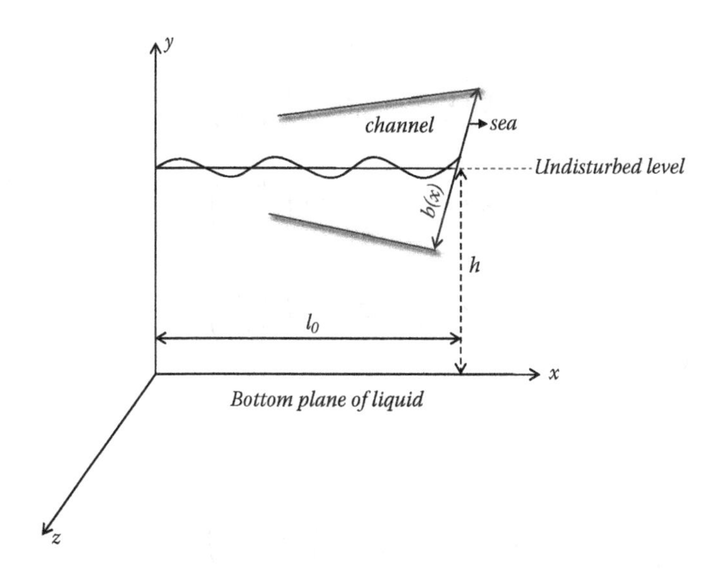

FIGURE 3.1
Tidal wave.

Since the breadth varies as the distance x from the origin, we have

$$b(x) = \lambda_1 x \tag{3.80}$$

Using (3.80) in (3.79), we get

$$\frac{1}{x}\frac{\partial}{\partial x}\left(x\frac{\partial \eta}{\partial x}\right) = \frac{1}{gh}\frac{\partial^2 \eta}{\partial t^2} \tag{3.81}$$

The boundary conditions imposed on $\eta(x, t)$ are

$$\frac{\partial \eta}{\partial x} = 0 \quad \text{at} \quad x = a \tag{3.82}$$

and

$$\eta = \eta_0 \cos \omega t \quad \text{at} \quad x = l_0 \tag{3.83}$$

Our task is to solve equation (3.81) under the boundary conditions (3.82) and (3.83). Let $L = \frac{1}{x}\frac{\partial}{\partial x}(x\frac{\partial}{\partial x})$ whose inverse form is $L^{-1} = L_1^{-1}[x^{-1}(L_1^{-1}x)]$. Then the equation (3.81) takes the form as

$$L\eta = \frac{1}{gh}\frac{\partial^2 \eta}{\partial t^2} \tag{3.84}$$

Following the regular decomposition procedure in the Section 3, we have

$$\eta_0(x, t) = a_0(t) + a_1(t) \log x \tag{3.85}$$

and

$$\eta_1(x, t) = \frac{1}{gh}L^{-1}\left(\frac{\partial^2 \eta_0}{\partial t^2}\right)$$

$$\eta_2(x, t) = \frac{1}{gh}L^{-1}\left(\frac{\partial^2 \eta_1}{\partial t^2}\right)$$

$$\eta_3(x, t) = \frac{1}{gh}L^{-1}\left(\frac{\partial^2 \eta_2}{\partial t^2}\right) \tag{3.86}$$

....

....

or

$$\eta_m(x, t) = \frac{1}{gh}L^{-1}\left(\frac{\partial^2 \eta_{m-1}}{\partial t^2}\right) \tag{3.87}$$

for $m = 0,\ 1,\ 2,$ etc., remembering that $\eta_{-1} = 0$ where $\eta(x, t) = \sum_{m=0}^{\infty} \lambda^m \eta_m$.

Putting (3.85) in (3.87), then performing integrations and finally adding all the components of η, we get

$$\eta(x,t) = \left[a_0 + a_0'' \frac{(x/\sqrt{gh})^2}{2^2} + a_0^{iv} \frac{(x/\sqrt{gh})^4}{2^2.4^2} + a_0^{vi} \frac{(x/\sqrt{gh})^6}{2^2.4^2.6^2} + \cdots\right]$$
$$+ \left[\left\{a_1 + a_1'' \frac{(x/\sqrt{gh})^2}{2^2} + a_1^{iv} \frac{(x/\sqrt{gh})^4}{2^2.4^2} + a_1^{vi} \frac{(x/\sqrt{gh})^6}{2^2.4^2.6^2} + \cdots\right\} \log x\right.$$
$$- \left\{a_1'' \frac{(x/\sqrt{gh})^2}{2^2}(1) + a_1^{iv} \frac{(x/\sqrt{gh})^4}{2^2.4^2}\left(1+\frac{1}{2}\right)\right.$$
$$\left.\left. + a_1^{vi} \frac{(x/\sqrt{gh})^6}{2^2.4^2.6^2}\left(1+\frac{1}{2}+\frac{1}{3}\right)\cdots\right\}\right] \tag{3.88}$$

To complete the solution of the problem we have to find out the functions $a_0(t)$ and $a_1(t)$. For the purpose we set the functions in the following forms:

$$\left.\begin{aligned} a_0(t) &= C_0 \cos \omega t \\ a_1(t) &= C_1 \cos \omega t \end{aligned}\right\} \tag{3.89}$$

where C_0 and C_1 are to be determined. Putting (3.89) in (3.88), we have

$$= \left[C_0 J_0(kx) + C_1 \left\{J_0(kx) \log x + \sum_{m=1}^{\infty} \frac{(-1)^{m-1} hm}{2^{2m}}\right\}\right] \cos \omega t$$

$$= [C_0 J_0(kx) + C_1 Y_0(kx)] \cos \omega t \tag{3.90}$$

where $k^2 = \omega^2/gh$. The constants involved in (3.90) can be determined from the boundary conditions. Satisfying the conditions (3.82) and (3.83), we find the expression for $\eta(x,t)$ as

$$\eta(x,t) = \eta_0 \frac{J_0(kx)Y_1(ka) - Y_0(kx)J_1(ka)}{J_0(kl_0)\,Y_1(ka) - Y_0(kl_0)\,J_1(ka)} \cos \omega t \tag{3.91}$$

which is the exact solution of the problem.

3.7 Temperature Distribution in an Infinitely Long Circular Cylinder

Consider an infinitely long circular cylinder of radius a with a constant initial temperature T_0. Let (r, θ, z) be the polar coordinates of any point P in the cylinder at any time t where r is the radical coordinate, θ is the rotational

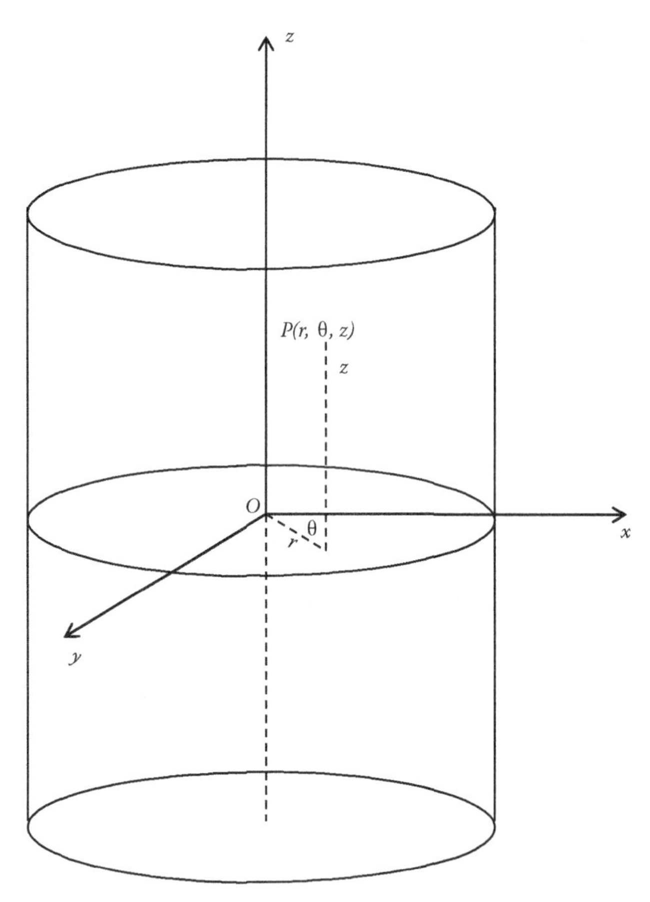

FIGURE 3.2
Temperature distribution in the cylinder.

coordinate and z is parallel to the axis of the cylinder (see Figure 3.2). It is assumed that the temperature is independent of θ and z, and hence, it is a function of r and t only. It is also assumed that the outer surface of the cylinder is insulated. If $T(r, t)$ is the temperature at P, then the partial differential equation that governs the distribution of temperature in the cylinder is

$$\frac{\partial^2 T}{\partial r^2} + \frac{1}{r}\frac{\partial T}{\partial r} = \frac{\partial T}{\partial t}, \quad 0 < r < a, t > 0 \tag{3.92}$$

The boundary and initial conditions imposed on T are

$$T(r, t) = 0 \quad \text{at} \quad r = a \tag{3.93}$$

for all t and

$$T\,(r, 0) = T_0 \tag{3.94}$$

Before we solve the equation (3.92), it is convenient to apply the Laplace transform of (3.92). Taking the Laplace transform of (3.92) and using the initial condition (3.94), we have an ordinary differential equation for $\bar{T}(r, s)$ as

$$\frac{d^2\bar{T}}{dr^2} + \frac{1}{r}\frac{d\bar{T}}{dr} - s\bar{T} = -T_0 \tag{3.95}$$

where $\bar{T}$ is the Laplace transform of T defined by

$$\bar{T}(r, s) = \int_0^\infty e^{-ts}T(r, t)dt \tag{3.96}$$

The corresponding boundary condition (3.93) by virtue of (3.96) takes the form

$$\bar{T}(a, s) = 0 \tag{3.97}$$

The equation (3.95) is a second-order ordinary differential equation that needs two conditions for its solution. One condition is the boundary condition given by (3.97), and the other condition states that T or $\bar{T}$ should be finite on the axis of the cylinder. Mathematically, we write,

$$\bar{T}(r, s) = \textit{finite on } r = 0 \tag{3.98}$$

For solving we write the equation (3.95) in the form

$$\frac{d^2U}{dr^2} + \frac{1}{r}\frac{dU}{dr} - sU = 0 \tag{3.99}$$

where

$$U = \bar{T} - \frac{T_0}{s} \tag{3.100}$$

We now follow the decomposition procedure and write for the purpose

$$U(r, s) = U_0(r) + L^{-1}(sU) \tag{3.101}$$

where $L^{-1} = L_1^{-1}[r^{-1}(L_1^{-1}r)]$ is the inverse operator of $L = \frac{1}{r}\frac{d}{dr}\left(r\frac{d}{dr}\right)$ and $U_0(r)$, the solution of the equation

$$\frac{1}{r}\frac{d}{dr}\left(r\frac{dU}{dr}\right) = 0 \tag{3.102}$$

is given by

$$U_0(r) = C_0 + C_1 \log r \tag{3.103}$$

with C_0 and C_1 being the integration constants to be determined later on.

We now decompose U into the following form:

$$U = \sum_{m=0}^{\infty} \lambda^m U_m \tag{3.104}$$

and write the parameterized form of (3.101) as

$$U(r,s) = U_0(r) + \lambda L^{-1}(sU) \tag{3.105}$$

Using (3.104) in (3.105) and then comparing like-power terms, we have the set of components U_1, U_2, etc., as

$$U_{m+1}(r,s) = L^{-1}(sU_m) \tag{3.106}$$

for $m = 0, 1, 2, 3,$ etc. Adding all the components in (3.103) and (3.106), and then using the sum in the expanded form of (3.104) where $\lambda = 1$, we have

$$U(r,s) = C_0 J_0(i\sqrt{s}r) + C_1 Y_0(i\sqrt{s}r) \tag{3.107}$$

which by virtue of (3.100) takes the form

$$\bar{T}(r,s) = C_0 J_0(i\sqrt{s}r) + C_1 Y_0(i\sqrt{s}r) + \frac{T_0}{s} \tag{3.108}$$

According to the condition (3.98), the solution (3.108) is bounded at $r = 0$, but $Y_0(i\sqrt{s}r)$ is unbounded at $r = 0$. Therefore, we choose $C_1 = 0$, and the solution (3.108) reduces to

$$\bar{T}(r,s) = C_0 J_0(i\sqrt{s}r) + \frac{T_0}{s} \tag{3.109}$$

Then we satisfy the boundary condition (3.97) by (3.109) for C_0 and find the resulting expression for $\bar{T}(r,s)$ to be

$$\bar{T}(r,s) = \frac{T_0}{s}\left[1 - \frac{J_0(i\sqrt{s}r)}{J_0(i\sqrt{s}a)}\right] \tag{3.110}$$

Using the complex inversion formula, we have

$$T(r,t) = T_0 - \frac{T_0}{2\pi i}\int_{\gamma-i\infty}^{\gamma+i\infty} \frac{e^{st} J_0(i\sqrt{s}r)}{s J_0\left(i\sqrt{s}\right)} ds \tag{3.111}$$

which, on evaluation with the help of calculus of residue, gives the required solution of the equation (3.92) as

$$T(r,t) = 2T_0\left[\sum_{n=1}^{\infty} e^{-\frac{\lambda_n^2}{a^2}t} \cdot \frac{J_0\left(\frac{\lambda_n}{\alpha}r\right)}{\lambda_n J_1\left(\lambda_n/a\right)}\right] \tag{3.112}$$

where the function $J_0\left(i\sqrt{s}a\right)$ has simple zeroes, that is, $i\sqrt{s}a = \lambda_n$ or $s = -\lambda_n^2/a^2$.

4

Navier–Stokes Equations in Cartesian Coordinates

4.1 Introduction

Fluid mechanics deals with the physical phenomenon of fluid flow by modelling this phenomenon of interest, generally, in the form of Navier–Stokes equations, which are nonlinear partial differential equations. Then the model needs an effective analysis to produce a result in accordance with observation and experiment, that is, the mathematical solution must be consistent with the physical reality. Therefore, the essentially nonlinear differential equations without simplifications, a customary in research in physics and mathematics, should be solvable.

Since the Navier–Stokes equations are nonlinear in character, it is very difficult to obtain analytic solutions of these equations, except in some special cases. Therefore, we take the help of linearization and restrictions, which reduce the problem to a mathematically tractable one. But the solution of the reduced problem is not consistent with the real world of physics and deviates much from the actual solution of the original problem. As a result, we take the advantage of traditional numerical techniques in order to get a sufficiently accurate result, and the methods result in massive computations.

In this chapter we have applied the decomposition method, originally developed by Adomian [1, 2], to two-dimensional Navier–Stokes equations in Cartesian coordinates and have developed the theory as far as possible. We have also considered some fluid flow problems from the boundary-layer theory of Schlichting [6] for clear illustration of the theory.

4.2 Equations of Motion

Consider two-dimensional flow of viscous incompressible fluid in the planes parallel to the reference plane $z = 0$. Let $u(x, y)$ and $v(x, y)$ be the components of velocity in the x and y directions, respectively. Then the momentum equations governing the flow field in the planes are given by

$$\frac{\partial u}{\partial t} + u\frac{\partial u}{\partial x} + v\frac{\partial u}{\partial y} = -\frac{1}{\rho}\frac{\partial p}{\partial x} + \nu L u \tag{4.1}$$

and

$$\frac{\partial v}{\partial t} + u\frac{\partial v}{\partial x} + v\frac{\partial v}{\partial y} = -\frac{1}{\rho}\frac{\partial p}{\partial y} + \nu L v \tag{4.2}$$

where p is the fluid pressure, ρ is the density, ν is the kinematic visocosity, and L is the Laplacian operator defined by

$$L = \frac{\partial^2}{\partial x^2} + \frac{\partial^2}{\partial y^2} \tag{4.3}$$

The equation of continuity is

$$\frac{\partial u}{\partial x} + \frac{\partial v}{\partial y} = 0 \tag{4.4}$$

If we introduce the stream function $\psi(x, y, t)$ defined by

$$u = \frac{\partial \psi}{\partial x}, \; v = -\frac{\partial \psi}{\partial y} \tag{4.5}$$

then the equation (4.4) is identically satisfied. Eliminating p between (4.1) and (4.2) and then using (4.5) we get the governing equation of motion in terms of $\psi(x, y, t)$ as

$$\frac{\partial}{\partial t}(L\psi) + \frac{\partial \psi}{\partial y} \cdot \frac{\partial}{\partial x}(L\psi) - \frac{\partial \psi}{\partial x} \cdot \frac{\partial}{\partial y}(L\psi) = \nu L^2 \psi \tag{4.6}$$

which we write for our purpose in the form

$$\frac{\partial^4 \psi}{\partial x^4} + \frac{\partial^4 \psi}{\partial y^4} = \nu^{-1}\frac{\partial}{\partial t}(L\psi) - 2\frac{\partial^4 \psi}{\partial x^2 \partial y^2} + \nu^{-1} N\psi \tag{4.7}$$

where the nonlinear term $N\psi$ is defined by

$$N\psi = \frac{\partial \psi}{\partial y} \cdot \frac{\partial}{\partial x}(L\psi) - \frac{\partial \psi}{\partial x} \cdot \frac{\partial}{\partial y}(L\psi) = \frac{\partial(L\psi, \psi)}{\partial(x, y)} \tag{4.8}$$

Equation (4.7) cannot be solved analytically except in some special cases because of the nonlinear character, but it can be solved numerically by means of traditional numerical techniques. The modern powerful method known as Adomian's decomposition method can provide analytical approximations to this nonlinear equation and is applied here in order to get the solution that demands to be parallel to any modern supercomputer.

4.2.1 Solution by Regular Decomposition

Let $L_1 = \frac{\partial^4}{\partial x^4}$ and $L_2 = \frac{\partial^4}{\partial y^4}$ be two fourth-order differential operators. Then the equation (4.7) is written as

$$L_1\psi + L_2\psi = \nu^{-1}\frac{\partial}{\partial t}(L\psi) - 2\frac{\partial^4\psi}{\partial x^2\partial y^2} + \nu^{-1}N\psi \tag{4.9}$$

We now proceed to apply decomposition procedure for different forms of the operator.

Case I

Operator $L_1 = \frac{\partial^4}{\partial x^4}$: The solution obtained by this operator is called the x-partial solution [2] of the equation (4.9). Then we write the equation (4.9), for solving $L_1\psi$, in the form

$$L_1\psi = -L_2\psi + \nu^{-1}\frac{\partial}{\partial t}(L\psi) - 2\frac{\partial^4\psi}{\partial x^2\partial y^2} + \nu^{-1}N\psi \tag{4.10}$$

Let L_1^{-1} be the inverse operator of L_1. It is a fourfold indefinite integral defined by

$$L_1^{-1} = \iiiint(\cdot)\,dx\,dx\,dx\,dx \tag{4.11}$$

Then we operate with the inverse operator L_1^{-1} on both sides of (4.10) and get the x-partial solution as

$$\psi(x, y, t) = \Phi(x, y, t) - L_1^{-1}(L_2\psi) + L_1^{-1}\left[\nu^{-1}\frac{\partial}{\partial t}(L\psi) - 2\frac{\partial^4\psi}{\partial x^2\partial y^2} + \nu^{-1}N\psi\right] \tag{4.12}$$

where Φ is the solution of homogeneous equation $L_1\psi = 0$, and it is given by

$$\Phi(x, y, t) = \xi_0(y, t) + \xi_1(y, t)x + \xi_2(y, t)\frac{x^2}{2} + \xi_3(y, t)\frac{x^3}{6} \tag{4.13}$$

Here $\xi_0(y, t)$, $\xi_1(y, t)$, $\xi_2(y, t)$, and $\xi_3(y, t)$ are the constants of integration to be determined from the prescribed boundary conditions imposed on Φ.

We now decompose ψ and $N\psi$ in the following forms [1]:

$$\psi(x, y, t) = \sum_{n=0}^{\infty}\lambda^n\psi_n(x, y, t) \tag{4.14}$$

and

$$N\psi = \sum_{n=0}^{\infty}\lambda^n A_n \tag{4.15}$$

The parameter λ has its usual meaning, and the A_n are Adomian's special polynomials [1]. The parameterized form [1] of (4.12) is

$$\psi(x, y, t) = \overline{\Phi}(x, y, t) - \lambda L_1^{-1}\left[L_2\psi - \nu^{-1}\frac{\partial}{\partial t}(L\psi) + 2\frac{\partial^4\psi}{\partial x^2\partial y^2} - \nu^{-1}N\psi\right] \tag{4.16}$$

Substituting (4.14) and (4.15) in (4.16) and equating like-power terms of λ from both sides of the resulting equation, we have

$$\begin{aligned} \psi_0(x, y, t) &= \Phi(x, y, t) \\ \psi_1(x, y, t) &= -L_1^{-1}\left[L_2\psi_0 - \nu^{-1}\frac{\partial}{\partial t}(L\psi_0) + 2\frac{\partial^4\psi_0}{\partial x^2\partial y^2} - \nu^{-1}A_0\right] \\ \psi_2(x, y, t) &= -L_1^{-1}\left[L_2\psi_1 - \nu^{-1}\frac{\partial}{\partial t}(L\psi_1) + 2\frac{\partial^4\psi_1}{\partial x^2\partial y^2} - \nu^{-1}A_1\right] \\ \cdots \quad \cdots \quad & \cdots \quad \cdots \quad \cdots \\ \cdots \quad \cdots \quad & \cdots \quad \cdots \quad \cdots \end{aligned} \tag{4.17}$$

The Adomian's polynomials A_0, A_1, etc., involved in (4.17) are defined in such a way that A_0 is a function of ψ_0, A_1 is a function of ψ_0, ψ_1, and so on. From (4.8) and (4.15), we have

$$\sum_{n=0}^{\infty}\lambda^n A_n = \frac{\partial(L\psi, \psi)}{\partial(x, y)} = \frac{\partial\psi}{\partial y}\cdot\frac{\partial}{\partial x}(L\psi) - \frac{\partial\psi}{\partial x}\cdot\frac{\partial}{\partial y}(L\psi) \tag{4.18}$$

Again using (4.14) in (4.18) and comparing the terms on both sides, we obtain the following set of expressions for the Adomian's polynomials:

$$\begin{aligned} A_0 &= \frac{\partial(L\psi_0, \psi_0)}{\partial(x, y)} \\ A_1 &= \frac{\partial(L\psi_0, \psi_1)}{\partial(x, y)} - \frac{\partial(L\psi_1, \psi_0)}{\partial(x, y)} \\ \cdots \quad & \cdots \quad \cdots \\ \cdots \quad & \cdots \quad \cdots \end{aligned} \tag{4.19}$$

The relation (4.14) together with (4.13), (4.17), and (4.19) gives the complete solution of the problem remembering that $\lambda = 1$.

Case II

Operator $L_2 = \frac{\partial^4}{\partial y^4}$: The procedure is exactly the same as that discussed in the Case I. The solution obtained by this analysis is called the y-partial solution of the problem. Now solving for $L_2\psi$ we have from (4.9)

$$L_2\psi = -L_1\psi + \nu^{-1}\frac{\partial}{\partial t}(L\psi) - 2\frac{\partial^4\psi}{\partial x^2\partial y^2} + \nu^{-1}N\psi \tag{4.20}$$

Then using the inverse operator L_2^{-1} defined by

$$L_2^{-1} = \iiiint (\cdot)\, dy\, dy\, dy\, dy \tag{4.21}$$

on both sides of (4.20), we get

$$\psi(x, y, t) = \overline{\Phi}(x, y, t) - L_2^{-1}\left[L_1\psi - \nu^{-1}\frac{\partial}{\partial t}(L\psi) + 2\frac{\partial^4\psi}{\partial x^2\partial y^2} - \nu^{-1}N\psi\right] \tag{4.22}$$

whose parameterized form is

$$\psi(x, y, t) = \overline{\Phi}(x, y, t) - \lambda L_2^{-1}\left[L_1\psi - \nu^{-1}\frac{\partial}{\partial t}(L\psi) + 2\frac{\partial^4\psi}{\partial x^2\partial y^2} - \nu^{-1}N\psi\right] \tag{4.23}$$

where $\overline{\Phi}(x, y, t)$ is the solution of homogeneous equation $L_2\psi = 0$, and it is given by

$$\overline{\Phi}(x, y, t) = \eta_0(x) + \eta_1(x)y + \eta_2(x)\frac{y^2}{2} + \eta_3(x)\frac{y^3}{6} \tag{4.24}$$

The constants $\eta_0(x)$, $\eta_1(x)$, $\eta_2(x)$, and $\eta_3(x)$ involved in the expression are the integrating constants, and these are to be evaluated from the given boundary conditions.

Now using (4.14) and (4.15) in (4.23) we get

$$\begin{aligned}
\psi_0(x, y, t) &= \overline{\Phi}(x, y, t)\\
\psi_1(x, y, t) &= -L_2^{-1}\left[L_1\psi_0 - \nu^{-1}\frac{\partial}{\partial t}(L\psi_0) + 2\frac{\partial^4\psi_0}{\partial x^2\partial y^2} - \nu^{-1}N\psi_0\right]\\
\psi_2(x, y, t) &= -L_2^{-1}\left[L_1\psi_1 - \nu^{-1}\frac{\partial}{\partial t}(L\psi_1) + 2\frac{\partial^4\psi_1}{\partial x^2\partial y^2} - \nu^{-1}N\psi_1\right]\\
&\cdots \quad \cdots \quad \cdots \quad \cdots\\
&\cdots \quad \cdots \quad \cdots \quad \cdots
\end{aligned} \tag{4.25}$$

The Adomian's special polynomials A_0, A_1, etc., involved in (4.25) are given by (4.19). The relations (4.14), (4.19), (4.24), and (4.25) together constitute the complete solution of the problem with the concept $\lambda = 1$.

Case III

Operator $L = \frac{\partial^4}{\partial x^4} + \frac{\partial^4}{\partial y^4}$: With the help of this operator the equation (4.7) takes the form

$$L\psi = \nu^{-1}\frac{\partial}{\partial t}(L\psi) - 2\frac{\partial^4\psi}{\partial x^2\partial y^2} + \nu^{-1}N\psi \tag{4.26}$$

If L^{-1} is the multidimensional inverse operator of L, then using this operator on both sides of (4.26) we obtain

$$\psi(x, y, t) = \bar{\psi}_0(x, y, t) + L^{-1}\left[\nu^{-1}\frac{\partial}{\partial t}(L\psi) - 2\frac{\partial^4 \psi}{\partial x^2 \partial y^2} + \nu^{-1}N\psi\right] \tag{4.27}$$

where $\psi_0(x, y, t)$ is the solution of Laplace's equation $L\psi = 0$.

We now proceed to find out the expression for L^{-1}. To do that, we consider the equation

$$\frac{\partial^4 \overline{\psi}}{\partial x^4} + \frac{\partial^4 \overline{\psi}}{\partial y^4} = 0 \tag{4.28}$$

which can be written as

$$L_1\overline{\psi} + L_2\overline{\psi} = 0 \tag{4.29}$$

Solving for $L_1\overline{\psi}$ and $L_2\overline{\psi}$, we have from (4.29)

$$L_1\overline{\psi} = -L_2\overline{\psi} \tag{4.30}$$

and

$$L_2\overline{\psi} = -L_1\overline{\psi} \tag{4.31}$$

Performing operations of the operators L_1^{-1} and L_2^{-1} on (4.30) and (4.31), respectively, we have

$$\overline{\psi}(x, y, t) = \overline{\psi}_1(x, y, t) - L_1^{-1}(L_2\overline{\psi}) \tag{4.32}$$

and

$$\overline{\psi}(x, y, t) = \overline{\psi}_2(x, y, t) - L_2^{-1}(L_1\overline{\psi}) \tag{4.33}$$

where $\overline{\psi}_1(x, y, t)$ and $\overline{\psi}_2(x, y, t)$ are the solutions of the equations $L_1\overline{\psi} = 0$ and $L_2\overline{\psi} = 0$. Adding (4.32) and (4.33) and then dividing by 2 we get

$$\overline{\psi}(x, y, t) = \overline{\psi}_0(x, y, t) - \frac{1}{2}\left[L_1^{-1}(L_2\overline{\psi}) + L_2^{-1}(L_1\overline{\psi})\right] \tag{4.34}$$

where $\overline{\psi}_0(x, y, t) = \frac{1}{2}(\overline{\psi}_1 + \overline{\psi}_2)$. Then we decompose $\overline{\psi}(x, y, t)$ into

$$\overline{\psi}(x, y, t) = \sum_{n=0}^{\infty}\lambda^n\overline{\psi}_n \tag{4.35}$$

and write the parameterized form of (4.34) as

$$\overline{\psi}(x, y, t) = \overline{\psi}_0(x, y, t) - \frac{\lambda}{2}\left[L_1^{-1}(L_2\overline{\psi}) + L_2^{-1}(L_1\overline{\psi})\right] \tag{4.36}$$

If we put (4.35) into (4.36) and compare the like-power terms of λ, we get

$$\begin{aligned}
\overline{\psi}_1(x, y, t) &= -\frac{1}{2}\left[L_1^{-1}L_2 + L_2^{-1}L_1\right]\overline{\psi}_0\\
\overline{\psi}_2(x, y, t) &= -\frac{1}{2}\left[L_1^{-1}L_2 + L_2^{-1}L_1\right]\overline{\psi}_1\\
\overline{\psi}_3(x, y, t) &= -\frac{1}{2}\left[L_1^{-1}L_2 + L_2^{-1}L_1\right]\overline{\psi}_2\\
\cdots\cdots &\quad \cdots\\
\cdots\cdots &\quad \cdots\\
\overline{\psi}_{n+1}(x, y, t) &= -\frac{1}{2}\left[L_1^{-1}L_2 + L_2^{-1}L_1\right]\overline{\psi}_n\\
\cdots\cdots &\quad \cdots\\
\cdots\cdots &\quad \cdots
\end{aligned} \tag{4.37}$$

which can be written as

$$\begin{aligned}
\overline{\psi}_1(x, y, t) &= \left(-\frac{1}{2}\right) S\overline{\psi}_0\\
\overline{\psi}_2(x, y, t) &= \left(-\frac{1}{2}\right)^2 S^2\overline{\psi}_0\\
\overline{\psi}_3(x, y, t) &= \left(-\frac{1}{2}\right)^3 S^3\overline{\psi}_0\\
\cdots\cdots &\quad \cdots\\
\cdots\cdots &\quad \cdots\\
\overline{\psi}_n(x, y, t) &= \left(-\frac{1}{2}\right)^n S^n\overline{\psi}_0\\
\cdots\cdots &\quad \cdots\\
\cdots\cdots &\quad \cdots
\end{aligned} \tag{4.38}$$

where

$$S = \left[L_1^{-1}L_2 + L_2^{-1}L_1\right]$$

and

$$\overline{\psi}_n(x, y, t) = \sum_{n=0}^{\infty}\overline{\psi}_n = \sum_{n=0}^{\infty}(-1)^n\frac{1}{2^n}\left[\overline{L}_1L_2 + L_2^{-1}L_1\right]^n\overline{\psi}_0 \tag{4.39}$$

Hence, the inverse $L^{-1} = (L_1 + L_2)^{-1}$ is identified as

$$L^{-1} = \sum_{n=0}^{\infty}(-1)^n\frac{1}{2^n}\left[L_1^{-1}L_2 + L_2^{-1}L_1\right]^n \tag{4.40}$$

We now return to the equation (4.27), whose parameterized form is

$$\psi(x, y, t) = \psi_0(x, y, t) + \lambda L^{-1}\left[\nu^{-1}\frac{\partial}{\partial t}(L\psi) - 2\frac{\partial^4\psi}{\partial x^2\partial y^2} + \nu^{-1}N\psi\right] \tag{4.41}$$

Substituting (4.14) and (4.15) in (4.41) and then equating like-power terms of λ, we have

$$\begin{aligned}
\psi_1(x, y, t) &= L^{-1}\left[\nu^{-1}\frac{\partial}{\partial t}(L\psi_0) - 2\frac{\partial^4\psi_0}{\partial x^2\partial y^2} + \nu^{-1}A_0\right] \\
\psi_2(x, y, t) &= L^{-1}\left[\nu^{-1}\frac{\partial}{\partial t}(L\psi_1) - 2\frac{\partial^4\psi_1}{\partial x^2\partial y^2} + \nu^{-1}A_1\right] \\
&\cdots \quad \cdots \quad \cdots \quad \cdots \\
&\cdots \quad \cdots \quad \cdots \quad \cdots \\
\psi_{n+1}(x, y, t) &= L^{-1}\left[\nu^{-1}\frac{\partial}{\partial t}(L\psi_n) - 2\frac{\partial^4\psi_n}{\partial x^2\partial y^2} + \nu^{-1}A_n\right] \\
&\cdots \quad \cdots \quad \cdots \quad \cdots \\
&\cdots \quad \cdots \quad \cdots \quad \cdots
\end{aligned} \tag{4.42}$$

where the Adomian's special polynomials A_0, A_1, etc., are given by (4.19).

If $\overline{\psi}_0$ is once obtained, then ψ_1 can be obtained in term of ψ_0, ψ_2 can be obtained in term of ψ_1, and so on. So all the components are computable and

$$\psi(x, y, t) = \sum_{n=0}^{\infty}\psi_n.$$

4.2.2 Solution by Modified Decomposition

The x-Partial Solution

Consider the equation (4.9) with $L_1 = \frac{\partial^4}{\partial x^4}$ and $L_2 = \frac{\partial^4}{\partial y^4}$. Solving for $L_1\psi$ and then operating with the inverse operator L_1^{-1} defined by (4.11), we have the expression (4.12) for ψ, and it is called the regular x-partial solution of (4.9). Then we follow the modified decomposition procedure [2] and write for the purpose

$$\psi(x, y, t) = \sum_{m=0}^{\infty}\sum_{n=0}^{\infty}\sum_{l=0}^{\infty} C_{m,n,l}\; x^m\; y^n\; t^l \tag{4.43}$$

and

$$N\psi = \sum_{m=0}^{\infty}\sum_{n=0}^{\infty}\sum_{l=0}^{\infty} B_{m,n,l}\; x^m\; y^n\; t^l \tag{4.44}$$

If we expand the right side of (4.43) in the ascending power of x, we have

$$\begin{aligned}
\psi(x, y, t) &= \sum_{m=0}^{\infty}\left[\sum_{n=0}^{\infty}\sum_{l=0}^{\infty} C_{m,n,l}\; y^n\; t^l\right]x^m = \left[\sum_{n=0}^{\infty}\sum_{l=0}^{\infty} C_{0,n,l}\; y^n\; t^l\right]x^0 \\
&+ \left[\sum_{n=0}^{\infty}\sum_{l=0}^{\infty} C_{1,n,l}\; y^n\; t^l\right]x + \left[\sum_{n=0}^{\infty}\sum_{l=0}^{\infty} C_{2,n,l}\; y^n\; t^l\right]x^2 + \cdots
\end{aligned} \tag{4.45}$$

Let the quantities in the brackets be defined by

$$
\begin{aligned}
a_0(y,t) &= \sum_{n=0}^{\infty}\sum_{l=0}^{\infty} C_{0,n,l}\; y^n\; t^l \\
a_1(y,t) &= \sum_{n=0}^{\infty}\sum_{l=0}^{\infty} C_{1,n,l}\; y^n\; t^l \\
&\cdots\cdots \quad \cdots \\
&\cdots\cdots \quad \cdots \\
a_m(y,t) &= \sum_{n=0}^{\infty}\sum_{l=0}^{\infty} C_{m,n,l}\; y^n\; t^l \\
&\cdots\cdots \quad \cdots
\end{aligned}
\tag{4.46}
$$

Then the relation (4.45), by means of (4.46), becomes

$$
\begin{aligned}
\psi(x,y,t) &= a_0(y,t)x^0 + a_1(y,t)x + a_2(y,t)x^2 + a_3(y,t)x^3 \\
&\quad + \cdots + a_m(y,t)x^m + \cdots = \sum_{m=0}^{\infty} a_m(y,t)x^m
\end{aligned}
\tag{4.47}
$$

Similarly, the relation (4.44) can be written as

$$
N\psi = \sum_{m=0}^{\infty} A_{m(y,t)}x^m \tag{4.48}
$$

where

$$
\begin{aligned}
A_0(y,t) &= \sum_{n=0}^{\infty}\sum_{l=0}^{\infty} B_{0,n,l}\; y^n\; t^l \\
A_1(y,t) &= \sum_{n=0}^{\infty}\sum_{l=0}^{\infty} B_{1,n,l}\; y^n\; t^l \\
&\cdots\cdots \quad \cdots \\
&\cdots\cdots \quad \cdots \\
A_m(y,t) &= \sum_{n=0}^{\infty}\sum_{l=0}^{\infty} B_{m,n,l}\; y^n\; t^l \\
&\cdots\cdots \quad \cdots
\end{aligned}
\tag{4.49}
$$

are the Adomian's special polynomials.

Substituting (4.47) and (4.48) in (4.12), we have

$$\sum_{m=0}^{\infty} a_m(y,t)x^m = \xi_0(y,t) + \xi_1(y,t)x + \xi_2(y,t)\frac{x^2}{2} + \xi_3(y,t)\frac{x^3}{6}$$
$$- L_1^{-1}\left[\frac{\partial^2 4}{\partial y^4}\sum_{m=0}^{\infty} a_m(y,t)x^m\right]$$
$$+ \nu^{-1}L_1^{-1}\left[\frac{\partial}{\partial t}\left\{\frac{\partial^2}{\partial x^2}\sum_{m=0}^{\infty} a_m(y,t)x^m + \frac{\partial^2}{\partial y^2}\sum_{m=0}^{\infty} a_m(y,t)x^m\right\}\right]$$
$$- 2L_1^{-1}\left[\frac{\partial^4}{\partial x^2 \partial y^2}\sum_{m=0}^{\infty} a_m(y,t)x^m\right] + \nu^{-1}L_1^{-1}\left[\sum_{m=0}^{\infty} A_m(y,t)x^m\right]$$
$$= \xi_0(y,t) + \xi_1(y,t)x + \xi_2(y,t)\frac{x^2}{2} + \xi_3(y,t)\frac{x^3}{6}$$
$$- L_1^{-1}\left[\sum_{m=0}^{\infty}\frac{\partial^4}{\partial y^4} a_m(y,t)x^m\right]$$
$$+ \nu^{-1}L_1^{-1}\left[\frac{\partial}{\partial t}\left\{\sum_{m=0}^{\infty}(m+1)(m+2)a_{m+2}(y,t)x^m\right.\right.$$
$$\left.\left.+ \sum_{m=0}^{\infty}\frac{\partial^2}{\partial y^2} a_m(y,t)x^m\right\}\right]$$
$$- 2L_1^{-1}\left[\sum_{m=0}^{\infty}(m+1)(m+2)\frac{\partial^2}{\partial y^2} a_{m+2}(y,t)x^m\right]$$
$$+ \nu^{-1}L_1^{-1}\left[\sum_{m=0}^{\infty} A_m(y,t)x^m\right] \tag{4.50}$$

Performing the integration on the right side of (4.50), we get

$$\sum_{m=0}^{\infty} a_m(y,t)x^m = \xi_0(y,t) + \xi_1(y,t)x + \xi_2(y,t)\frac{x^2}{2} + \xi_3(y,t)\frac{x^3}{6}$$
$$- \sum_{m=0}^{\infty}\frac{\frac{\partial^4}{\partial y^4} a_m(y,t)}{(m+1)(m+2)(m+3)(m+4)}x^{m+4}$$
$$+ \nu^{-1}\frac{\partial}{\partial t}\sum_{m=0}^{\infty}\left[\frac{(m+1)(m+2)a_{m+2}(y,t) + \frac{\partial^2}{\partial y^2} a_m(y,t)}{(m+1)(m+2)(m+3)(m+4)}\right]x^{m+4}$$

$$- 2\sum_{m=0}^{\infty} \frac{(m+1)(m+2)\frac{\partial^2}{\partial y^2}a_{m+2}(y,t)}{(m+1)(m+2)(m+3)(m+4)} x^{m+4}$$

$$+ \nu^{-1}\sum_{m=0}^{\infty} \frac{A_m(y,t)}{(m+1)(m+2)(m+3)(m+4)} x^{m+4}$$

We now replace m by $m - 4$ on the right side of the above relation and get

$$\sum_{m=0}^{\infty} a_m(y,t)x^m = \xi_0(y,t) + \xi_1(y,t)x + \xi_2(y,t)\frac{x^2}{2} + \xi_3(y,t)\frac{x^3}{6}$$

$$- \sum_{m=4}^{\infty} \frac{\frac{\partial^4}{\partial y^4}a_{m-4}(y,t)}{m(m-1)(m-2)(m-3)} x^m$$

$$+ \nu^{-1}\sum_{m=4}^{\infty} \frac{(m-2)(m-3)\,(\partial/\partial t)\,a_{m-2}(y,t) + \left(\frac{\partial^3}{\partial t\partial y^2}\right)a_{m-4}(y,t)}{m(m-1)(m-2)(m-3)} x^m$$

$$- 2\sum_{m=4}^{\infty} \frac{(m-2)(m-3)\frac{\partial^2}{\partial y^2}a_{m-2}(y,t)}{m(m-1)(m-2)(m-3)} x^m$$

$$+ \nu^{-1}\sum_{m=4}^{\infty} \frac{A_{m-4}(y,t)}{m(m-1)(m-2)(m-3)} x^m$$

Equating the coefficients of like-power terms of x, we have

$$\begin{aligned} \xi_0(y,t) &= a_0(y,t) \\ \xi_1(y,t) &= a_1(y,t) \\ \frac{1}{2}\xi_2(y,t) &= a_2(y,t) \\ \frac{1}{6}\xi_3(y,t) &= a_3(y,t) \end{aligned} \tag{4.51}$$

and for $m \geq 4$ the recurrence relation for the coefficients

$$\begin{aligned} &m(m-1)(m-2)(m-3)a_m(y,t) \\ &\quad = -\frac{\partial^4}{\partial y^4}a_{m-4}(y,t) + \nu^{-1}\left[(m-2)(m-3)\frac{\partial}{\partial t}a_{m-2}(y,t) + \frac{\partial^3}{\partial t\partial y^2}a_{m-4}(y,t)\right] \\ &\qquad -2(m-2)(m-3)\frac{\partial^2}{\partial y^2}a_{m-2}(y,t) + \nu^{-1}A_{m-4}(y,t) \end{aligned} \tag{4.52}$$

We next proceed to find out the expressions for the Adomian's polynomials. To do that, we substitute (4.47) and (4.48) in (4.8) to get

$$\sum_{m=0}^{\infty} A_m(y,t)x^m = \left[\frac{\partial}{\partial y}\sum_{m=0}^{\infty} a_m(y,t)x^m\right]\left[\frac{\partial}{\partial x}\left\{L\sum_{m=0}^{\infty} a_m(y,t)x^m\right\}\right] - \left[\frac{\partial}{\partial x}\sum_{m=0}^{\infty} a_m(y,t)x^m\right]\left[\frac{\partial}{\partial y}\left\{L\sum_{m=0}^{\infty} a_m(y,t)x^m\right\}\right] \tag{4.53}$$

Considering the first product on the right side of (4.53), we have

$$\begin{aligned}
&\left[\frac{\partial}{\partial y}\sum_{m=0}^{\infty} a_m(y,t)x^m\right]\left[\frac{\partial}{\partial x}\left\{L\sum_{m=0}^{\infty} a_m(y,t)x^m\right\}\right] \\
&= \left[\sum_{m=0}^{\infty}\frac{\partial}{\partial y}a_m(y,t)x^m\right]\left[\frac{\partial}{\partial x}\left\{\frac{\partial^2}{\partial x}\sum_{m=0}^{\infty} a_m(y,t)x^m + \frac{\partial^2}{\partial y^2}\sum_{m=0}^{\infty} a_m(y,t)x^m\right\}\right] \\
&= \left[\sum_{m=0}^{\infty}\frac{\partial}{\partial y}a_m(y,t)x^m\right]\left[\sum_{m=0}^{\infty}(m+1)(m+2)(m+3)a_{m+3}(y,t)x^m\right. \\
&\quad \left. +\sum_{m=0}^{\infty}(m+1)\frac{\partial^2}{\partial y^2}a_{m+1}(y,t)x^m\right] \\
&= \left[\sum_{m=0}^{\infty}\frac{\partial}{\partial y}a_m(y,t)x^m\right]\left[\sum_{m=0}^{\infty}(m+1)(m+2)(m+3)a_{m+3}(y,t)x^m\right] \\
&\quad + \left[\sum_{m=0}^{\infty}\frac{\partial}{\partial y}a_m(y,t)x^m\right]\left[\sum_{m=0}^{\infty}(m+1)\frac{\partial^2}{\partial y^2}a_{m+1}(y,t)x^m\right]
\end{aligned}$$

Carrying out the Cauchy product [2] of infinite series, we have

$$\begin{aligned}
&\left[\frac{\partial}{\partial y}\sum_{m=0}^{\infty} a_m(y,t)\right]\left[\frac{\partial}{\partial x}\left\{L\sum_{m=0}^{\infty} a_m(y,t)x^m\right\}\right] \\
&= \sum_{m=0}^{\infty}\left[\sum_{n=0}^{m}\frac{\partial}{\partial y}a_{m-n}(y,t)(n+1)(n+2)(n+3)a_{n+3}(y,t)\right]x^m \\
&\quad +\sum_{m=0}^{\infty}\left[\sum_{n=0}^{m}\frac{\partial}{\partial y}a_{m-n}(y,t)(n+1)\frac{\partial^2}{\partial y^2}a_{n+1}(y,t)\right]x^m \\
&= \sum_{m=0}^{\infty}\left[\sum_{n=0}^{m}\frac{\partial}{\partial y}a_{m-n}(y,t)(n+1)\right. \\
&\quad \left.\times\left\{(n+2)(n+3)a_{n+3}(y,t) + \frac{\partial^2}{\partial y^2}a_{n+1}(y,t)\right\}\right]x^m
\end{aligned} \tag{4.54}$$

Considering the second product on the right side of (4.53), we have

$$\left[\frac{\partial}{\partial x}\sum_{m=0}^{\infty}a_m(y,t)x^m\right]\left[\frac{\partial}{\partial y}\left\{L\sum_{m=0}^{\infty}a_m(y,t)x^m\right\}\right]$$

$$=\left[\sum_{m=0}^{\infty}(m+1)a_{m+1}(y,t)x^m\right]$$

$$\times\left[\frac{\partial}{\partial y}\left\{\frac{\partial^2}{\partial x^2}\sum_{m=0}^{\infty}a_m(y,t)x^m+\frac{\partial^2}{\partial y^2}\sum_{m=0}^{\infty}a_m(y,t)x^m\right\}\right]$$

$$=\left[\sum_{m=0}^{\infty}(m+1)a_{m+1}(y,t)x^m\right]$$

$$\times\left[\sum_{m=0}^{\infty}(m+1)(m+2)\frac{\partial}{\partial y}a_{m+2}(y,t)x^m+\sum_{m=0}^{\infty}\frac{\partial^3}{\partial y^3}a_m(y,t)x^m\right]$$

$$=\left[\sum_{m=0}^{\infty}(m+1)a_{m+1}(y,t)x^m\right]\left[\sum_{m=0}^{\infty}(m+1)(m+2)\frac{\partial}{\partial y}a_{m+2}(y,t)x^m\right]$$

$$+\left[\sum_{m=0}^{\infty}(m+1)a_{m+1}(y,t)x^m\right]\left[\sum_{m=0}^{\infty}\frac{\partial^3}{\partial y^3}a_m(y,t)x^m\right]$$

Using again the Cauchy product of infinite series, we have

$$\left[\frac{\partial}{\partial x}\sum_{m=0}^{\infty}a_m(y,t)x^m\right]\left[\frac{\partial}{\partial y}\left\{L\sum_{m=0}^{\infty}a_m(y,t)x^m\right\}\right]$$

$$=\sum_{m=0}^{\infty}\left[\sum_{n=0}^{m}(m-n+1)a_{m-n+1}(y,t)(n+1)(n+2)\frac{\partial}{\partial y}a_{n+2}(y,t)\right]x^m$$

$$+\sum_{m=0}^{\infty}\left[\sum_{n=0}^{m}(m-n+1)a_{m-n+1}(y,t)\frac{\partial^3}{\partial y^3}a_n(y,t)\right]x^m$$

$$=\sum_{m=0}^{\infty}\left[\sum_{n=0}^{m}(m-n+1)a_{m-n+1}(y,t)\right.$$

$$\left.\times\left\{(n+1)(n+2)\frac{\partial}{\partial y}a_{n+2}(y,t)+\frac{\partial^3}{\partial y^3}a_n(y,t)\right\}\right]x^m \tag{4.55}$$

Substituting (4.54) and (4.55) in (4.53), we get

$$\sum_{m=0}^{\infty} A_m(y,t)x^m = \sum_{m=0}^{\infty}\left[\sum_{n=0}^{m}\frac{\partial}{\partial y}a_{m-n}(y,t)(n+1)\left\{(n+2)(n+3)a_{n+3}(y,t) + \frac{\partial^2}{\partial y^2}a_{n+1}(y,t)\right\}\right]x^m - \sum_{m=0}^{\infty}\left[\sum_{n=0}^{m}(m-n+1)a_{m-n+1}(y,t)\left\{(n+1)(n+2)\frac{\partial}{\partial y}a_{n+2}(y,t) + \frac{\partial^3}{\partial y^3}a_n(y,t)\right\}\right]x^m$$

Comparing both sides of the above relation, we have

$$A_m(y,t) = \sum_{n=0}^{m}\frac{\partial}{\partial y}a_{m-n}(y,t)(n+1)\left[(n+2)(n+3)a_{n+3}(y,t) + \frac{\partial^2}{\partial y^2}a_{n+1}(y,t)\right] - \sum_{n=0}^{m}(m-n+1)a_{m-n+1}(y,t) \times \left[(n+1)(n+2)\frac{\partial}{\partial y}a_{n+2}(y,t) + \frac{\partial^3}{\partial y^3}a_n(y,t)\right] \tag{4.56}$$

which gives the set of expressions for the Adomian's special polynomials for different values of m. The final solution is now given by $\psi(x,y,t) = \sum_{m=0}^{\infty} a_m(y,t)x^m$.

The y-Partial Solution

For the y-partial solution, we solve the equation (4.9) for $L_2\psi$ to get

$$L_2\psi = -L_1\psi + \nu^{-1}\frac{\partial}{\partial t}(L\psi) - 2\frac{\partial^4\psi}{\partial x^2\partial y^2} + \nu^{-1}N\psi$$

Then we operate with the inverse operator L_2^{-1} defined by (4.21) on both sides of the above equation and get

$$\psi(x,y,t) = \eta_0(x,t) + \eta_1(x,t)y + \eta_2(x,t)\frac{y^2}{2} + \eta_3(x,t)\frac{y^3}{6} + L_2^{-1}\left[-L_1\psi + \nu^{-1}\frac{\partial}{\partial t}(L\psi) - 2\frac{\partial^4\psi}{\partial x^2\partial y^2} + \nu^{-1}N\psi\right] \tag{4.57}$$

We now proceed to apply the modified decomposition procedure and write for the purpose

$$\psi(x,y,t) = \sum_{m=0}^{\infty} a_m(x,t)y^m \tag{4.58}$$

and

$$N\psi = \sum_{m=0}^{\infty} A_m(x,t)y^m \tag{4.59}$$

Substituting (4.58) and (4.59) in (4.57), we have

$$\begin{aligned}
\sum_{m=0}^{\infty} a_m(x,t)y^m &= \eta_0(x,t) + \eta_1(x,t)y + \eta_2(x,t)\frac{y^2}{2} + \eta_3(x,t)\frac{y^3}{6} \\
&\quad - L_2^{-1}\left[\frac{\partial^4}{\partial x^4}\sum_{m=0}^{\infty} a_m(x,t)y^m\right] \\
&\quad + \nu^{-1}L_2^{-1}\left[\frac{\partial}{\partial t}\left\{\frac{\partial^2}{\partial x^2}\sum_{m=0}^{\infty} a_m(x,t)y^m + \frac{\partial^2}{\partial y^2}\sum_{m=0}^{\infty} a_m(x,t)y^m\right\}\right] \\
&\quad - 2L_2^{-1}\left[\frac{\partial^4}{\partial x^2\partial y^2}\sum_{m=0}^{\infty} a_m(x,t)y^m\right] + \nu^{-1}L_2^{-1}\left[\sum_{m=0}^{\infty} A_m(x,t)y^m\right] \\
&= \eta_0(x,t) + \eta_1(x,t)y + \eta_2(x,t)\frac{y^2}{2} + \eta_3(x,t)\frac{y^3}{6} \\
&\quad - L_2^{-1}\left[\sum_{m=0}^{\infty}\frac{\partial^4}{\partial x^4} a_m(x,t)y^m\right] + \nu^{-1}L_2^{-1}\left[\sum_{m=0}^{\infty}\frac{\partial^3}{\partial t\partial x^2} a_m(x,t)y^m\right. \\
&\quad \left. + \sum_{m=0}^{\infty}(m+1)(m+2)\frac{\partial}{\partial t}a_{m+2}(x,t)y^m\right] \\
&\quad - 2L_2^{-1}\left[\sum_{m=0}^{\infty}(m+1)(m+2)\frac{\partial^2}{\partial x^2}a_{m+2}(x,t)y^m\right] \\
&\quad + \nu^{-1}L_2^{-1}\left[\sum_{m=0}^{\infty} A_m(x,t)y^m\right]
\end{aligned} \tag{4.60}$$

Performing the integrations on the right side of (4.60), we have

$$\begin{aligned}
\sum_{m=0}^{\infty} a_m(x,t)y^m &= \eta_0(x,t) + \eta_1(x,t)y + \eta_2(x,t)\frac{y^2}{2} + \eta_3(x,t)\frac{y^3}{6} \\
&\quad - \sum_{m=0}^{\infty}\frac{(\partial^4/\partial x^4)a_m(x,t)}{(m+1)(m+2)(m+3)(m+4)}y^{m+4} \\
&\quad + \nu^{-1}\left[\sum_{m=0}^{\infty}\frac{\frac{\partial^3}{\partial t\partial x^2}a_m(x,t)y^{m+4}}{(m+1)(m+2)(m+3)(m+4)}\right. \\
&\quad \left. + \sum_{m=0}^{\infty}\frac{(m+1)(m+2)(\partial/\partial t)a_{m+2}(x,t)}{(m+1)(m+2)(m+3)(m+4)}y^{m+4}\right]
\end{aligned}$$

$$- 2\sum_{m=0}^{\infty}\frac{(m+1)(m+2)(\partial^2/\partial x^2)a_{m+2}(x,t)}{(m+1)(m+2)(m+3)(m+4)}y^{m+4}$$

$$+ \nu^{-1}\sum_{m=0}^{\infty}\frac{A_m(x,t)y^{m+4}}{(m+1)(m+2)(m+3)(m+4)} \tag{4.61}$$

Replacing m by $m-4$ on the right side of the above relation, we get

$$\sum_{m=0}^{\infty}a_m(x,t)y^m = \eta_0(x,t) + \eta_1(x,t)y + \eta_2(x,t)\frac{y^2}{2} + \eta_3(x,t)\frac{y^3}{6}$$

$$-\sum_{m=4}^{\infty}\frac{(\partial^4/\partial x^4)a_{m-4}(x,t)}{m(m-1)(m-2)(m-3)}y^m$$

$$+\nu^{-1}\sum_{m=4}^{\infty}\frac{(\partial^3/\partial t\partial x^2)a_{m-4}(x,t)+(m-2)(m-3)(\partial/\partial t)a_{m-2}(x,t)}{m(m-1)(m-2)(m-3)}y^m$$

$$-2\sum_{m=4}^{\infty}\frac{(m-2)(m-3)(\partial^2/\partial x^2)a_{m-2}(x,t)}{m(m-1)(m-2)(m-3)}y^m$$

$$+\nu^{-1}\sum_{m=4}^{\infty}\frac{A_{m-4}(x,t)y^m}{m(m-1)(m-2)(m-3)} \tag{4.62}$$

By equating the coefficients of like-power terms of y, we have

$$\begin{aligned}\eta_0(x,t) &= a_0(x,t)\\ \eta_1(x,t) &= a_1(x,t)\\ \frac{1}{2}\eta_2(x,t) &= a_2(x,t)\\ \frac{1}{6}\eta_3(x,t) &= a_3(x,t)\end{aligned} \tag{4.63}$$

and for $n \geq 4$, the recurrence relation for the coefficients

$$m(m-1)(m-2)(m-3)a_m(x,t)$$

$$= -\frac{\partial^4}{\partial x^4}a_{m-4}(x,t) + \nu^{-1}\left[(m-2)(m-3)\frac{\partial}{\partial t}a_{m-2}(y,t) + \frac{\partial^3}{\partial t\partial y^2}a_{m-4}(y,t)\right]$$

$$- 2(m-2)(m-3)\frac{\partial^2}{\partial x^2}a_{m-2}(x,t) + \nu^{-1}A_{m-4}(x,t) \tag{4.64}$$

from which we can have the coefficients for different values of $m \geq 4$.

Our next step is to find out the Adomian's special polynomials. To do that, we consider the equation (4.8), and putting (4.58) and (4.59) in it, we have

$$\begin{aligned}\sum_{m=0}^{\infty} A_m(x,t)y^m &= \left[\frac{\partial}{\partial y}\sum_{m=0}^{\infty} a_m(x,t)y^m\right]\\ &\quad\times\left[\frac{\partial}{\partial x}\left\{\frac{\partial^2}{\partial x^2}\sum_{m=0}^{\infty} a_m(x,t)y^m+\frac{\partial^2}{\partial y^2}\sum_{m=0}^{\infty} a_m(x,t)y^m\right\}\right]\\ &\quad-\left[\frac{\partial}{\partial x}\sum_{m=0}^{\infty} a_m(x,t)y^m\right]\\ &\quad\times\left[\frac{\partial}{\partial y}\left\{\frac{\partial^2}{\partial x^2}\sum_{m=0}^{\infty} a_m(x,t)y^m+\frac{\partial^2}{\partial y^2}\sum_{m=0}^{\infty} a_m(x,t)y^m\right\}\right]\\ &=\left[\sum_{m=0}^{\infty}(m+1)a_{m+1}(x,t)y^m\right]\left[\sum_{m=0}^{\infty}(\partial^3/\partial x^3)a_m(x,t)y^m\right.\\ &\quad\left.+\sum_{m=0}^{\infty}(m+1)(m+2)(\partial/\partial x)a_{m+2}(x,t)y^m\right]\\ &\quad-\left[\sum_{m=0}^{\infty}(\partial/\partial x)a_m(x,t)y^m\right]\left[\sum_{m=0}^{\infty}(m+1)(\partial^2/\partial x^2)a_{m+1}(x,t)y^m\right.\\ &\quad\left.+\sum_{m=0}^{\infty}(m+1)(m+2)(m+3)a_{m+3}(x,t)y^m\right]\end{aligned}\tag{4.65}$$

Considering the first product of the quantities in the brackets on the right side of (4.65) and then using the Cauchy product of infinite series, we get

$$\begin{aligned}&\left[\sum_{m=0}^{\infty}(m+1)a_{m+1}(x,t)y^m\right]\left[\sum_{m=0}^{\infty}(\partial^3/\partial x^3)a_m(x,t)y^m\right.\\ &\qquad\left.+\sum_{m=0}^{\infty}(m+1)(m+2)(\partial/\partial x)a_{m+2}(x,t)y^m\right]\\ &\quad=\left[\sum_{m=0}^{\infty}(m+1)a_{m+1}(x,t)y^m\right]\left[\sum_{m=0}^{\infty}(\partial^3/\partial x^3)a_m(x,t)y^m\right]\\ &\qquad+\left[\sum_{m=0}^{\infty}(m+1)a_{m+1}(x,t)y^m\right]\left[\sum_{m=0}^{\infty}(m+1)(m+2)(\partial/\partial x)a_{m+2}(x,t)y^m\right]\end{aligned}$$

$$= \sum_{m=0}^{\infty} \left[\sum_{n=0}^{m} (m-n+1) a_{m-n+1}(x,t)(\partial^3/\partial x^3) a_n(x,t) \right] y^m$$
$$+ \sum_{m=0}^{\infty} \left[\sum_{n=0}^{m} (m-n+1) a_{m-n+1}(x,t)(n+1)(n+2)(\partial/\partial x) a_{n+2}(x,t) \right] y^m$$
$$= \sum_{m=0}^{\infty} \left[\sum_{n=0}^{m} (m-n+1) a_{m-n+1}(x,t)(\partial^3/\partial x^3) a_n(x,t) \right.$$
$$\left. + \sum_{n=0}^{m} (m-n+1) a_{m-n+1}(x,t)(n+1)(n+2)(\partial/\partial x) a_{n+2}(x,t) \right] y^m$$
$$= \sum_{m=0}^{\infty} \left[\sum_{n=0}^{m} (m-n+1) a_{m-n+1}(x,t) \{(\partial^3/\partial x^3) a_n(x,t) \right.$$
$$\left. + (n+1)(n+2)(\partial/\partial x) a_{n+2}(x,t)\} \right] y^m \tag{4.66}$$

Considering again the second bracket product on the right side of (4.65) and then carrying out the Cauchy product [2] of infinite series, we have

$$\left[\sum_{m=0}^{\infty} (\partial/\partial x) a_m(x,t) y^m \right] \left[\sum_{m=0}^{\infty} (m+1)(\partial^2/\partial x^2) a_{m+1}(x,t) y^m \right.$$
$$\left. + \sum_{m=0}^{\infty} (m+1)(m+2)(m+3) a_{m+3}(x,t) y^m \right]$$
$$= \left[\sum_{m=0}^{\infty} (\partial/\partial x) a_m(x,t) y^m \right] \left[\sum_{m=0}^{\infty} (m+1)(\partial^2/\partial x^2) a_{m+1}(x,t) y^m \right]$$
$$+ \left[\sum_{m=0}^{\infty} (\partial/\partial x) a_m(x,t) y^m \right] \left[\sum_{m=0}^{\infty} (m+1)(m+2)(m+3) a_{m+3}(x,t) y^m \right]$$
$$= \sum_{m=0}^{\infty} \left[\sum_{n=0}^{m} (\partial/\partial x) a_{m-n}(x,t)(n+1)(\partial^2/\partial x^2) a_{n+1}(x,t) \right] y^m$$
$$+ \sum_{m=0}^{\infty} \left[\sum_{n=0}^{m} (\partial/\partial x) a_{m-n}(x,t)(n+1)(n+2)(n+3) a_{n+3}(x,t) \right] y^m$$
$$= \sum_{m=0}^{\infty} \left[\sum_{n=0}^{m} (\partial/\partial x) a_{m-n}(x,t)(n+1) \{(\partial^2/\partial x^2) a_{n+1}(x,t) \right.$$
$$\left. + (n+2)(n+3) a_{n+3}(x,t)\} \right] y^m \tag{4.67}$$

Now using (4.66) and (4.67) in (4.65) and then equating the coefficients of y^m from both sides of it, we get the expression for $A_m(x, t)$ as

$$\begin{aligned} A_m(x,t) = {} & \sum_{n=0}^{m}(m-n+1)a_{m-n+1}(x,t)[(\partial^3/\partial x^3)a_n(x,t) \\ & + (n+1)(n+2)(\partial/\partial x)a_{n+2}(x,t)] - \sum_{n=0}^{m}(\partial/\partial x)a_{m-n}(x,t)(n+1) \\ & \times \{(\partial^2/\partial x^2)a_{n+1}(x,t) + (n+2)(n+3)a_{n+2}(x,t)\} \end{aligned} \quad (4.68)$$

from which we can have the Adomian's polynomials for different values of m, and the final solution is now given by $\psi(x, y, t) = \sum_{m=0}^{\infty} a_m(x, t)y^m$.

In the next sections, some physical problems that Haldar [3] has studied by means of modified decomposition methods will be considered in order to make the theories clear. In these sections, some fluid flow problems on the basis of these theories will be discussed. It is shown that the solution in the case of modified decomposition can be obtained more easily than that of the regular decomposition technique. From this point of view, it can be inferred that the modified decomposition procedure is more powerful than the regular decomposition method.

4.3 Steady Laminar Flow of Viscous Fluid through a Tube of an Elliptic Cross Section

4.3.1 Equation of Motion

Consider the steady laminar flow of viscous incompressible fluid through a tube of an elliptic cross section. The z-axis is assumed along the axis of the tube, and the flow of fluid is parallel to this axis. The velocity components parallel to x- and y-axes are all zero except the velocity component w, which is parallel to the tube axis. This component is also a function of x and y. Under the above assumption, the mass-conservation equation is satisfied identically, and the Navier–Stokes equations take the following form:

$$\frac{\partial^2 W}{\partial x^2} + \frac{\partial^2 W}{\partial y^2} = -K \quad (4.69)$$

where $K = -\frac{1}{\mu}\frac{dp}{dz}$, p and μ being the pressure and viscosity of the fluid.

The boundary condition is

$$W = 0 \quad \text{on} \quad \frac{x^2}{a^2} + \frac{y^2}{b^2} = 1 \quad (4.70)$$

The equation (4.69) is a Poission's equation, and it is difficult to solve this equation easily. The decomposition method can be applied here directly to find out the solution of it very easily. The equation (4.69) with the boundary condition (4.70) constitutes the mathematical model of the problem.

4.3.2 Solution by Regular Decomposition

The x-Partial Solution

Let $L_x = \frac{\partial^2}{\partial x^2}$ and $L_y = \frac{\partial^2}{\partial y^2}$ be the two partial differential operators. Then the partial form of (4.69) is

$$L_x W + L_y W = -K \tag{4.71}$$

Solving for $L_x W$, we have

$$L_x W = -K - L_y W \tag{4.72}$$

If $L_x^{-1} = \int\int(\cdot)dxdx$ is the inverse operator of L_x, then we have, by means of the operation of L_x^{-1} on both sides of (4.72),

$$W(x, y) = \underline{\Phi}_0(x, y) - L_x^{-1}[L_y W] \tag{4.73}$$

where Φ_0 is the solution of

$$\frac{\partial^2 W}{\partial x^2} = -K \tag{4.74}$$

and it is given by

$$\Phi_0(x, y) = a_0(y) + a_1(y)x - K\frac{x^2}{2} \tag{4.75}$$

We now decompose $W(x, y)$ into

$$W(x, y) = \sum_{m=0}^{\infty} \lambda^m W_m \tag{4.76}$$

Then the parameterized form of (4.73) is

$$W(x, y) = \Phi_0(x, y) - \lambda L_x^{-1}\left[L_y W\right] \tag{4.77}$$

where λ is used for grouping the terms of different orders; it is not a perturbation parameter.

Substituting (4.76) in (4.77), we get

$$\sum_{m=0}^{\infty} \lambda^m W_m = \Phi_0(x, y) - \lambda L_x^{-1}\left[L_y \sum_{m=0}^{\infty} \lambda^m W_m\right]$$

Equating the coefficients of like-powers of λ from both sides of the above relation, we get

$$\begin{aligned}
W_0(x, y) &= \Phi_0(x, y) \\
W_1(x, y) &= -L_x^{-1}\left[L_y W_0\right] \\
W_2(x, y) &= -L_x^{-1}\left[L_y W_1\right] \\
\cdots \quad &\cdots \quad \cdots \\
W_{m+1}(x, y) &= -L_x^{-1}\left[L_y W_m\right] \\
\cdots \quad &\cdots \quad \cdots
\end{aligned} \tag{4.78}$$

Using (4.75) in (4.78) and performing integrations, we have the following set of components:

$$
\begin{aligned}
W_0(x, y) &= a_0(y) + a_1(y)x - \frac{Kx^2}{2} \\
W_1(x, y) &= -\left[a_{0,2}(y)\frac{x^2}{2!} + a_{1,2}(y)\frac{x^3}{3!}\right] \\
W_2(x, y) &= \left[a_{0,4}(y)\frac{x^4}{4!} + a_{1,4}(y)\frac{x^5}{5!}\right] \\
W_3(x, y) &= -\left[a_{0,6}(y)\frac{x^6}{6!} + a_{1,6}(y)\frac{x^7}{7!}\right] \\
&\cdots \quad \cdots \quad \cdots \\
&\cdots \quad \cdots \quad \cdots \\
W_{m+1}(x, y) &= (-1)^{m+1}\left[a_{0,(2m+2)}(y)\frac{x^{2m+2}}{(2m+2)^1} + a_{1,(2m+2)}(y)\frac{x^{2m+3}}{(2m+3)!}\right] \\
&\cdots \quad \cdots \quad \cdots
\end{aligned}
\tag{4.79}
$$

where the suffix of each of the functions $a_0(y)$ and $a_1(y)$ denotes the order of differentiation with respect to y.

We now substitute (4.79) in the expanded form of (4.76), remembering that $\lambda = 1$, and get

$$
\begin{aligned}
W(x, y) = &\left[a_0(y) - a_{0,2}(y)\frac{x^2}{2!} + a_{0,4}(y)\frac{x^4}{4!} - a_{0,6}(y)\frac{x^6}{6!} + .\right] \\
&+ \left[a_1(y) - a_{1,2}(y)\frac{x^3}{3!} + a_{1,4}(y)\frac{x^5}{5!} - a_{1,6}(y)\frac{x^7}{7!} + .\right] - \frac{Kx^2}{2}
\end{aligned}
\tag{4.80}
$$

Applying the boundary condition (4.70), we obtain

$$
\begin{aligned}
0 = &\left[a_0(y) - a_{0,2}(y)\frac{x^2}{2!} + a_{0,4}(y)\frac{x^4}{4!} - a_{0,6}(y)\frac{x^6}{6!} + \cdot\right] \\
&+ \left[a_1(y)x - a_{1,2}(y)\frac{x^3}{3!} + a_{1,4}(y)\frac{x^5}{5!} - a_{1,6}(y)\frac{x^7}{7!} + \cdot\right] - \frac{Kx^2}{2}
\end{aligned}
\tag{4.81}
$$

Since the velocity $W(x, y)$ vanishes on the ellipse $\frac{x^2}{a^2} + \frac{y^2}{b^2} = 1$, the relation (4.81) represents an ellipse, and we write

$$
\begin{aligned}
a_{0,4}(y) = a_{0,6}(y) = \cdots = 0 \text{ (each)} \\
a_1(y) = a_{1,2}(y) = \cdots = 0 \text{ (each)}
\end{aligned}
\tag{4.82}
$$

By virtue of (4.82), the relation (4.81) takes the form

$$
a_0(y) - a_{0,2}(y)\frac{x^2}{2} - \frac{Kx^2}{2} = 0 \tag{4.83}
$$

Since the relation (4.83) represents an ellipse, it should therefore contain x^2-term and y^2-term together with a constant term. For this purpose the function $a_0(y)$ in the following form is assumed:

$$a_0(y) = \alpha y^2 + \beta \tag{4.84}$$

where α and β are constants to be determined later on. Substituting (4.84) in (4.83), we get

$$\left(\frac{K}{2} + \alpha\right) x^2 - \alpha y^2 = \beta \tag{4.85}$$

Relation (4.85) represents an ellipse if we put

$$\left.\begin{aligned} \beta \Big/ \left(\frac{K}{2} + \alpha\right) &= a^2 \\ -\beta/\alpha &= b^2 \end{aligned}\right\} \tag{4.86}$$

which, on solving for α and β, gives

$$\begin{aligned} \alpha &= -\frac{Ka^2}{2(a^2 + b^2)} \\ \beta &= \frac{Ka^2b^2}{2(a^2 + b^2)} \end{aligned} \tag{4.87}$$

Therefore, using (4.87) in (4.84), we have the expression for $a_0(y)$ as

$$a_0(y) = \frac{Ka^2b^2}{2(a^2 + b^2)} - \frac{Ka^2}{2(a^2 + b^2)} y^2 \tag{4.88}$$

Now the expression (4.80) for $W(x, y)$, by virtue of (4.82), becomes

$$W(x, y) = a_0(y) - a_{0,2}\frac{x^2}{2} - \frac{Kx^2}{2} \tag{4.89}$$

Combining (4.88) and (4.89), we have the expression for the velocity $W(x, y)$ as

$$W(x, y) = \frac{Ka^2b^2}{2(a^2 + b^2)} \left(1 - \frac{x^2}{a^2} - \frac{y^2}{b^2}\right) \tag{4.90}$$

which is the exact solution of the problem.

The y-Partial Solution

For the y-partial solution, we solve the equation (4.71) for $L_y W$ and get

$$L_y W = -K - L_x W \tag{4.91}$$

Then operating both sides of this equation with the inverse operator L_y^{-1} defined by $L_y^{-1} = \int\int(\cdot)dydy$, we obtain

$$W(x, y) = \overline{\Phi}(x, y) - L_y^{-1}(L_x W) \tag{4.92}$$

where $\overline{\Phi}(x, y)$ is the solution of

$$\frac{\partial^2 W}{\partial y^2} = -K \tag{4.93}$$

and it is given by

$$\overline{\Phi}(x, y) = b_0(x) + b_1(x)y - \frac{Ky^2}{2} \tag{4.94}$$

$b_0(x)$ and $b_1(x)$ being the integrating constants to be evaluated later on.

We now decompose $W(x, y)$ into

$$W(x, y) = \sum_{m=0}^{\infty} \lambda^m W_m \tag{4.95}$$

and then write the parameterized representation of (4.92) as

$$W(x, y) = \overline{\Phi}(x, y) - \lambda L_y^{-1}(L_x W) \tag{4.96}$$

Substitution of (4.95) in (4.96) gives

$$\sum_{m=0}^{\infty} \lambda^m W_m = \overline{\Phi}(x, y) - \lambda L_y^{-1} \left[L_x \sum_{m=0}^{\infty} \lambda^m W_m \right] \tag{4.97}$$

which on comparison leads to the following set of component functions:

$$\begin{aligned} W_0(x, y) &= \overline{\Phi}(x, y) \\ W_1(x, y) &= -L_y^{-1} [L_x W_0] \\ W_2(x, y) &= -L_y^{-1} [L_x W_1] \\ &\cdots \quad \cdots \quad \cdots \\ &\cdots \quad \cdots \quad \cdots \\ W_{m+1}(x, y) &= -L_y^{-1} [L_x W_m] \\ &\cdots \quad \cdots \quad \cdots \\ &\cdots \quad \cdots \quad \cdots \end{aligned} \tag{4.98}$$

We use (4.94) in (4.98) and then perform integrations to obtain

$$\begin{aligned} W_0(x, y) &= b_0(x) + b_1(x)y - \frac{Ky^2}{2} \\ W_1(x, y) &= -\left[b_{0,2}(x)\frac{y^2}{2!} + b_{1,2}(x)\frac{y^3}{3!} \right] \\ W_2(x, y) &= \left[b_{0,4}(x)\frac{y^4}{4!} + b_{1,4}(x)\frac{y^5}{5!} \right] \\ &\cdots \quad \cdots \quad \cdots \\ &\cdots \quad \cdots \quad \cdots \\ W_{m+1}(x, y) &= (-1)^{m+1} \left[b_{0,(2m+2)}(x)\frac{y^{2m+2}}{(2m+2)} + b_{1,(2m+2)}(x)\frac{y^{2m+3}}{(2m+3)!} \right] \\ &\cdots \quad \cdots \quad \cdots \end{aligned} \tag{4.99}$$

where the suffix of each of the functions $b_0(x)$ and $b_1(x)$ denotes the order of differentiation. Adding all the components in (4.99) and using the expanded form of (4.95) for $\lambda = 1$, we have

$$\begin{aligned} W(x, y) = &\left[b_0(x) - b_{0,2}(x)\frac{y^2}{2!} + b_{0,4}(x)\frac{y^4}{4!} - \cdots\right] \\ &+ \left[b_1(x)y - b_{1,2}(x)\frac{y^3}{3!} + b_{1,4}(x)\frac{y^5}{5!} - \cdots\right] - \frac{Ky^2}{2} \end{aligned} \tag{4.100}$$

Applying boundary condition (4.70) in (4.100), we get

$$\begin{aligned} 0 = &\left[b_0(x) - b_{0,2}(x)\frac{y^2}{2!} + b_{0,4}(x)\frac{y^4}{4!} - \cdots\right] \\ &+ \left[b_1(x)y - b_{1,2}(x)\frac{y^3}{3!} + b_{1,4}(x)\frac{y^5}{5!} - \cdots\right] - \frac{Ky^2}{2} \end{aligned} \tag{4.101}$$

Since the velocity $W(x, y)$ vanishes on the ellipse, the relation (4.101) should be an ellipse. For this reason we set

$$\begin{aligned} b_{0,4}(x) = b_{0,6}(x) = \cdots = 0 \text{ (each)} \\ b_1(x) = b_{1,2}(x) = \cdots = 0 \text{ (each)} \end{aligned} \tag{4.102}$$

and the relation (4.101) takes the form

$$b_0(x) - b_{0,2}(x)\frac{y^2}{2} - \frac{Ky^2}{2} = 0 \tag{4.103}$$

As the relation (4.103) represents an ellipse, the function $b_0(x)$ must contain the x^2-term and a constant term. For this purpose we assume the function $b_0(x)$ in the form

$$b_0(x) = \gamma x^2 + \delta \tag{4.104}$$

where γ and δ are constants to be evaluated.

Combining (4.103) and (4.104), we get

$$\frac{x^2}{(-\delta/\gamma)} + \frac{y^2}{2\delta/(2\gamma + K)} = 1 \tag{4.105}$$

which represents an ellipse if we put

$$\begin{aligned} -\delta/\gamma = a^2 \\ \delta/\left(\gamma + \frac{K}{2}\right) = b^2 \end{aligned} \tag{4.106}$$

Solving (4.106) for γ and δ, we get

$$\begin{aligned} \gamma = -\frac{Kb^2}{2(a^2 + b^2)} \\ \delta = \frac{Ka^2b^2}{2(a^2 + b^2)} \end{aligned} \tag{4.107}$$

and hence the relation (4.104) takes the form

$$b_0(x) = \frac{Ka^2b^2}{2(a^2+b^2)} - \frac{Kb^2}{2(a^2+b^2)}\,x^2 \tag{4.108}$$

By virtue of the relations (4.102), the expression (4.100) becomes

$$W(x,y) = b_0(x) - b_{0,2}(x)\frac{y^2}{2} - \frac{Ky^2}{2} \tag{4.109}$$

and using (4.108) in (4.109), the velocity $W(x, y)$ is given by

$$W(x,y) = \frac{Ka^2b^2}{2(a^2+b^2)}\left(1 - \frac{x^2}{a^2} - \frac{y^2}{b^2}\right)$$

which is an exact solution of the problem.

4.3.3 Solution by Modified Decomposition

The x-Partial Solution

Once again beginning with the operator equation $L_xW + L_yW = -K$, where $L_x\frac{\partial^2}{\partial x^2}$ and $L_y = \frac{\partial^2}{\partial y^2}$, and solving it for the x-partial solution followed by application of the regular decomposition procedure, one arrives at

$$W(x,y) = \xi_0(y) + \xi_1(y)x - L_x^{-1}\left[L_y W(x,y) + K\right] \tag{4.110}$$

where L_x^{-1} is the inverse operator of L_x defined earlier.

Now following the modified decomposition technique, we write for the purpose

$$W(x,y) = \sum_{m=0}^{\infty} a_m(y)x^m \tag{4.111}$$

$$K = \sum_{m=0}^{\infty} K_m x^m \tag{4.112}$$

Substituting (4.111) and (4.112) in (4.110), we get

$$\begin{aligned}\sum_{m=0}^{\infty} a_m(y)x^m &= \xi_0(y) + \xi_1(y)x - \iint\left[\frac{\partial^2}{\partial y^2}\sum_{m=0}^{\infty} a_m(y)x^m + \sum_{m=0}^{\infty} K_m x^m\right]dxdx \\ &= \xi_0(y) + \xi_1(y)x \\ &\quad - \left[\sum_{m=0}^{\infty}(\partial^2/\partial y^2)a_m(y)\frac{x^{m+2}}{(m+1)(m+2)} + \sum_{m=0}^{\infty} K_m\frac{x^{m+2}}{(m+1)(m+2)}\right]\end{aligned} \tag{4.113}$$

Replacing m by $(m-2)$ on the right side of (4.113), we obtain

$$\sum_{m=0}^{\infty} a_m(y)x^m = \xi_0(y) + \xi_1(y)x - \left[\sum_{m=2}^{\infty} \frac{(\partial^2/\partial y^2)a_{m-2(y)} + K_{m-2}}{m(m-1)} x^m\right] \tag{4.114}$$

We now compare the coefficients of like-power terms of x on both sides of (4.114) to get

$$\left.\begin{aligned} \xi_0(y) &= a_0(y) \\ \xi_1(y) &= a_1(y) \end{aligned}\right\} \tag{4.115}$$

and the recurrence relation for $m \geq 2$

$$a_m(y) = -\frac{(\partial^2/\partial y^2)a_{m-2}(y) + K_{m-2}}{m(m-1)} \tag{4.116}$$

which determines the other coefficients.

According to the boundary condition (4.70), the velocity W vanishes on the boundary of the elliptic tube. Therefore, we have from (4.111)

$$\sum_{m=0}^{\infty} a_m(y)x^m = 0 \tag{4.117}$$

If this relation represents an ellipse, then we can write

$$a_1(y) = a_3(y) = a_4(y) = \cdots = 0 \text{ (each)} \tag{4.118}$$

and the relation (4.117) reduces to

$$a_0(y) + a_2(y)x^2 = 0 \tag{4.119}$$

By virtue of (4.118), the expression for the velocity W becomes

$$W(x, y) = a_0(y) + a_2(y)x^2 \tag{4.120}$$

Putting $m = 2$ in (4.116), we get

$$a_2(y) = -\frac{1}{2}\left[\frac{\partial^2}{\partial y^2}a_0(y) + K_0\right] \tag{4.121}$$

From the relation (4.112), we have $K_0 = K$ and $K_1 = K_2 = \cdots = 0$. Then the relations (4.119) and (4.120) take the forms by virtue of (4.121):

$$a_0(y) - \left[\frac{\partial^2}{\partial y^2}a_0(y) + K\right]\frac{x^2}{2} = 0 \tag{4.122}$$

and

$$W_0(x, y) = a_0(y) - \left[\frac{\partial^2}{\partial y^2}a_0(y) + K\right]\frac{x^2}{2} \tag{4.123}$$

Since the relation (4.122) represents an ellipse, it should contain the x^2-term, the y^2-term, and a constant term. For this reason we set the function $a_0(y)$ in the following form:

$$a_0(y) = Ly^2 + M \tag{4.124}$$

Putting (4.124) in (4.122), we get

$$\frac{x^2}{2M/(2L+K)} + \frac{y^2}{(-M/L)} = 1$$

which gives

$$\left.\begin{aligned} 2M &= (2L+K)a^2 \\ M &= -Lb^2 \end{aligned}\right\} \tag{4.125}$$

Solving (4.125) for L and M, we have

$$\begin{aligned} L &= -\frac{Ka^2}{2(a^2+b^2)} \\ M &= \frac{Ka^2b^2}{2(a^2+b^2)} \end{aligned} \tag{4.126}$$

and the expression for $a_0(y)$ is given by

$$a_0(y) = \frac{Ka^2b^2}{2(a^2+b^2)} - \frac{Ka^2}{2(a^2+b^2)}y^2 \tag{4.127}$$

Using (4.127) in (4.123), we have the velocity component $W(x, y)$ as

$$W(x, y) = \frac{Ka^2b^2}{2(a^2+b^2)}\left[1 - \frac{x^2}{a^2} - \frac{y^2}{b^2}\right]$$

which is the exact solution of the problem. The theories can also be applied to the problems of fluid flow through the tubes of triangular and rectangular cross sections.

The y-Partial Solution

Here we consider also the operator equation (4.71) and solve it for the y-partial solution by means of the regular decomposition technique. The solution is found to be

$$W(x, y) = \eta_0(x) + \eta_1(x)y - L_y^{-1}[L_x W + K] \tag{4.128}$$

where $L_y^{-1} = \iint(\cdot)dydy$ and $\eta_0(x)$, $\eta_1(x)$ are integrating constants.

Now applying the modified decomposition procedure to the regular decomposition solution (4.128), we write

$$W(x, y) = \sum_{m=0}^{\infty} b_m(x)y^m \tag{4.129}$$

and

$$K = \sum_{m=0}^{\infty} K_m y^m \tag{4.130}$$

Substitution of (4.129) and (4.130) in (4.128) gives after integrations

$$\sum_{m=0}^{\infty} b_m(x) y^m = \eta_0(x) + \eta_1(x) y - \left[\sum_{m=0}^{\infty} \frac{(\partial^2/\partial x^2) b_m(x) + K_m}{(m+1)(m+2)} y^{m+2} \right]$$

which leads to the following form after changing m by $(m-2)$ on its right side:

$$\sum_{m=0}^{\infty} b_m(x) y^m = \eta_0(x) + \eta_1(x) y - \sum_{m=2}^{\infty} \frac{(\partial^2/\partial x^2) b_{m-2}(x) + K_{m-2}}{m(m-1)} y^m \tag{4.131}$$

Then equating the coefficients of like-power terms of y from both sides of (4.131), we get

$$\begin{aligned} \eta_0(x) &= b_0(x) \\ \eta_1(x) &= b_1(x) \end{aligned} \tag{4.132}$$

and the recurrence relation for $m \geq 2$

$$b_m(x) = -\frac{(\partial^2/\partial x^2) a_{m-2}(x) + K_{m-2}}{m(m-1)} \tag{4.133}$$

which gives for $m = 2$

$$b_2(x) = -\frac{1}{2} \left[\frac{\partial^2}{\partial x^2} b_0(x) + K_0 \right] \tag{4.134}$$

Applying the boundary condition (4.70) we have from (4.129)

$$\sum_{m=0}^{\infty} b_m(x) y^m = 0 \tag{4.135}$$

If this relation represents an ellipse, then

$$b_1(x) = b_3(x) = \cdots = 0 \tag{4.136}$$

and the relation (4.135) becomes

$$b_0(x) + b_2 y^2 = 0 \tag{4.137}$$

By virtue of (4.134), the relation (4.137) takes another form

$$b_0(x) - \left[\frac{\partial^2}{\partial y^2} b_0(x) + K \right] \frac{y^2}{2} = 0 \tag{4.138}$$

where $K_0 = K$ by (4.130).

Since the relation (4.138) is an ellipse, then we assume $b_0(x)$ in the following form:

$$b_0(x) = Ax^2 + B \tag{4.139}$$

Using (4.139) in (4.138), we get

$$\frac{x^2}{(-B/A)} + \frac{y^2}{2B/(2A+K)} = 1$$

which gives

$$B = -Aa^2$$
$$2B = (2A+K)\, b^2$$

Then we solve the above equations for A and B to get

$$A = -\frac{Kb^2}{2(a^2+b^2)}$$
$$B = \frac{Ka^2b^2}{2(a^2+b^2)} \tag{4.140}$$

and the expression (4.139) for the function $b_0(x)$ is found to be

$$b_0(x) = \frac{Ka^2b^2}{2(a^2+b^2)} - \frac{Kb^2}{2(a^2+b^2)}\, x^2 \tag{4.141}$$

By virtue of the relation (4.136), the expression (4.129) becomes

$$W(x, y) = b_0(x) + b_2(x)y^2 \tag{4.142}$$

Using (4.134) and (4.141) in (4.142), we get

$$W(x, y) = \frac{Ka^2b^2}{2(a^2+b^2)}\left[1 - \frac{x^2}{a^2} - \frac{y^2}{b^2}\right]$$

which is the same result as the previous one.

4.4 Stokes's First Problem: The Suddenly Accelerated Plane Wall

4.4.1 Equations of Motion

Let us consider a flat plate extending to large distances in the x- and z-directions. Let us also consider an incompressible viscous fluid over a half plane $y = 0$. The fluid that is in contact with the flat plate is infinite in extent and is also at rest at time $t < 0$. At time $t = 0$ the plate is suddenly set in motion with a constant velocity U in the x-direction, and this sudden movement of the plate produces a two-dimensional parallel motion of the fluid over it. Selecting the x-axis along the plate in the direction of U, we have the simplified Navier–Stokes equation

$$\frac{\partial u}{\partial t} = \nu \frac{\partial^2 u}{\partial y^2} \tag{4.143}$$

subject to the boundary conditions

$$\left.\begin{array}{llll} t \leq 0, & u = 0 & \text{for} & \text{all } y \\ t > 0, & u = U & \text{for} & y = 0 \\ & u = 0 & \text{for} & y = \propto \end{array}\right\} \tag{4.144}$$

The equation (4.143) is a partial differential equation, and this can be converted into an ordinary differential equation by the following transformations:

$$\left.\begin{array}{l} \eta = \dfrac{y}{2\sqrt{\nu t}} \\ u = Uf(\eta) \end{array}\right\} \tag{4.145}$$

Using (4.145), the equation (4.143) becomes

$$\frac{d^2 f}{d\eta^2} + 2\eta \frac{df}{d\eta} = 0 \tag{4.146}$$

and the corresponding boundary conditions are

$$f = 1 \quad \text{at} \quad \eta = 0 \tag{4.147}$$

$$f = 0 \quad \text{at} \quad \eta = \propto \tag{4.148}$$

4.4.2 Solution by Regular Decomposition

We proceed to solve the equation (4.146) under the boundary conditions (4.147) and (4.148) by means of the regular decomposition method. To do that, we let $L = \frac{d^2}{d\eta^2}$ whose inverse form is $L^{-1} = \int\int(\cdot)d\eta\, d\eta$. Then the equation (4.146) becomes

$$Lf + 2\eta \frac{df}{d\eta} = 0$$

which, on solving for Lf, gives

$$Lf = -2\eta \frac{df}{d\eta} \tag{4.149}$$

If we operate both sides of (4.149) with the inverse operator L^{-1}, then we get

$$f(\eta) = f_0 - L^{-1}\left[2\eta\frac{df}{d\eta}\right] \tag{4.150}$$

where the solution of $Lf = 0$ is found to be

$$f_0 = \xi_0 + \xi_1\eta \tag{4.151}$$

Here the constants of integration, ξ_0 and ξ_1, are evaluated by the boundary conditions (4.147) and (4.148). Satisfying (4.147) by (4.151), we get $\xi_0 = 1$, and the expression (4.151) for f_0 becomes

$$f_0 = 1 + \xi_1\eta \tag{4.152}$$

We now decompose $f(\eta)$ into

$$f(\eta) = \sum_{m=0}^{\infty}\lambda^m fm \tag{4.153}$$

where λ has its usual meanings. The parameterized representation of (4.150) is

$$f(\eta) = f_0 - \lambda L^{-1}\left[2\eta\frac{df}{d\eta}\right] \tag{4.154}$$

Using (4.153) in (4.154), we get

$$\sum_{m=0}^{\infty}\lambda^m f_m = f_0 - \lambda L^{-1}\left[2\eta\frac{d}{d\eta}\sum_{m=0}^{\infty}\lambda^m f_m\right]$$

which gives

$$\left.\begin{aligned}
f_1 &= -2L^{-1}\left[\eta\frac{d}{d\eta}f_0\right]\\
f_2 &= -2L^{-1}\left[\eta\frac{d}{d\eta}f_1\right]\\
f_3 &= -2L^{-1}\left[\eta\frac{d}{d\eta}f_2\right]\\
&\cdots\quad\cdots\cdots\quad\cdots\\
&\cdots\quad\cdots\cdots\quad\cdots\\
f_{m+1} &= -2L^{-1}\left[\eta\frac{d}{d\eta}f_m\right]\\
&\cdots\quad\cdots\cdots\quad\cdots
\end{aligned}\right\} \tag{4.155}$$

Then we use (4.152) in (4.155) and get the velocity components as

$$\left.\begin{aligned} f_1 &= -2.\xi_1\frac{\eta^3}{3!} \\ f_2 &= \xi_1 2^2.1.3.\frac{\eta^5}{5!} \\ f_3 &= -\xi_1 2^3 1.3.5.\frac{\eta^7}{7!} \\ \cdots &\quad \cdots\cdots \\ \cdots &\quad \cdots\cdots \end{aligned}\right\} \tag{4.156}$$

Adding all the components given in (4.152) and (4.156) and then using the expanded form of (4.153), we get

$$\begin{aligned} f(\eta) &= 1 + \xi_1\left[\eta - 2.\frac{\eta^3}{3!} + 2^2.1.3.\frac{\eta^5}{5!} - 2^3.1.3.5.\frac{\eta^7}{7!} + \cdots\right] \\ &= 1 + \xi_1\int_0^{\eta} e^{-\eta^2} d_\eta \end{aligned} \tag{4.157}$$

We now apply the boundary condition (4.148) and get $\xi_1 = -2 \mid \sqrt{\pi}$. Putting ξ_1 in (4.157), we have the expression for $f(\eta)$ as

$$\begin{aligned} f(\eta) &= 1 - \frac{2}{\sqrt{\pi}}\int_0^{\eta} e^{-\eta^2} d_\eta \\ &= 1 - erf_\eta \end{aligned} \tag{4.158}$$

which is a closed-form solution of the problem.

4.4.3 Solution by Modified Decomposition

We know that the decomposition method can be applied to any type of differential equation, linear or nonlinear, ordinary or partial, with constant or variable coefficients. Here the equation (4.146) is an ordinary differential equation with variable coefficients, and we shall apply the modified decomposition technique to this equation.

For this equation we let $2\eta = \alpha$ and $L = \frac{d^2}{d\eta^2}$ whose inverse L^{-1} is a twofold indefinite integral defined by $L^{-1} = \int\int(\cdot)d_\eta d_\eta$. Then the equation (4.146) becomes

$$Lf + \alpha\frac{df}{d\eta} = 0 \tag{4.159}$$

Solving (4.159) for f by the application of the ordinary decomposition technique, we obtain

$$f(\eta) = f_0 - L^{-1}\left[\alpha\frac{df}{d\eta}\right] \tag{4.160}$$

where

$$f_0 = \beta_0 + \beta_1\eta \tag{4.161}$$

Satisfying the boundary condition (4.147) by (4.161), we have $\beta_0 = 1$, and the expression for f_0 is found to be

$$f_0 = 1 + \beta_1\eta \tag{4.162}$$

We now follow the modified decomposition procedure and write for the purpose

$$\left.\begin{aligned} f(\eta) &= \sum_{m=0}^{\infty} a_m\eta^m \\ \frac{df}{d\eta} &= \sum_{m=0}^{\infty}(m+1)a_{m+1}\eta^m \end{aligned}\right\} \tag{4.163}$$

$$2\eta = \alpha = \sum_{m=0}^{\infty}\alpha_m\eta^m \tag{4.164}$$

Using (4.162), (4.163), and (4.164) in (4.160), we have

$$\sum_{m=0}^{\infty}a_m\eta^m = 1 + \beta_1\eta - \iint\left[\left\{\sum_{m=0}^{\infty}\alpha_m\eta^m\right\}\left\{\sum_{m=0}^{\infty}(m+1)a_{m+1}\eta^m\right\}\right]d_\eta d_\eta$$

Carrying out the Cauchy product of infinite series on the right side, we have

$$\sum_{m=0}^{\infty}a_m\eta^m = 1 + \beta_1\eta - \iint\sum_{m=0}^{\infty}\left[\sum_{n=0}^{m}\alpha_{m-n}(n+1)a_{n+1}\right]\eta^m d_\eta d_\eta$$

Performing integrations, we get

$$\sum_{m=0}^{\infty}a_m\eta^m = 1 + \beta_1\eta - \sum_{m=0}^{\infty}\left[\sum_{n=0}^{m}\alpha_{m-n}(n+1)a_{n+1}\right]\frac{\eta^{m+2}}{(m+1)(m+2)}$$

Replacing m by $(m-2)$ on the right side, we obtain

$$\sum_{m=0}^{\infty}a_m\eta^m = 1 + \beta_1\eta - \sum_{m=2}^{\infty}\left[\sum_{n=0}^{m-2}\alpha_{m-n-2}(n+1)a_{n+1}\right]\frac{\eta^m}{m(m-1)}$$

Hence, the coefficients are identified by

$$\left.\begin{aligned} a_0 &= 1 \\ a_1 &= \beta_1 \end{aligned}\right\} \tag{4.165}$$

and the recurrence relation for $m \geq 2$

$$a_m = -\sum_{n=0}^{m-2} \alpha_{m-n-2}(n+1)a_{n+1}/m(m-1) \tag{4.166}$$

From (4.164), we get

$$\alpha_1 = 2 \tag{4.167}$$

and

$$\alpha_0 = \alpha_2 = \ldots = 0 \text{ (each)} \tag{4.168}$$

Again, from (4.166), we obtain

$$\left.\begin{aligned} a_2 &= a_4 = a_6 \cdots = 0 \text{ (each)} \\ a_3 &= -\frac{2a_1}{3!}, \quad a_5 = a_1 2^2.1.3.\frac{1}{5!} \\ a_7 &= a_1.2^3.1.3.5.\frac{1}{7!}, \text{ etc.} \end{aligned}\right\} \tag{4.169}$$

Using (4.165) and (4.169), we have from the first relation of (4.163)

$$\begin{aligned} f(\eta) &= 1 + a_1\left[\eta - 2\frac{\eta^3}{3!} + 2^2.1.3.\frac{\eta^5}{5!} - 2^3.1.3.5.\frac{\eta^7}{7!} + \cdots\right] \\ &= 1 + a_1 \int_0^n e^{-\eta^2} d_\eta \end{aligned} \tag{4.170}$$

If we satisfy the boundary condition (4.148) by (4.170), we get $a_1 = -2/\sqrt{\pi}$, and the expression for $f(\eta)$ is found to be

$$\begin{aligned} f(\eta) &= 1 - \frac{2}{\sqrt{\pi}} \int_0^\eta e^{-\eta^2} d_\eta \\ &= 1 - erf\,\eta \end{aligned} \tag{4.171}$$

which is the exact solution of the problem.

4.5 Stokes's Second Problem: The Flow Near an Oscillating Flat Plate

4.5.1 Equations of Motion

In this section we discuss the flow of viscous incompressible fluid about an infinite plane wall that executes oscillations in its own plane. This problem was first studied by G. Stokes [5] and later by Lord Rayleigh [4]. We assume that the fluid is infinite in extent and the pressure is constant everywhere. Denoting the x-axis in the direction of motion and the y-axis perpendicular to the plane wall, the simplified Navier–Stokes equation that governs the flow field is

$$\frac{\partial u}{\partial t} = \nu \frac{\partial^2 u}{\partial y^2} \tag{4.172}$$

Because of the no-slip condition at the wall, the fluid velocity at it must be equal to that of the wall. We suppose that the motion is given by

$$u(y, t) = \text{Ucosnt} \quad \text{at} \quad y = 0 \tag{4.173}$$

and the other boundary condition is

$$u(y, t) = 0 \quad \text{at} \quad y = \infty \tag{4.174}$$

If we make use of the complex variable, we write

$$u(y, t) = f(y)e^{int} \tag{4.175}$$

whose real part is the solution of the problem. By virtue of this transformation, the mathematical model described by (4.172), (4.173), and (4.174) becomes

$$\frac{d^2 f}{dy^2} = \alpha f \tag{4.176}$$

$$f = U \quad \text{at} \quad y = 0 \tag{4.177}$$

$$f = 0 \quad \text{at} \quad y = \infty \tag{4.178}$$

where $\alpha = in/\nu$.

4.5.2 Solution by Regular Decomposition

Let $L = \frac{d^2}{dy^2}$, and the equation (4.176) becomes $Lf = \alpha f$. Then we follow the regular decomposition procedure to get

$$f(y) = \xi_0 + \xi_1 y + L^{-1}(\alpha f) \tag{4.179}$$

whose parameterized form is

$$f(y) = \xi_0 + \xi_1 y + \lambda L^{-1}(\alpha f) \tag{4.180}$$

when we decompose f into

$$f(y) = \sum_{m=0}^{\infty} \lambda^m f_m \tag{4.181}$$

where L^{-1} is a twofold indefinite integral defined by $L^{-1} = \iint(\cdot)dydr$. Putting (4.181) in (4.180) and equating the coefficients of like-power terms of λ, we get

$$\begin{aligned}
f_0(y) &= \xi_0 + \xi_1 y \\
f_1(y) &= \alpha L^{-1} f_0 = \alpha \left[\xi_0 \frac{y^2}{2!} + \xi_1 \frac{y^3}{3!}\right] \\
f_2(y) &= \alpha L^{-1} f_1 = \alpha^2 \left[\xi_0 \frac{y^4}{4!} + \xi_1 \frac{y^5}{5!}\right] \\
f_3(y) &= \alpha L^{-1} f_2 = \alpha^3 \left[\xi_0 \frac{y^6}{6!} + \xi_1 \frac{y^7}{7!}\right] \\
&\cdots \quad \cdots \quad \cdots \\
&\cdots \quad \cdots \quad \cdots \\
f_{n+1}(y) &= \alpha L^{-1} f_n = \alpha^{n+1} \left[\xi_0 \frac{y^{2n+2}}{(2n+2)!} + \xi_1 \frac{y^{2n+3}}{(2n+3)!}\right] \\
&\cdots \quad \cdots \quad \cdots \quad \cdots \\
&\cdots \quad \cdots \quad \cdots \quad \cdots
\end{aligned} \tag{4.182}$$

Adding all the component terms and then using the expanded form of (4.181), we have

$$\begin{aligned}
f(y) &= \xi_0 \left[1 + \alpha \frac{y^2}{2!} + \frac{\alpha^2 y^4}{4!} + \alpha^3 \frac{y^6}{6!} + \cdots\right] \\
&\quad + \xi_1 \left[y + \frac{\alpha y^3}{3!} + \frac{\alpha^2 y^5}{5!} + \cdots\right] \\
&= \xi_0 \left[1 + \frac{(\sqrt{\alpha} y)^2}{2!} + \frac{(\sqrt{\alpha} y)^4}{4!} + \frac{(\sqrt{\alpha} y)^6}{6!} + \cdots\right] \\
&\quad + \frac{\xi_1}{\sqrt{\alpha}} \left[(\sqrt{\alpha} y) + \frac{(\sqrt{\alpha} y)^3}{3!} + \frac{(\sqrt{\alpha} y)^5}{5!} + \cdots\right] \\
&= \frac{\xi_0}{2}\left(e^{\sqrt{\alpha} y} + e^{-\sqrt{\alpha} y}\right) + \frac{\xi_1}{2\sqrt{\alpha}}\left(e^{\sqrt{\alpha} y} - e^{-\sqrt{\alpha} y}\right) \\
&= \frac{1}{2}\left(\xi_0 + \frac{\xi_1}{\sqrt{\alpha}}\right) e^{\sqrt{\alpha} y} + \frac{1}{2}\left(\xi_0 - \frac{\xi_1}{\sqrt{\alpha}}\right) e^{-\sqrt{\alpha} y}
\end{aligned} \tag{4.183}$$

Satisfying the boundary condition (4.177) by (4.183), we get $\xi_0 = U$, and the expression for $f(y)$ becomes

$$f(y) = \frac{1}{2}\left(U + \frac{\xi_1}{\sqrt{\alpha}}\right) e^{\sqrt{\alpha} y} + \frac{1}{2}\left(U - \frac{\xi_1}{\sqrt{\alpha}}\right) e^{-\sqrt{\alpha} y} \tag{4.184}$$

Since $f(y)$ is finite at infinity according to the boundary condition (4.178), we have from (4.184) $U + \frac{\xi_1}{\sqrt{\alpha}} = 0$, that is, $\xi_1 = -U\sqrt{\alpha}$, and the expression (4.184) takes the form

$$f(y) = Ue^{-\sqrt{\alpha}y} = Ue^{-\sqrt{n/2\nu}y}e^{-\sqrt{n/2\nu}y} \tag{4.185}$$

Therefore, from (4.175) we have, using (4.185),

$$\begin{aligned} u(y) &= \text{ real } \left\{ Ue^{-\sqrt{n/2\nu y}}.e^{i\left[n+-\sqrt{n/2\nu}\right]y} \right\} \\ &= Ue^{-\sqrt{n/2\nu.y}}.\cos\left(nt - \sqrt{\frac{n}{2\nu}}.y\right) \end{aligned} \tag{4.186}$$

which is the exact solution of the problem.

4.5.3 Solution by Modified Decomposition

For modified decomposition, we write the solution in the form given by

$$f(y) = \sum_{m=0}^{\infty} a_m y^m \tag{4.187}$$

Substituting (4.187) in the regular decomposition solution (4.179), we get

$$\sum_{m=0}^{\infty} a_m y^m = \xi_0 + \xi_1 y + \iint \left[\propto \sum_{m=0}^{\infty} a_m y^m \right] d_y d_y$$

Carrying out integrations, we obtain

$$\sum_{m=0}^{\infty} a_m y^m = \xi_0 + \xi_1 y + \sum_{m=0}^{\infty} \propto a_m \frac{y^{m+2}}{(m+1)(m+2)}$$

Replacement of m by $(m-2)$ on the right side gives

$$\sum_{m=0}^{\infty} a_m y^m = \xi_0 + \xi_1 y + \sum_{m=2}^{\infty} \alpha a_{m-2} \frac{y^m}{m(m-1)}$$

The coefficients are identified by

$$\xi_0 = a_0, \xi_1 = a_1 \tag{4.188}$$

and for $m \geq 2$, the recurrence relation

$$a_m = \frac{\alpha a_{m-2}}{m(m-1)} \tag{4.189}$$

Putting $m = 2, 3, 4$, etc., we have from (4.189)

$$a_2 = \frac{\alpha a_0}{2!}, a_3 = \frac{\alpha a_1}{3!}$$
$$a_4 = \frac{\alpha^2 a_0}{4!}, a_5 = \frac{\alpha^2 a_1}{5!} \tag{4.190}$$
$$a_6 = \frac{\alpha^3 a_0}{6!}, \text{ etc.}$$

Using (4.190) in (4.187), we have

$$\begin{aligned} f(y) &= a_0\left[1+\frac{\alpha y^2}{2!}+\frac{\alpha^2 y^4}{4!}+\cdots\right]+a_1\left[y+\frac{\alpha y^3}{3!}+\frac{\alpha^2 y^5}{5!}+\cdots\right] \\ &= a_0\left[1+\frac{(\sqrt{\alpha}y)^2}{2!}+\frac{(\sqrt{\alpha}y)^4}{4!}+\cdots\right] \\ &+ \frac{a_1}{\sqrt{\alpha}}\left[(\sqrt{\alpha}y)+\frac{(\sqrt{\alpha}y)^3}{3!}+\frac{(\sqrt{\alpha}y)^5}{5!}+\cdots\right] \\ &= \frac{a_0}{2}\left(e^{\sqrt{\alpha}y}+e^{-\sqrt{\alpha}y}\right)+\frac{a_1}{2\sqrt{\alpha}}\left(e^{\sqrt{\alpha}y}-e^{-\sqrt{\alpha}y}\right) \\ &= \frac{1}{2}\left(a_0+\frac{a_1}{\sqrt{\alpha}}\right)e^{\sqrt{\alpha}y}+\frac{1}{2}\left(a_0-\frac{a_1}{\sqrt{\alpha}}\right)e^{-\sqrt{\alpha}y} \end{aligned} \tag{4.191}$$

Satisfying the boundary conditions (4.177) and (4.178), we have $a_0 = U$ and $a_1 = -U\sqrt{\alpha}$. Then the expression for $f(y)$ is found to be

$$f(y) = Ue^{-\sqrt{\alpha}y} \tag{4.192}$$

Combining (4.175) and (4.192), we get the expression for $u(y, t)$ as

$$\begin{aligned} u(y,t) &= \text{real of } Ue^{-\sqrt{\alpha}y}.e^{int} \\ &= \text{real of } Ue^{-\sqrt{n/2\nu}y}.e^{i\left(nt-\sqrt{\frac{n}{2\nu}}.y\right)} \\ &= Ue^{-\sqrt{\frac{n}{2\nu}}.y}\cos\left(nt-\sqrt{\frac{n}{2\nu}}.y\right) \end{aligned} \tag{4.193}$$

which is the exact solution.

4.6 Unsteady Flow of Viscous Incompressible Fluid between Two Parallel Plates

4.6.1 Equations of Motion

In this section we consider the flow of viscous incompressible fluid between two parallel plates separated by a distance d. The plates extend to infinity in

x- and z-directions, and the y-axis is perpendicular to them. The fluid in the space between the plates is at rest, and the lower plate $y = 0$ at time $t = 0$ is suddenly set in motion with the velocity U in the x direction. This sudden motion of the lower plate, as a result, generates a two-dimensional motion of fluid in the space between the plates. The simplified Navier–Stokes equation in this case is

$$\frac{\partial u}{\partial t} = \nu \frac{\partial^2 u}{\partial y^2}$$

where u is the fluid velocity and ν is the kinematic viscosity.

The boundary conditions are

$$u = U \quad \text{at} \quad y = 0$$

and

$$u = 0 \quad \text{at} \quad y = d$$

To solve the partial differential equation given above, we find it convenient to convert this equation to an ordinary differential equation by the following transformations:

$$\left.\begin{aligned} \eta &= \tfrac{y}{2\sqrt{\nu t}} \\ u &= Uf(\eta) \end{aligned}\right\} \tag{4.194}$$

The transformed differential equation is

$$\frac{d^2 f}{d\eta^2} + 2\eta \frac{df}{d\eta} = 0 \tag{4.195}$$

and the corresponding boundary conditions are

$$f(\eta) = 1 \quad \text{at} \quad \eta = 0 \tag{4.196}$$

$$f(\eta) = 0 \quad \text{at} \quad \eta = \frac{d}{2\sqrt{\nu t}} = \eta_0 \text{ (say)} \tag{4.197}$$

The equations ranging from (4.195) to (4.197) constitute the mathematical model of the problem.

4.6.2 Solution by Regular Decomposition

Let $L = \frac{d^2}{d\eta^2}$ whose inverse form L^{-1} was defined earlier. Then the operator form of (4.195) is $Lf = -2\eta \frac{df}{d\eta}$. Operating both sides of this equation by L^{-1}, we get

$$f(\eta) = a_0 + a_1 \eta - 2L^{-1}\left(\eta \frac{df}{d\eta}\right) \tag{4.198}$$

where a_0 and a_1 are integrating constants to be evaluated later on.

We now decompose f into

$$f(\eta) = \sum_{m=0}^{\infty} \lambda^m f_m \tag{4.199}$$

Then the parameterized representation of (4.198) is

$$f(\eta) = a_0 + a_1\eta - \lambda L^{-1}\left(2\eta \frac{df}{d\eta}\right) \tag{4.200}$$

Substitution of (4.199) in (4.200) and then comparison of like-power terms on both sides of the resulting expression gives

$$\left.\begin{aligned} f_0(\eta) &= a_0 + a_1\eta \\ f_1(\eta) &= -a_1 2\frac{2.\eta^3}{3!} \\ f_2(\eta) &= a_1 2^2.1.3\frac{\eta^5}{5!} \\ \cdots & \quad \cdots \quad \cdots \\ \cdots & \quad \cdots \quad \cdots \end{aligned}\right\} \tag{4.201}$$

Using (4.201) in the expanded form of (4.199) for $\lambda = 1$, we have

$$\begin{aligned} f(\eta) &= a_0 + a_1\left[\eta - 2.\frac{\eta^3}{3!} + 2^2.1.3.\frac{\eta^5}{5!} - \cdots\right] \\ &= a_0 + a_1\int_0^{\eta} e^{-\eta^2} d_\eta \end{aligned} \tag{4.202}$$

Satisfying the boundary conditions (4.196) and (4.197), we obtain

$$\begin{aligned} a_0 &= 1 \\ a_1 &= -1/\left[\int_0^{\eta_0} e^{-\eta^2} d_\eta\right] \end{aligned} \tag{4.203}$$

Putting the values of a_0 and a_1 from (4.203) in (4.202), we find the expression for $f(\eta)$ as

$$f(\eta) = 1 - \left[\int_0^{\eta} e^{-\eta^2} d_\eta / \int_0^{\eta_0} e^{-\eta^2} d_\eta\right] \tag{4.204}$$

Combining (4.194) and (4.204), we get

$$\begin{aligned} u(y,t) &= U\left[1 - \frac{\int_0^{\eta} e^{-\eta^2} d_\eta}{\int_0^{\eta_0} e^{-\eta^2} d_\eta}\right] \\ \eta &= y/2\sqrt{\nu t} \end{aligned} \tag{4.205}$$

which is a closed-form solution of the problem.

4.6.3 Solution by Modified Decomposition

In this section we find out the solution by means of the modified decomposition method. Since the equation (4.195) is an ordinary differential equation with variable coefficient 2η, we consider again the regular decomposition solution (4.198) of the previous section, and then following the modified decomposition technique, we write

$$\begin{aligned} f(\eta) &= \sum_{m=0}^{\infty} a_m \eta^m \\ \frac{df}{d\eta} &= \sum_{m=0}^{\infty} (m+1) a_{m+1} \eta^m \\ 2\eta = \alpha &= \sum_{m=0}^{\infty} \alpha_m \, \eta^m \end{aligned} \tag{4.206}$$

Substituting (4.206) in (4.198), we have

$$\sum_{m=0}^{\infty} a_m \eta^m = a_0 + a_1 \eta - L^{-1} \left[\sum_{m=0}^{\infty} \alpha_m \eta^m \right] \left[\sum_{m=0}^{\infty} a_{m+1} \eta^m \right]$$

Using the Cauchy product of infinite series and then replacing m by $(m-2)$ on the right side of the above relation, we obtain the recurrence relation for $m \geq 2$ as

$$a_m = -\frac{\sum_{n=0}^{\infty} \alpha_{m-n} (n+1) a_{n+1}}{m(m-1)} \tag{4.207}$$

From the third relation of (4.206) we get

$$\begin{aligned} \alpha_1 &= 2, \\ \alpha_0 &= \alpha_2 = \cdots = 0 \text{ (each)} \end{aligned} \tag{4.208}$$

Using (4.208), we have from (4.207)

$$\begin{aligned} a_2 &= a_4 = a_6 = \cdots = 0 \text{ (each)} \\ a_3 &= -\frac{2a_1}{3!}, \, a_5 = a_1 2^2.1.3.\frac{1}{5!} \\ a_7 &= a_1 2^3.1.3.5.\frac{1}{7!}, \text{ etc.} \end{aligned} \tag{4.209}$$

By virtue of (4.209), the expanded form of $f(\eta) = \sum_{m=0}^{\infty} a_m \eta^m$ is found to be

$$\begin{aligned} f(\eta) &= a_0 + a_1 \left[\eta - 2\frac{\eta^3}{3!} + 2^2.1.3.\frac{\eta^5}{5!} - 2^3.1.3.5.\frac{\eta^7}{7!} + \cdots \right] \\ &= a_0 + a_1 \int_0^{\eta} e^{-\eta^2} d_{\eta} \end{aligned} \tag{4.210}$$

where a_0 and a_1 are the constants. Satisfying the boundary conditions (4.196) and (4.197), we have $a_0 = 1$ and $a_1 = -1/\int_0^{\eta_0} e^{-\eta^2} d_\eta$. Hence, the expression for $f(\eta)$ is

$$f(\eta) = 1 - \frac{\int_0^{\eta} e^{-\eta^2} d_\eta}{\int_0^{\eta_0} e^{-\eta^2} d_\eta} \tag{4.211}$$

Therefore, the desired velocity component is given by

$$u(y,t) = U\left[1 - \frac{\int_0^{\eta} e^{-\eta^2} d_\eta}{\int_0^{\eta_0} e^{-\eta^2} d_\eta}\right]$$
$$\eta = y/2\sqrt{\nu t} \tag{4.212}$$

4.7 Pulsatile Flow between Two Parallel Plates

4.7.1 Equations of Motion

Consider an oscillatory flow of viscous incompressible fluid in the space between two fixed parallel plates separated by a distance $2d$. The plates are extended to infinity in the x- and z-directions, and the y-axis is perpendicular to the plates. We also consider the oscillatory pressure gradient in the x-direction. This generates a two-dimensional flow, and the Navier–Stokes equations reduce to

$$\frac{\partial u}{\partial t} = -\frac{1}{\rho}.\frac{\partial p}{\partial x} + \nu \frac{\partial^2 u}{\partial y^2} \tag{4.213}$$

The boundary conditions are

$$u(d,t) = u(-d,t) = 0 \tag{4.214}$$

If p_0 is the magnitude of the pressure gradient, which is constant, then the pressure gradient is assumed in the following form:

$$\frac{\partial p}{\partial x} = p_0 \cos nt \tag{4.215}$$

We should keep in mind that velocity and the pressure gradient are oscillatory in nature. Therefore, we suppose that

$$u(y,t) = R\left[f(y)e^{int}\right] \tag{4.216}$$

and

$$\frac{\partial p}{\partial x} = R\left[p_0 e^{int}\right] \tag{4.217}$$

where R denotes the real parts of the expressions. Substitution of (4.216) and (4.217) in (4.213) gives

$$\frac{d^2 f}{dy^2} - \alpha f = K \tag{4.218}$$

where

$$\alpha = in/\nu, \quad K = p_0/\rho\nu \tag{4.219}$$

The corresponding boundary conditions (4.214), by means of (4.216), take the forms

$$f(d) = f(-d) = 0 \tag{4.220}$$

The equation (4.218), together with the boundary conditions (4.220), constitutes the mathematical model of the problem.

4.7.2 Solution by Regular Decomposition

To solve the equation (4.218) subject to boundary conditions (4.220), we use the regular decomposition method and write its solution as

$$f(y) = f_0(y) + L^{-1}[\alpha f] \tag{4.221}$$

where f_0, the solution of the equation $\frac{d^2 g}{dy^2} = K$, is given by

$$f_0(y) = \xi_0 + \xi_1 y + K\frac{y^2}{2!} \tag{4.222}$$

The L^{-1} defined by $L^{-1} = \int\int(\cdot)dydy$ is the inverse operator of the linear operator $L = \frac{d^2}{dy^2}$, and the integrating constants ξ_0 and ξ_1 are to be evaluated from the boundary conditions.

We now decompose $f(y)$ into $f(y) = \sum_{m=0}^{\infty}\lambda^m f_m$, and putting this decomposed form into the parameterized form of (4.221), we have, on comparison of the terms,

$$\begin{aligned}
f_0(y) &= \xi_0 + \xi_1 y + \frac{Ky^2}{2!} \\
f_1(y) &= L^{-1}[\alpha f_0] = \xi_0\frac{(\sqrt{\alpha}y)^2}{2!} + \frac{\xi_1}{\sqrt{\alpha}}\frac{(\sqrt{\alpha}y)^3}{3!} + \frac{K}{\alpha}\frac{(\sqrt{\alpha}y)^4}{4!} \\
f_2(y) &= L^{-1}[\alpha f_1] = \xi_0\frac{(\sqrt{\alpha}y)^4}{4!} + \frac{\xi_1}{\sqrt{\alpha}}\frac{(\sqrt{\alpha}y)^5}{5!} + \frac{K}{\alpha}\frac{(\sqrt{\alpha}y)^6}{6!} \\
&\cdots \quad \cdots \quad \cdots \quad \cdots \quad \cdots \\
&\cdots \quad \cdots \quad \cdots \quad \cdots \quad \cdots
\end{aligned} \tag{4.223}$$

Adding all the components given in (4.223) and then using the decomposed form of $f(y)$, we get

$$f(y) = \left(\xi_0 + \frac{K}{\alpha}\right)\left[1 + \frac{(\sqrt{\alpha}y)^2}{2!} + \frac{(\sqrt{\alpha}y)^4}{4!} + \frac{(\sqrt{\alpha}y)^6}{6!} + \cdots\right]$$
$$+ \frac{\xi_1}{\sqrt{\alpha}}\left[(\sqrt{\alpha}y) + \frac{(\sqrt{\alpha}y)^3}{3!} + \frac{(\sqrt{\alpha}y)^5}{5!} + \cdots\right] - \frac{K}{\alpha}$$
$$= A\cos h\left(\sqrt{\alpha}y\right) + B\sin h\left(\sqrt{\alpha}y\right) - \frac{K}{\alpha} \quad (4.224)$$

where $A = \xi_0 + \frac{K}{\alpha}$ and $B = \xi_1/\sqrt{\alpha}$.

Satisfying the boundary conditions (4.220), we have $A = K/\alpha \cos h(\sqrt{\alpha}d)$, $B = 0$, and the expression (4.224) becomes

$$f(y) = -\frac{K}{\alpha}\left[1 - \frac{\cos h(\sqrt{\alpha}y)}{\cos h(\sqrt{\alpha}d)}\right] \quad (4.225)$$

Hence, the velocity distribution of the fluid is given by (4.216), that is,

$$u(y,t) = -R\left[\frac{K}{\alpha}e^{int}\left\{1 - \frac{\cos h(\sqrt{\alpha}y)}{\cos h(\sqrt{\alpha}d)}\right\}\right] \quad (4.226)$$

which gives the exact solution of the problem.

4.7.3 Solution by Modified Decomposition

For solving the equation (4.218) under the boundary condition (4.220) by means of the modified decomposition technique, we first write down its regular decomposition solution in the form given by

$$f(y) = \eta_0 + \eta_1 y + L^{-1}[\alpha f + K] \quad (4.227)$$

where η_0, η_1 are integrating constants and L^{-1} is the inverse operator of $L = \frac{d^2}{dy^2}$ defined earlier.

Then following the modified decomposition procedure, we write

$$f(y) = \sum_{m=0}^{\infty} a_m y^m \quad (4.228)$$

$$g = \sum_{m=0}^{\infty} g_m y^m = K \quad (4.229)$$

Putting (4.228) and (4.229) in (4.227), then carrying out integrations on the right side and finally replacing m by $(m-2)$, we get

$$\sum_{m=0}^{\infty} a_m y^m = \eta_0 + \eta_1 y + \sum_{m=2}^{\infty} \frac{g_{m-2} + \alpha a_{m-2}}{m(m-1)}$$

which gives, on comparison,

$$\left.\begin{aligned} \eta_0 &= a_0 \\ \eta_1 &= a_1 \end{aligned}\right\} \tag{4.230}$$

and

$$a_m = \frac{g_{m-2} + \alpha a_{m-2}}{m(m-1)}, m \geq 2 \tag{4.231}$$

From (4.229), we obtain

$$g_0 = K, g_1 = g_2 \cdots = 0 \text{ (each)} \tag{4.232}$$

Again, from (4.231), we have

$$\begin{aligned} a_2 &= \frac{K}{2!} + \frac{\alpha a_0}{2!}, \quad a_3 = \frac{\alpha a_1}{3!} \\ a_4 &= \frac{K\alpha}{4!} + \frac{\alpha^2 a_0}{4!}, \quad a_5 = \frac{\alpha^2 a_1}{5!} \\ a_6 &= \frac{K\alpha^2}{6!} + \frac{\alpha^3 a_0}{6!}, \text{ etc.} \end{aligned} \tag{4.233}$$

The use of (4.233) in (4.228) leads to

$$f(y) = C \cos h(\sqrt{\alpha} y) + D \sin h(\sqrt{\alpha} y) - \frac{K}{\alpha} \tag{4.234}$$

where $C = a_0 + \frac{K}{\alpha}$ and $D = a_1/\sqrt{\alpha}$. Satisfying the boundary conditions (4.220), we have $C = K/\alpha \cos h(\sqrt{\alpha} d)$, $D = 0$. The expression for $f(y)$ is found to be

$$f(y) = -\frac{K}{\alpha}\left[1 - \frac{\cos h(\sqrt{\alpha} y)}{\cos h(\sqrt{\alpha} d)}\right] \tag{4.235}$$

Hence, the velocity distribution is given by (4.214), that is,

$$u(y,t) = -R\left[\frac{K}{\alpha} e^{int}\left\{1 - \frac{\cos h(\sqrt{\alpha} y)}{\cos h(\sqrt{\alpha} d)}\right\}\right] \tag{4.236}$$

which is a closed-form solution of the problem.

References

1. Adomian, G.: *Nonlinear Stochastic Systems Theory and Application to Physics*, Kluwer Academic Publishers (1989).
2. Adomian, G.: *Solving Frontier Problems of Physics: The Decomposition Method*, Kluwer Academic Publishers (1994).

3. Haldar, K.: Some Exact Solutions of Linear Fluid Flow Problems by Modified Decomposition Method, *Bull. Cal. Math. Soc.*, 100(3), 283–300 (2008).
4. Rayleigh, L.: On the Motion of Solid Bodies Through Viscous Liquid, *Phil. Mag.*, *21*, 697–711 (1911).
5. Stokes, G. G.: On the Effect of the Internal Friction of Fluids on the Motion of Pendulum, *Cambr. Phil. Trans.*, *IX*, 8 (1851).
6. Schlichting, H.: *Boundary-Layer Theory*, McGraw-Hill (1968).

5

Navier–Stokes Equations in Cylindrical Polar Coordinates

5.1 Introduction

Fundamental problems regarding the flow of fluid in a circular tube frequently exist in physics, chemistry, biology, medicine, and engineering. These problems are nonlinear in character and can be analyzed by means of momentum balance. Their most important feature is that they can be described by the Navier–Stokes equations in cylindrical polar coordinates.

When the fluid enters a circular tube from an external device, the velocity profile at the entry of the tube is flat. Immediately after the entry, the velocity profile adjacent to the wall is affected by the friction exerted by the surface of the tube, on the flowing fluid. As the fluid moves further along the tube, this flat portion gradually diminishes, and it ultimately disappears at a point where the flow is fully developed with an asymptotic parabolic velocity profile. The distance between the entry of the tube and this point is called *entry length*. To summarize, one can say that the flow develops and assumes a parabolic velocity profile beyond this entry length.

Theoretically, the length of the circular pipe should be infinite when a steady laminar flow with constant fluid density is considered. Their measure is to avoid the end effects.

5.2 Equations of Motion

Consider an axially symmetrical flow of viscous incompressible fluid. To study this type of flow, we used the equations of motion in cylindrical coordinates (r, θ, z), where the z-axis is taken as the axis of symmetry. Let u, v, and w be the components of velocity, where u is the radial velocity component perpendicular to the symmetrical axis, v is the rotational component, and w is the velocity component parallel to the z-axis. For the axisymmetric case, the value of v is taken to be zero i.e., $v = 0$, and the motion is considered to be two-dimensional, so that u, w, and the pressure p are independent of θ.

Under such assumptions the simplified Navier–Stokes equations are

$$\frac{\partial u}{\partial t}+u\frac{\partial u}{\partial r}+w\frac{\partial u}{\partial z}=-\frac{1}{\rho}\frac{\partial p}{\partial r}+\upsilon\left[\frac{\partial^2 u}{\partial r^2}+\frac{1}{r}\frac{\partial u}{\partial r}-\frac{u}{r^2}+\frac{\partial^2 u}{\partial z^2}\right] \tag{5.1}$$

$$\frac{\partial w}{\partial t}+u\frac{\partial w}{\partial r}+w\frac{\partial w}{\partial z}=-\frac{1}{\rho}\frac{\partial p}{\partial z}+\nu\left[\frac{\partial^2 w}{\partial r^2}+\frac{1}{r}\frac{\partial w}{\partial r}+\frac{\partial^2 w}{\partial z^2}\right] \tag{5.2}$$

The equation of continuity is

$$\frac{1}{r}\frac{\partial}{\partial r}(ru)+\frac{\partial w}{\partial z}=0 \tag{5.3}$$

If the stream function ψ defined by

$$u=-\frac{1}{r}\frac{\partial\psi}{\partial z},\qquad w=\frac{1}{r}\frac{\partial\psi}{\partial r}, \tag{5.4}$$

is introduced, then the continuity equation is identically satisfied. Substituting (5.4) in (5.1) and (5.2), and then eliminating p, we obtain a fourth-order partial differential equation for ψ, as shown below:

$$\frac{\partial}{\partial t}(D^2\psi)+\frac{1}{r}J-\frac{2}{r^2}\frac{\partial\psi}{\partial z}\cdot D^2\psi=\nu D^4\psi \tag{5.5}$$

where the operator D^2 and the Jacobian are defined by

$$D^2=\frac{\partial^2}{\partial r^2}-\frac{1}{r}\frac{\partial}{\partial r}+\frac{\partial^2}{\partial z^2} \tag{5.6}$$

and

$$J=\frac{\partial(D^2\psi,\psi)}{\partial(r,z)}=\begin{vmatrix}\frac{\partial}{\partial r}(D^2\psi) & \frac{\partial\psi}{\partial r}\\ \frac{\partial}{\partial z}(D^2\psi) & \frac{\partial\psi}{\partial z}\end{vmatrix} \tag{5.7}$$

The equation (5.5) is a nonlinear partial differential equation in ψ, and it is not always possible to obtain an analytic solution of this equation. There are many traditional numerical methods that are followed to solve this equation numerically, but these methods result in massive numerical computations. The modern powerful method known as the decomposition method developed by Adomian [1, 2, 3, 4] has been applied here to obtain analytical approximations to this fourth-order nonlinear partial differential equation without restrictive assumptions and simplifications.

5.2.1 Solution by Regular Decomposition

Let

$$L=\frac{\partial^2}{\partial r^2}-\frac{1}{r}\frac{\partial}{\partial r}=r\frac{\partial}{\partial r}\left(\frac{1}{r}\frac{\partial}{\partial r}\right) \tag{5.8}$$

Then the operator D^2 takes the form

$$D^2 = L + \frac{\partial^2}{\partial z^2} \tag{5.9}$$

Using (5.9), we can write the equation (5.5) as

$$L^2\psi = \frac{\partial}{\partial t}(D^2\psi) - \frac{\partial^4\psi}{\partial z^4} - 2\frac{\partial^2}{\partial z^2}(L\psi) + N\psi \tag{5.10}$$

where $N\psi$ is the nonlinear term defined by

$$N\psi = \frac{1}{r}J - \frac{2}{r^2}\cdot\frac{\partial\psi}{\partial z}\cdot D^2\psi \tag{5.11}$$

The solution ψ_0 involves four integration constants to be evaluated from the prescribed boundary conditions.

If L^{-1} is the inverse operator of L, then operating with this operator on both sides of (5.10), we have

$$\psi(r, z, t)\psi = \psi_0(r, z, t) + L^{-2}\left[\frac{\partial}{\partial t}(D^2\psi) - \frac{\partial^4\psi}{\partial z^4} - 2\frac{\partial^2}{\partial z^2}(L\psi) + N\psi\right] \tag{5.12}$$

where ψ_0 is the solution of

$$L^2\psi = 0 \tag{5.13}$$

The solution ψ_0 involves four constants of integration to be evaluated from the prescribed boundary conditions.

Let ψ and the nonlinear term $N\psi$ be decomposed into the following forms:

$$\psi(r, z, t) = \sum_{m=0}^{\infty}\lambda^m\psi_m \tag{5.14}$$

$$N\psi = \sum_{m=0}^{\infty}\lambda^m A_m \tag{5.15}$$

where A_m is the Adomian's special polynomials to be discussed later. The parameter λ used here has the usual meanings. Then we write the equation (5.12) in the parameterized form

$$\psi(r, z, t) = \psi_0(r, z, t) + \lambda L^{-2}\left[\frac{\partial}{\partial t}(D^2\psi) - \frac{\partial^4\psi}{\partial z^4} - 2\frac{\partial^2}{\partial z^2}(L\psi) + N\psi\right] \tag{5.16}$$

Substituting (5.14) and (5.15) into (5.16), and then equating like-power terms of λ, we have

$$\begin{aligned}
\psi_1(r,z,t) &= L^{-2}\left[\frac{\partial}{\partial t}(D^2\psi_0) - \frac{\partial^4\psi_0}{\partial z^4} - 2\frac{\partial^2}{\partial z^2}(L\psi_0) + A_0\right]\\
\psi_2(r,z,t) &= L^{-2}\left[\frac{\partial}{\partial t}(D^2\psi_1) - \frac{\partial^4\psi_1}{\partial z^4} - 2\frac{\partial^2}{\partial z^2}(L\psi_1) + A_1\right]\\
&\cdots \quad \cdots \quad \cdots \quad \cdots\\
&\cdots \quad \cdots \quad \cdots \quad \cdots\\
\psi_{m+1}(r,z,t) &= L^{-2}\left[\frac{\partial}{\partial t}(D^2\psi_m) - \frac{\partial^4\psi_m}{\partial z^4} - 2\frac{\partial^2}{\partial z^2}(L\psi_m) + A_m\right]\\
&\cdots \quad \cdots \quad \cdots \quad \cdots\\
&\cdots \quad \cdots \quad \cdots \quad \cdots
\end{aligned} \tag{5.17}$$

The polynomials $A_0, A_1, \ldots, A_m$ are Adomian's polynomials. These are defined in such a way that each A_m depends on $\psi_0, \psi_1, \ldots, \psi_n$, that is, $A_0 = A_0(\psi_0)$, $A_1 = A_1(\psi_0, \psi_1)$, $A_2 = A_2(\psi_0, \psi_1, \psi_2)$, etc.

For determining each A_m, we equate (5.11) and (5.15) to get

$$\sum_{m=0}^{\infty}\lambda^m A_m = \frac{1}{r}J - \frac{2}{r^2}\cdot\frac{\partial\psi}{\partial z}\cdot D^2\psi \tag{5.18}$$

Using (5.14) in (5.18) and then comparing the coefficients of like-power terms of λ on both sides of the resulting equation, we get

$$\begin{aligned}
A_0 &= \frac{1}{r}\frac{\partial(D^2\psi_0, \psi_0)}{\partial(r,z)} - \frac{2}{r^2}\cdot\frac{\partial\psi_0}{\partial z}\cdot D^2\psi_0\\
A_1 &= \frac{1}{r}\left[\frac{\partial(D^2\psi_1, \psi_1)}{\partial(r,z)} + \frac{2}{r^2}\cdot\frac{\partial(D^2\psi_0, \psi_0)}{\partial(r,z)}\right]\\
&\quad - \frac{2}{r^2}\left[\frac{\partial\psi_0}{\partial z}\cdot D^2\psi_1 + \frac{\partial\psi_1}{\partial z}\cdot D^2\psi_0\right]\\
&\cdots \quad \cdots \quad \cdots \quad \cdots\\
&\cdots \quad \cdots \quad \cdots \quad \cdots
\end{aligned} \tag{5.19}$$

Finally, we proceed to find out the inverse operator L^{-1} of the operator L, and to do this, we consider the equation

$$L\psi = F \tag{5.20}$$

which can be written as

$$r\frac{\partial}{\partial r}\left(\frac{1}{r}\cdot\frac{\partial\psi}{\partial r}\right) = F \tag{5.21}$$

Solving this equation for ψ, we have

$$\psi = \left[L_1^{-1}r\left(L_1^{-1}r^{-1}\right)\right]F \tag{5.22}$$

remembering that the boundary condition terms vanish and L_1^{-1} is a onefold indefinite integral, that is, $L_1^{-1} = \int(\cdot)dr$. The expression (5.22) shows that the inverse operator L^{-1} has been identified as

$$L^{-1} = \left[L_1^{-1}r\left(L_1^{-1}r^{-1}\right)\right] \tag{5.23}$$

and hence we get

$$L^{-2} = L_1^{-1}\left[rL_1^{-1}\left\{r^{-1}L_1^{-1}\left(rL_1^{-1}r^{-1}\right)\right\}\right] \tag{5.24}$$

Thus we see that once ψ_0 is determined, ψ_1 is computable. Again, if ψ_0 and ψ_1 are determined, ψ_2 is computable and so on. Then the final solution is given by

$$\psi(r, z, t) = \sum_{m=0}^{\infty} \psi_m(r, z, t) \tag{5.25}$$

remembering that $\lambda = 1$.

5.2.2 Solution by Double Decomposition

For the solution of the double decomposition procedure, we begin with the equation (5.10), that is,

$$L^2\psi = \frac{\partial}{\partial t}(D^2\psi) - \frac{\partial^4\psi}{\partial z^4} - 2\frac{\partial^2}{\partial z^2}(L\psi) + N\psi$$

where the nonlinear term $N\psi$ is given by (5.11). The operators L and D^2 are defined by (5.8) and (5.9).

Then following the single (regular) decomposition procedure, we write the solution as

$$\psi(r, z, t) = \psi_0(r, z, t) + L^{-2}\left[\frac{\partial}{\partial t}(D^2\psi) - \frac{\partial^4\psi}{\partial z^4} - 2\frac{\partial^2}{\partial z^2}(L\psi) + N\psi\right]$$

where ψ_0 is the solution of the homogeneous equation $L^2\psi = 0$, and it is given by

$$\psi(r, z, t)\psi = \frac{1}{16}\alpha(z, t)r^4 + \beta(z, t)L_1^{-1}r\log r + \frac{1}{2}\gamma(z, t)r^2 + \delta(z, t) \tag{5.26}$$

Here $L_1^{-1} = \int(\cdot)dr$ and $\alpha(z, t)$, $\beta(z, t)$, $\gamma(z, t)$, $\delta(z, t)$ are the constants of integration to be determined by the prescribed boundary conditions.

One systematic way of using the decomposition method is to introduce a formal counting parameter λ to write the solution $\psi(r, z, t)$ in the following form:

$$\psi(r, z, t) = \psi_0(r, z, t) + \lambda L_1^{-1}\left[\frac{\partial}{\partial t}(D^2\psi) - \frac{\partial^4\psi}{\partial z^4} - 2\frac{\partial^2}{\partial z^2}(L\psi) + N\psi\right] \tag{5.27}$$

The equation (5.27) is called a *parameterized equation*, and the parameter λ inserted here is not a perturbation parameter; it is used only for grouping the terms.

Then ψ and the nonlinear term $N\psi$ are decomposed into the following parameterized forms:

$$\psi(r, z, t) = \sum_{m=0}^{\infty} \lambda^m \psi_m \tag{5.28}$$

$$N\psi = \sum_{m=0}^{\infty} \lambda^m A_m \tag{5.29}$$

where the A_m is the Adomian's special polynomials given by (5.19).

If we again take the parameterized decomposition of ψ_0 given by

$$\psi_0(r, z, t) = \sum_{m=0}^{\infty} \lambda^m \psi_{0,m} \tag{5.30}$$

we mean the double decomposition of the solution ψ. Substituting (5.28), (5.29), and (5.30) in (5.27)

$$\sum_{m=0}^{\infty} \lambda^m \psi_m = \sum_{m=0}^{\infty} \lambda^m \psi_{0,m} + \lambda L^{-2} \left[\frac{\partial}{\partial t} \left(D^2 \sum_{m=0}^{\infty} \lambda^m \psi_m \right) - \frac{\partial^4}{\partial z^4} \sum_{m=0}^{\infty} \lambda^m \psi_m \right.$$
$$\left. - 2\frac{\partial^2}{\partial z^2} \left(L \sum_{m=0}^{\infty} \lambda^m \psi_m \right) + \sum_{m=0}^{\infty} \lambda^m A_m \right]$$

Equating the coefficient of like-power terms of λ from both sides of the above equation, we get

$$\psi_0(r, z, t) = \psi_{0,0}$$

$$\psi_1(r, z, t) = \psi_{0,1} + L^{-2} \left[\frac{\partial}{\partial t} \left(D^2 \psi_0 \right) - \frac{\partial^4 \psi_0}{\partial z^4} - 2\frac{\partial^2}{\partial z^2} (L\psi_0) + A_0 \right]$$

$$\psi_2(r, z, t) = \psi_{0,2} + L^{-2} \left[\frac{\partial}{\partial t} \left(D^2 \psi_1 \right) - \frac{\partial^4 \psi_1}{\partial z^4} - 2\frac{\partial^2}{\partial z^2} (L\psi_1) + A_1 \right]$$

$$\cdots \quad \cdots \quad \cdots \quad \cdots \quad \cdots \quad \cdots \tag{5.31}$$

$$\cdots \quad \cdots \quad \cdots \quad \cdots \quad \cdots \quad \cdots$$

$$\psi_{m+1}(r, z, t) = \psi_{0(m+1)} + L^{-2} \left[\frac{\partial}{\partial t} \left(D^2 \psi_m \right) - \frac{\partial^4 \psi_m}{\partial z^4} - 2\frac{\partial^2}{\partial z^2} (L\psi_m) + A_1 \right]$$

$$\cdots \quad \cdots \quad \cdots \quad \cdots \quad \cdots \quad \cdots$$

$$\cdots \quad \cdots \quad \cdots \quad \cdots \quad \cdots \quad \cdots$$

Since the constants of integration in the solution ψ_0 are $\alpha(z,t)$, $\beta(z,t)$, $\gamma(z,t)$, and $\delta(z,t)$, the parameterized decomposition forms of these constants are

$$\begin{aligned}\alpha(z,t) &= \sum_{m=0}^{\infty}\lambda^m\alpha_m\\ \beta(z,t) &= \sum_{m=0}^{\infty}\lambda^m\beta_m\\ \gamma(z,t) &= \sum_{m=0}^{\infty}\lambda^m\gamma_m\\ \delta(z,t) &= \sum_{m=0}^{\infty}\lambda^m\delta_m\end{aligned} \tag{5.32}$$

Then we substitute (5.30) and (5.32) in (5.26) to get

$$\sum_{m=0}^{\infty}\lambda^m\psi_{0m} = \frac{r^4}{16}\sum_{m=0}^{\infty}\lambda^m\alpha_m + L_1^{-1}\log r\sum_{m=0}^{\infty}\lambda^m\beta_m + \frac{r^2}{2}\sum_{m=0}^{\infty}\lambda^m\gamma_m + \sum_{m=0}^{\infty}\lambda^m\delta_m$$

Comparison of like-power terms of λ on both sides of the above relation gives

$$\begin{aligned}\psi_{0,0}(r,z,t) &= \frac{1}{16}\alpha_0 r^4 + L_1^{-1}r\log r\beta_0 + \frac{1}{2}\gamma_0 r^2 + \delta_0\\ \psi_{0,1}(r,z,t) &= \frac{1}{16}\alpha_1 r^4 + L_1^{-1}r\log r\beta_1 + \frac{1}{2}\gamma_1 r^2 + \delta_1\\ \psi_{0,2}(r,z,t) &= \frac{1}{16}\alpha_2 r^4 + L_1^{-1}r\log r\beta_2 + \frac{1}{2}\gamma_2 r^2 + \delta_2\\ &\cdots \quad \cdots \quad \cdots \quad \cdots \quad \cdots\\ &\cdots \quad \cdots \quad \cdots \quad \cdots \quad \cdots\\ \psi_{0,m}(r,z,t) &= \frac{1}{16}\alpha_m r^4 + L_1^{-1}r\log r\beta_m + \frac{1}{2}\gamma_m r^2 + \delta_m\\ &\cdots \quad \cdots \quad \cdots \quad \cdots \quad \cdots\\ &\cdots \quad \cdots \quad \cdots \quad \cdots \quad \cdots\end{aligned} \tag{5.33}$$

In this analysis the solution is obtained, as the components of ψ are determined. But each component contains some constants of integration that are to be evaluated by their specific conditions.

5.2.3 Solution by Modified Decomposition

For the solution of modified decomposition, we use the following transformation:

$$y = \log r \tag{5.34}$$

Then the equation (5.10) becomes

$$\begin{aligned}\frac{\partial^4\psi}{\partial y^4} &= 4\frac{\partial^3\psi}{\partial y^3} - 12\frac{\partial^2\psi}{\partial y^2} + 16\frac{\partial\psi}{\partial y}\\ &\quad + e^{-4y}\left[\frac{\partial}{\partial t}\left(D^2\psi\right) - \frac{\partial^4\psi}{\partial z^4} - 2\frac{\partial^2}{\partial z^2}(L\psi) + N\psi\right]\\ &= 4\frac{\partial^3\psi}{\partial y^3} - 12\frac{\partial^2\psi}{\partial y^2} + 16\frac{\partial\psi}{\partial y}\\ &\quad + e^{-6y}\frac{\partial}{\partial t}\left(\frac{\partial^2\psi}{\partial y^2} - 2\frac{\partial\psi}{\partial y}\right) + e^{-4y}\frac{\partial^3\psi}{\partial t\partial z^2}\\ &\quad - e^{-4y}\frac{\partial^4\psi}{\partial z^4} - 2e^{-6y}\frac{\partial^2}{\partial z^2}\left(\frac{\partial^2\psi}{\partial y^2} - 2\frac{\partial\psi}{\partial y}\right) + e^{-4y}N\psi \end{aligned}\tag{5.35}$$

Let $L_1 = \frac{\partial^4}{\partial y^4}$ be a fourth-order linear operator. Its inverse form is a fourfold indefinite integrals, and it is defined by

$$L_1^{-1} = \iiiint(\cdot)\,dy\,dy\,dy\,dy \tag{5.36}$$

Using this inverse form on both sides of (5.35), we get the regular decomposition as

$$\begin{aligned}\psi(y,z,t) &= \overline{\psi}_y(y,z,t) + L_1^{-1}\left[4\frac{\partial^3\psi}{\partial y^3} - 12\frac{\partial^2\psi}{\partial y^2} + 16\frac{\partial\psi}{\partial y}\right.\\ &\quad + e^{-6y}\frac{\partial}{\partial t}\left(\frac{\partial^2\psi}{\partial y^2} - 2\frac{\partial\psi}{\partial y}\right) + e^{-4y}\frac{\partial^3\psi}{\partial t\partial z^2}\\ &\quad \left. - 2e^{-6y}\frac{\partial}{\partial z^2}\left(\frac{\partial^2\psi}{\partial y^2} - 2\frac{\partial\psi}{\partial y}\right) - e^{-4y}\frac{\partial^4\psi}{\partial z^4} + e^{-4y}N\psi\right]\end{aligned}\tag{5.37}$$

where ψ_y is the solution of $\frac{\partial^4\psi}{\partial y^4} = 0$. This solution involves four constants of integration that are to be evaluated from the prescribed boundary conditions, and it is given by

$$\overline{\psi}_y(y,z,t) = \xi_0(z,t) + \xi_1(z,t)y + \xi_2(z,t)\frac{y^2}{2} + \xi_3(z,t)\frac{y^3}{6} \tag{5.38}$$

We now proceed to apply the modified decomposition procedure to (5.37). Since the equation (5.37) is a fourth-order nonlinear partial differential equation with variable coefficients, we write for the purpose

$$\begin{aligned}\psi(y,z,t) &= \sum_{m=0}^{\infty}\sum_{n=0}^{\infty}\sum_{l=0}^{\infty} C_{m,n,1}Y^m z^n t^l\\ &= \sum_{m=0}^{\infty}\left[\sum_{n=0}^{\infty}\sum_{l=0}^{\infty} C_{m,n,l}z^n t^l\right]y^m = \sum_{m=0}^{\infty} a_m(z,t)y^m\end{aligned}\tag{5.39}$$

$$N\psi = \sum_{m=0}^{\infty}\sum_{n=0}^{\infty}\sum_{l=0}^{\infty} A_{m,n,t} Y^m z^n t^l = \sum_{m=0}^{\infty}\left[\sum_{n=0}^{\infty}\sum_{l=0}^{\infty} A_{m,n,l} z^n t^l\right] y^m$$

$$= \sum_{m=0}^{\infty} A_m(z, t) y^m \tag{5.40}$$

$$e^{-6y} = g\sum_{m=0}^{\infty} g_m y^m \tag{5.41}$$

$$e^{-4y} = h\sum_{m=0}^{\infty} h_m y^m \tag{5.42}$$

where

$$a_m(z, t) = \sum_{n=0}^{\infty}\sum_{l=0}^{\infty} C_{m,n,l} z^n t^l \tag{5.43}$$

$$A_m(z, t) = \sum_{n=0}^{\infty}\sum_{l=0}^{\infty} A_{m,n,l} z^n t^l \tag{5.44}$$

Substituting the relations from (5.38) to (5.42) in (5.37), we get

$$\sum_{m=0}^{\infty} a_m(z, t) y^m = \xi_0(z, t) + \xi_1(z, t) y + \xi_2(z, t)\frac{y^2}{2} + \xi_3(z, t)\frac{y^3}{16}$$

$$+ L_1^{-1}\left[4\frac{\partial^3}{\partial y^3}\sum_{m=0}^{\infty} a_m(z, t) y^m - 12\frac{\partial^2}{\partial y^2}\sum_{m=0}^{\infty} a_m(z, t) y^m + 16\frac{\partial}{\partial y}\sum_{m=0}^{\infty} a_m(z, t) y^m\right]$$

$$+ L_1^{-1}\left[\sum_{m=0}^{\infty} g_m y^m\right]\left[\frac{\partial}{\partial t}\left\{\frac{\partial^2}{\partial y^2}\sum_{m=0}^{\infty} a_m(z, t) y^m - 2\frac{\partial}{\partial y}\sum_{m=0}^{\infty} a_m(z, t) y^m\right\}\right]$$

$$- 2L_1^{-1}\left[\sum_{m=0}^{\infty} g_m y^m\right]\left[\frac{\partial^2}{\partial z^2}\left\{\frac{\partial^2}{\partial y^2}\sum_{m=0}^{\infty} a_m(z, t) y^m - 2\frac{\partial}{\partial y}\sum_{m=0}^{\infty} a_m(z, t) y^m\right\}\right]$$

$$+ L_1^{-1}\left[\sum_{m=0}^{\infty} h_m y^m\right]\left[\frac{\partial}{\partial t}\left\{\frac{\partial^2}{\partial z^2}\sum_{m=0}^{\infty} a_m(z, t) y^m\right\}\right]$$

$$- L_1^{-1}\left[\sum_{m=0}^{\infty} h_m y^m\right]\left[\frac{\partial^4}{\partial z^4}\sum_{m=0}^{\infty} a_m(z, t) y^m\right]$$

$$+ L_1^{-1}\left[\sum_{m=0}^{\infty} h_m y^m\right]\left[\sum_{m=0}^{\infty} A_m(z, t) y^m\right]$$

$$\sum_{m=0}^{\infty} a_m(z,t)y^m = \xi_0(x,t) + \xi_1(z,t)y + \xi_2(z,t)\frac{y^2}{2} + \xi_3(z,t)\frac{y^3}{16}$$
$$+ L_1^{-1}\left[\sum_{m=0}^{\infty} 4(m+1)(m+2)(m+3)a_{m+3}(z,t)y^m\right]$$
$$- L_1^{-1}\left[\sum_{m=0}^{\infty} 12(m+1)(m+2)a_{m+2}(z,t)y^m\right]$$
$$+ L_1^{-1}\left[\sum_{m=0}^{\infty} 16(m+1)a_{m+1}(z,t)y^m\right]$$
$$+ L_1^{-1}\left[\left\{\sum_{m=0}^{\infty} g_m y^m\right\}\left\{\sum_{m=0}^{\infty} (m+1)(m+2)(\partial/\partial t)a_{m+2}(z,t)y^m\right\}\right]$$
$$- L_1^{-1}\left[\left\{\sum_{m=0}^{\infty} g_m y^m\right\}\left\{\sum_{m=0}^{\infty} 2(m+1)(\partial/\partial t)a_{m+1}(z,t)y^m\right\}\right]$$
$$- L_1^{-1}\left[\left\{\sum_{m=0}^{\infty} g_m y^m\right\}\left\{\sum_{m=0}^{\infty} 2(m+1)(m+2)(\partial^2/\partial z^2)a_{m+2}(z,t)y^m\right\}\right]$$
$$+ L_1^{-1}\left[\left\{\sum_{m=0}^{\infty} g_m y^m\right\}\left\{\sum_{m=0}^{\infty} 4(m+1)(\partial^2/\partial z^2)a_{m+1}(z,t)y^m\right\}\right]$$
$$+ L_1^{-1}\left[\left\{\sum_{m=0}^{\infty} h_m y^m\right\}\left\{\sum_{m=0}^{\infty} (\partial^3/\partial t\partial z^2)a_m(z,t)y^m\right\}\right]$$
$$- L_1^{-1}\left[\left\{\sum_{m=0}^{\infty} h_m y^m\right\}\left\{\sum_{m=0}^{\infty} (\partial^4/\partial z^4)a_m(z,t)y^m\right\}\right]$$
$$+ L_1^{-1}\left[\left\{\sum_{m=0}^{\infty} h_m y^m\right\}\left\{\sum_{m=0}^{\infty} A_m(z,t)y^m\right\}\right]$$

Applying the Cauchy product of infinite series, we have

$$\sum_{m=0}^{\infty} a_m(z,t)y^m = \xi_0(z,t) + \xi_1(z,t)y + \xi_2(z,t)\frac{y^2}{2} + \xi_3(z,t)\frac{y^3}{6}$$
$$+ L_1^{-1}\left[\sum_{m=0}^{\infty} 4(m+1)(m+2)(m+3)a_{m+3}(z,t)y^m\right]$$
$$- L_1^{-1}\left[\sum_{m=0}^{\infty} 12(m+1)(m+2)a_{m+2}(z,t)y^m\right]$$

$$+ L_1^{-1}\left[\sum_{m=0}^{\infty} 16(m+1)a_{m+1}(z,t)y^m\right]$$

$$+ L_1^{-1}\left[\sum_{m=0}^{\infty}\left\{\sum_{n=0}^{m} g_{m-n}(n+1)(n+2)(\partial/\partial t)a_{n+2}(z,t)\right\}y^m\right]$$

$$- L_1^{-1}\left[\sum_{m=0}^{\infty}\left\{\sum_{n=0}^{m} 2g_{m-n}(n+1)(\partial/\partial t)a_{n+1}(z,t)\right\}y^m\right]$$

$$- L_1^{-1}\left[\sum_{m=0}^{\infty}\left\{\sum_{n=0}^{m} 2g_{m-n}(n+1)(n+2)(\partial^2/\partial z^2)a_{n+2}(z,t)\right\}y^m\right]$$

$$+ L_1^{-1}\left[\sum_{m=0}^{\infty}\left\{\sum_{n=0}^{m} 4g_{m-n}(n+1)(\partial^2/\partial z^2)a_{n+1}(z,t)\right\}Y^m\right]$$

$$+ L_1^{-1}\left[\sum_{m=0}^{\infty}\left\{\sum_{n=0}^{m} h_{m-n}(\partial^3/\partial t\partial z^2)a_n(z,t)\right\}y^m\right]$$

$$- L_1^{-1}\left[\sum_{m=0}^{\infty}\left\{\sum_{n=0}^{m} h_{m-n}(\partial^4/\partial z^4)a_n(z,t)y^m\right\}y^m\right]$$

$$+ L_1^{-1}\left[\sum_{m=0}^{\infty}\left\{\sum_{n=0}^{m} h_{m-n}A_n(z,t)\right\}y^m\right]$$

Integrating and then replacing m by $(m-4)$ on the right side of the above relation, we assign the following equation at

$$\sum_{m=0}^{\infty} a_m(z,t)y^m = \xi_0(z,t) + \xi_1(z,t)y + \xi_2(z,t)\frac{y^2}{2} + \xi_3(z,t)\frac{y^3}{16}$$

$$+ \sum_{m=4}^{\infty}[\{4(m-1)(m-2)(m-3)a_{m-1}(z,t)\}$$

$$- \{12(m-2)(m-3)a_{m-2}(z,t)\}$$

$$+ 16(m-3)a_{m-3}(z,t)$$

$$+ \sum_{n=0}^{m-4} g_{m-n-4}(n+1)(n+2)(\partial/\partial t)a_{n+2}(z,t)$$

$$- \sum_{n=0}^{m-4} 2g_{m-n-4}(n+1)(\partial/\partial t)a_{n+1}(z,t)$$

$$- \sum_{n=0}^{m-4} 2g_{m-n-4}(n+1)(n+2)(\partial^2/\partial z^2)a_{n+2}(z,t)$$

$$+\sum_{n=0}^{m-4} 4g_{m-n-4}(n+1)(\partial^2/\partial z^2)a_{n+1}(z,t)$$

$$+\sum_{n=0}^{m-4} h_{m-n-4}(\partial^3/\partial t\partial z^2)a_n(z,t)$$

$$-\sum_{n=0}^{m-4} h_{m-n-4}(\partial^4/\partial z^4)a_n(z,t)$$

$$\left.+\sum_{n=0}^{m-4} h_{m-n-4}A_n(z,t)\right] y^m/m(m-1)(m-2)(m-3)$$

The coefficients are identified by

$$\left.\begin{aligned} a_0(z,t) &= \xi_0(z,t) \\ a_1(z,t) &= \xi_1(z,t) \\ a_2(z,t) &= \tfrac{1}{2}\xi_2(z,t) \\ a_3(z,t) &= \tfrac{1}{16}\xi_3(z,t) \end{aligned}\right\} \tag{5.45}$$

and the recurrence relation for $m \geq 4$

$$\begin{aligned} &m(m-1)(m-2)(m-3)a_m(z,t) \\ &\quad = 4(m-1)(m-2)(m-3)a_{m-1}(z,t) - 12(m-2)(m-3)a_{m-2}(z,t) \\ &\qquad + 16(m-3)a_{m-3}(z,t) + \sum_{n=0}^{m-4} g_{m-n-4}(n+1)(n+2)(\partial/\partial t)a_{n+2}(z,t) \\ &\qquad - \sum_{n=0}^{m-4} g_{m-n-4}(n+1)(\partial/\partial t)a_{n+1}(z,t) \\ &\qquad - \sum_{n=0}^{m-4} g_{m-n-4}(n+1)(n+2)(\partial^2/\partial z^2)a_{n+2}(z,t) \\ &\qquad + \sum_{n=0}^{m-4} 4g_{m-n-4}(n+1)(\partial^2/\partial z^2)a_{n+1}(z,t) + \sum_{n=0}^{m-4} h_{m-n-4}(\partial^3/\partial t\partial z^2)a_n(z,t) \\ &\qquad - \sum_{n=0}^{m-4} h_{m-n-4}(\partial^4/\partial z^4)a_n(z,t) + \sum_{n=0}^{m-4} h_{m-n-4}A_n(z,t) \end{aligned} \tag{5.46}$$

which gives other coefficients for different values of $m \geq 4$.

Determination of A_m

Using the transformation (5.34) in (5.11), we have

$$\begin{aligned} N\psi &= e^{-4y}\frac{\partial\psi}{\partial z}\left(\frac{\partial^3\psi}{\partial y^3} - 4\frac{\partial^2\psi}{\partial y^2} + 4\frac{\partial\psi}{\partial y}\right) + e^{-2y}\frac{\partial\psi}{\partial z}.\frac{\partial^3\psi}{\partial y\partial z^2} \\ &- e^{-4y}\frac{\partial\psi}{\partial y}\left(\frac{\partial^3\psi}{\partial y^2\partial z} - 2\frac{\partial^2\psi}{\partial y\partial z}\right) - e^{-2y}\frac{\partial\psi}{\partial y}.\frac{\partial^3\psi}{\partial z^3} \\ &- 2e^{-4y}.\frac{\partial\psi}{\partial z}\left(\frac{\partial^2\psi}{\partial y^2} - 2\frac{\partial\psi}{\partial y}\right) - 2e^{-2y}\frac{\partial\psi}{\partial z}.\frac{\partial^2\psi}{\partial z^2} \end{aligned} \tag{5.47}$$

Substituting the relations from (5.39) to (5.42) and the relation given by

$$e^{-2y} = k = \sum_{m=0}^{\infty} k_m y^m \tag{5.48}$$

in (5.47), we get

$$\begin{aligned} N\psi = &\left[\sum_{m=0}^{\infty} k_m y^m\right]\left[\sum_{m=0}^{\infty}(\partial/\partial z)a_m(z,t)y^m\right] \\ &\times \left[\sum_{m=0}^{\infty}(m+1)(m+2)(m+3)a_{m+3}(z,t)y^m\right] \\ &- 4\left[\sum_{m=0}^{\infty} h_m y^m\right]\left[\sum_{m=0}^{\infty}(\partial/\partial z)a_m(z,t)y^m\right]\left[\sum_{m=0}^{\infty}(m+1)(m+2)a_{m+2}(z,t)y^m\right] \\ &+ 4\left[\sum_{m=0}^{\infty} h_m y^m\right]\left[\sum_{m=0}^{\infty}(\partial/\partial z)a_m(z,t)y^m\right]\left[\sum_{m=0}^{\infty}(m+1)a_{m+1}(z,t)y^m\right] \\ &- \left[\sum_{m=0}^{\infty} h_m y^m\right]\left[\sum_{m=0}^{\infty}(m+1)a_{m+1}(z,t)y^m\right] \\ &\times \left[\sum_{m=0}^{\infty}(m+1)(m+2)(\partial/\partial z)a_{m+2}(z,t)y^m\right] \\ &+ 2\left[\sum_{m=0}^{\infty} h_m y^m\right]\left[\sum_{m=0}^{\infty}(m+1)a_{m+1}(z,t)y^m\right]\left[\sum_{m=0}^{\infty}(m+1)(\partial/\partial z)a_{m+1}(z,t)y^m\right] \\ &- 2\left[\sum_{m=0}^{\infty} h_m y^m\right]\left[\sum_{m=0}^{\infty}(\partial/\partial z)a_m(z,t)y^m\right]\left[\sum_{m=0}^{\infty}(m+1)(m+2)a_{m+2}(z,t)y^m\right] \\ &+ 4\left[\sum_{m=0}^{\infty} h_m y^m\right]\left[\sum_{m=0}^{\infty}(\partial/\partial z)a_m(z,t)y^m\right]\left[\sum_{m=0}^{\infty}(m+1)a_{m+1}(z,t)y^m\right] \end{aligned}$$

$$+\left[\sum_{m=0}^{\infty}k_m y^m\right]\left[\sum_{m=0}^{\infty}(\partial/\partial z)a_m(z,t)y^m\right]\left[\sum_{m=0}^{\infty}(m+1)(\partial^2/\partial z^2)a_{m+1}(z,t)y^m\right]$$

$$-\left[\sum_{m=0}^{\infty}k_m y^m\right]\left[\sum_{m=0}^{\infty}(m+1)(z,t)y^m\right]\left[\sum_{m=0}^{\infty}(\partial^3/\partial z^3)a_m(z,t)y^m\right]$$

$$-2\left[\sum_{m=0}^{\infty}k_m y^m\right]\left[\sum_{m=0}^{\infty}(\partial/\partial z)a_m(z,t)y^m\right]\left[\sum_{m=0}^{\infty}(\partial^2/\partial z^2)a_m(z,t)y^m\right] \quad (5.49)$$

Carrying out the Cauchy product of infinite series, we obtain

$$\begin{aligned}
N\psi =& \sum_{m=0}^{\infty}\left[\sum_{i=0}^{m}\left\{\sum_{n=0}^{m-i}h_{m-n-i}(\partial/\partial z)a_n(z,t)(i+1)(i+2)(i+3)a_{i+3}(z,t)\right\}\right]y^m \\
&- 4\sum_{m=0}^{\infty}\left[\sum_{i=0}^{m}\left\{\sum_{n=0}^{m-i}h_{m-n-i}(\partial/\partial z)a_n(z,t)(i+1)(i+2)(i+3)a_{i+3}(z,t)\right\}\right]y^m \\
&+ 4\sum_{m=0}^{\infty}\left[\sum_{i=0}^{m}\left\{\sum_{n=0}^{m-i}h_{m-n-i}(\partial/\partial z)a_n(z,t)(i+1)a_{i+1}(z,t)\right\}\right]y^m \\
&- \sum_{m=0}^{\infty}\left[\sum_{i=0}^{m}\left\{\sum_{n=0}^{m-i}h_{m-n-i}(n+1)a_{n+1}(z,t)(i+1)(i+2)(\partial/\partial z)a_{i+2}(z,t)\right\}\right]y^m \\
&+ 2\sum_{m=0}^{\infty}\left[\sum_{i=0}^{m}\left\{\sum_{n=0}^{m-i}h_{m-n-i}(n+1)a_{n+1}(z,t)(i+1)(\partial/\partial z)a_{i+1}(z,t)\right\}\right]y^m \\
&- 2\sum_{m=0}^{\infty}\left[\sum_{i=0}^{m}\left\{\sum_{n=0}^{m-i}h_{m-n-i}(\partial/\partial z)a_n(z,t)(i+1)(i+2)a_{i+2}(z,t)\right\}\right]y^m \\
&+ 4\sum_{m=0}^{\infty}\left[\sum_{i=0}^{m}\left\{\sum_{n=0}^{m-i}h_{m-n-i}(\partial/\partial z)a_n(z,t)(i+1)a_{i+2}(z,t)\right\}\right]y^m \\
&+ \sum_{m=0}^{\infty}\left[\sum_{i=0}^{m}\left\{\sum_{n=0}^{m-i}k_{m-n-i}(\partial/\partial z)a_n(z,t)(i+1)(\partial^2/\partial z^2)a_{i+1}(z,t)\right\}\right]y^m \\
&- \sum_{m=0}^{\infty}\left[\sum_{i=0}^{m}\left\{\sum_{n=0}^{m-i}k_{m-n-i}(n+1)a_{n+1}(z,t)(\partial^3/\partial z^3)a_i(z,t)\right\}\right]y^m \\
&- 2\sum_{m=0}^{\infty}\left[\sum_{i=0}^{m}\left\{\sum_{n=0}^{m-i}k_{m-n-i}(\partial/\partial z)a_n(z,t)(\partial^2/\partial z^2)a_i(z,t)\right\}\right]y^m
\end{aligned} \quad (5.50)$$

Using (4.40) and (4.50), and comparing the coefficients from both sides, we obtain

$$\begin{aligned}
A_m(z,t) = \sum_{i=0}^{m}\Bigg[&\sum_{n=0}^{m-i} h_{m-n-i}(\partial/\partial z)a_n(z,t)\cdot(i+1)(i+2)(i+3)a_{i+3}(z,t)\\
&-4\sum_{n=0}^{m-i} h_{m-n-i}(\partial/\partial z)a_n(z,t)\cdot(i+1)(i+2)a_{i+2}(z,t)\\
&+4\sum_{n=0}^{m-i} h_{m-n-i}(\partial/\partial z)a_n(z,t)\cdot(i+1)a_{i+1}(z,t)\\
&-\sum_{n=0}^{m-i} h_{m-n-i}(n+1)a_{n+1}(z,t)(i+1)(i+2)(\partial/\partial z)a_{i+2}(z,t)\\
&+2\sum_{n=0}^{m-i} h_{m-n-i}(n+1)a_{n+1}(z,t)\cdot(i+1)(\partial/\partial z)a_{i+1}(z,t)\\
&-2\sum_{n=0}^{m-i} h_{m-n-i}(\partial/\partial z)a_n(z,t)\cdot(i+1)(i+2)a_{i+2}(z,t)\\
&+u\sum_{n=0}^{m-i} h_{m-n-i}(\partial/\partial z)a_n(z,t)\cdot(i+1)a_{i+1}(z,t)\\
&+\sum_{n=0}^{m-i} k_{m-n-i}(\partial/\partial z)a_n(z,t)\cdot(i+1)(\partial^2/\partial z^2)a_{i+1}(z,t)\\
&-\sum_{n=0}^{m-i} k_{m-n-i}(n+1)a_{n+1}(z,t)\cdot(\partial^3/\partial z^3)a_i(z,t)\\
&-2\sum_{n=0}^{m-i} k_{m-n-i}(\partial/\partial z)a_n(z,t)\cdot(\partial^2/\partial z^2)a_i(z,t)\Bigg]
\end{aligned} \tag{5.51}$$

which gives the expressions for the Adomian's special polynomials for different values of m.

For a clear illustration, we discuss some physical problems in the next sections on the basis of the decomposition method.

5.3 Hagen–Poiseuille Theory: The Steady Laminar Flow of Fluid through a Circular Tube

Consider the steady laminar flow of viscous incompressible fluid through a circular pipe of infinite length, uniform cross section, and radius a Figure 5.1.

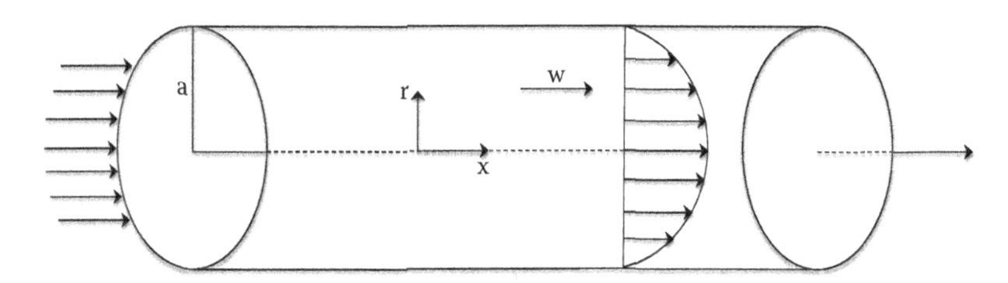

FIGURE 5.1
Hagen–Poiseuille flow through a tube.

To study this type of flow, we find it convenient to use the cylindrical polar coordinates (r, θ, z) where z is taken as the axis of the cylinder. It is assumed that there is no rotational motion in the tube, that is, the tangential velocity component is negligible. Furthermore, the variation of the radial velocity component is very very small in comparison to the axial velocity component. So it is also negligible.

If (u, v, w) are the velocity components along the radial, tangential, and axial directions, respectively, then $u = v = 0$ and the continuity equation, under the above assumptions, reduces to $(\partial w/\partial z) = 0$, which shows that w is a function of radial coordinate r only. The simplified Navier–Stokes equations are

$$\frac{\partial p}{\partial r} = 0, \quad \frac{\partial p}{\partial \theta} = 0 \tag{5.52}$$

which show that pressure p is a function of z only and

$$\frac{d^2 w}{\partial r^2} + \frac{1}{r}\frac{dw}{dr} = \frac{1}{\mu}\frac{dp}{dz} \tag{5.53}$$

The boundary conditions imposed on w are

$$w(a) = \left(\frac{dw}{dr}\right)_{r=0} = 0 \tag{5.54}$$

Now we have to solve the equation (5.53) under the boundary conditions (5.54) by means of the decomposition method. For solving this problem, we use the transformation defined by $y = \log r$ and write the equation (5.53) as

$$\frac{d^2 w}{dy^2} = \frac{1}{\mu}\frac{dp}{dz}e^{2y} \tag{5.55}$$

Let $Ly = \frac{d^2}{dy^2}$ be the linear second-order differential operator, and its inverse form is defined by $L_y^{-1} = \int\int(\cdot)dydy$, which is a twofold indefinite integral. Operating with this inverse operator on both sides of (5.5), we have the single (regular) decomposition solution as

$$w(y) = \alpha + \beta y + L_y^{-1} g \tag{5.56}$$

where

$$g = \overline{K}e^{2y} \tag{5.57}$$

and

$$\overline{K} = \frac{1}{\mu}\frac{dp}{dz} \tag{5.58}$$

Then we apply the modified decomposition procedure to (5.56) and write

$$w(y) = \sum_{m=0}^{\infty} \eta_m y^m \tag{5.59}$$

$$\overline{K}e^{2y} = g = \sum_{m=0}^{\infty} g_m y^m \tag{5.60}$$

We substitute (5.59) and (5.60) in (5.56) to get

$$\sum_{m=0}^{\infty} \eta_m y^m = \alpha + \beta y + L_y^{-1}\sum_{m=0}^{\infty} g_m y^m \tag{5.61}$$

Performing integrations and then replacing m by $(m-2)$ on the right side of (5.61), we have

$$\sum_{m=0}^{\infty} \eta_m y^m = \alpha + \beta y + \sum_{m=2}^{\infty} \frac{g_{m-2}}{m(m-1)} y^m \tag{5.62}$$

Equating the coefficient of different terms from both sides of (5.52), we obtain

$$\eta_0 = \alpha, \quad \eta_1 = \beta \tag{5.63}$$

and the recurrence relations for $m \geq 2$

$$\eta_m = \frac{g_{m-2}}{m(m-1)} \tag{5.64}$$

Now from (5.60) we get

$$g_0 + g_1 y + g_2 y^2 + g_3 y^3 + \cdots = \overline{K}\left[1 + \frac{(2y)}{1!} + \frac{(2y)^2}{2!} + \frac{(2y)^3}{3!} + \cdots\right]$$

Comparing the terms on both sides of the above relation, we have

$$g_0 = \overline{K}, \; g_1 = \frac{2\overline{K}}{1!}, \; g_2 = \frac{2^2\overline{K}}{2!}, \; g_3 = \frac{2^3\overline{K}}{3!}, \quad \text{etc.} \tag{5.65}$$

Again from (5.64), we obtain using (5.65)

$$\begin{aligned} \eta_2 &= \frac{g_0}{2!} = \frac{\overline{K}}{2!} \\ \eta_3 &= \frac{g_1}{3!} = \frac{2\overline{K}}{3!} \\ \eta_4 &= \frac{g_2}{3.4} = \frac{2^2\overline{K}}{4!} \\ \eta_5 &= \frac{g_3}{4.5} = \frac{2^3\overline{K}}{5!}, \quad \text{etc.} \end{aligned} \tag{5.66}$$

Using (5.66), we write the solution (5.59) as

$$w(y) = \left(\eta_0 - \frac{\overline{K}}{4}\right) + \left(\eta_1 - \frac{\overline{K}}{2}\right) y + \frac{\overline{K}}{4} e^{2y}$$

$$w(r) = \left(\eta_0 - \frac{\overline{K}}{4}\right) + \left(\eta_1 - \frac{\overline{K}}{2}\right) \log r + \frac{\overline{K}}{4} r^2 \tag{5.67}$$

Satisfying the boundary conditions by (5.67), we get

$$\eta_0 - \frac{\overline{K}}{4} = -\frac{\overline{K}a^2}{4}, \quad \eta_1 - \frac{\overline{K}}{2} = 0$$

and the resulting closed form solution of the problem is

$$w(r) = -\frac{\overline{K}}{4}(a^2 - r^2) \tag{5.68}$$

5.4 Couette Flow: Steady Laminar Flow between Two Concentric Rotating Circular Cylinders

5.4.1 Equations of Motions

Consider two infinitely long concentric circular cylinders of inner and outer radii r_1 and r_2, respectively, rotating with corresponding angular velocities, w_1 and w_2 Figure 5.2. The annular region between the cylinders is filled with viscous incompressible fluid. Then the cylinders induce steady, laminar, axisymmetric tangential motion in the fluid. To study this type of motion, we find

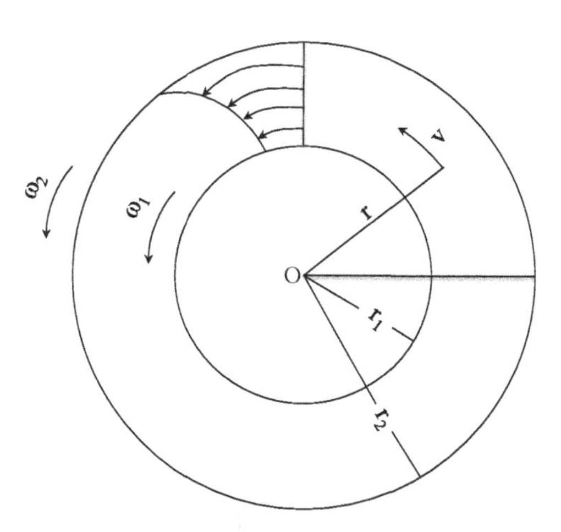

FIGURE 5.2
Velocity Distribution between two concentric rotating cylinders.

it convenient to use the cylindrical polar coordinates (r, θ, z), where z is taken as the axis of the cylinders. Let (u, v, w) be the radial, tangential, and axial components of the velocity. Since the motion is purely tangential, $u = w = 0$ and, therefore, the continuity equation reduces to $\partial v/\partial\theta = 0$, which implies that v is a function of r and z only. Again, the cylinders being infinitely long, the flow will not depend on z, and hence v is a function of r only.

Since the velocity component v is a function of r and the motion is axially symmetric, it follows that the pressure p must be a function of r. Under these above assumptions, the Navier–Stokes equations reduce to forms

$$\frac{dp}{dr} = \rho\frac{v^2}{r} \tag{5.69}$$

and

$$\frac{d^2v}{dr^2} + \frac{1}{r}\frac{dv}{dr} - \frac{v}{r^2} = 0 \tag{5.70}$$

The relation (5.69) states that the centripetal force $\rho\frac{v^2}{r}$ and the centrifugal force $\frac{d\beta}{dr}$ are equal in magnitude but opposite in sign and balance with each other so that the fluid particle tends to move in the tangential direction to its circular path. The tangential velocity component v can be obtained by solving the equation (5.70), which governs the flow field in the fluid between the cylinders.

The corresponding boundary conditions are

$$v = w_1r_1 \quad \text{at} \quad r = r_1 \tag{5.71}$$

and

$$v = w_2r_2 \quad \text{at} \quad r = r_2 \tag{5.72}$$

The differential equation (5.70) subject to the boundary conditions (5.71) and (5.72) constitutes the mathematical model of the problem.

5.4.2 Solution by Regular Decomposition

To solve the equation (5.70) by regular decomposition, we write

$$L = \frac{d^2}{dr^2} + \frac{1}{r}\frac{d}{dr} = \frac{1}{r}\frac{d}{dr}\left(r\frac{d}{dr}\right)$$

whose inverse form is defined by $L^{-1} = L_1^{-1}\left[r^{-1}\left(L_1^{-1}r\right)\right]$ where $L_1^{-1} = \int(\cdot)dr$. Then the operator form of (5.70) is

$$Lv = \frac{v}{r^2} \tag{5.73}$$

Using the inverse form L^{-1} on both sides of (5.73), we get

$$v(r) = v_0(r) + L^{-1}(v/r^2) \tag{5.74}$$

where $v_0(r)$, the solution of $Lv = 0$, is given by

$$v_0(r) = \xi_0 + \xi_1 \log r \tag{5.75}$$

ξ_0 and ξ_1 being the integration constants to be evaluated later on.

We now decompose $v(r)$ into

$$v(r) = \sum_{m=0}^{\infty} \lambda^m v_m \tag{5.76}$$

and the parameterized form of (5.74) is

$$v(r) = v_0(r) + \lambda L^{-1}(v/r^2) \tag{5.77}$$

where λ has the usual meanings. Substitution of (5.76) in (5.77) gives

$$\sum_{m=0}^{\infty} \lambda^m v_m = v_0(r) + \lambda L^{-1} \sum_{m=0}^{\infty} \lambda^m (v_m/r^2) \tag{5.78}$$

Comparing like-power terms of λ from both sides of (5.78), we have the following set of component terms:

$$\begin{aligned}
v_0(r) &= \xi_0 + \xi_1 \log r \\
v_1(r) &= L^{-1}\frac{v_0}{r^2} = \xi_0 \frac{(\log r)^2}{2!} + \xi_1 \frac{(\log r)^3}{3!} \\
v_2(r) &= L^{-1}(v_1/r^2) = \xi_0 \frac{(\log r)^4}{4!} + \xi_1 \frac{(\log r)^5}{5!} \\
\cdots &\quad \cdots \quad \cdots \quad \cdots \\
\cdots &\quad \cdots \quad \cdots \quad \cdots \\
v_{m+1}(r) &= L^{-1}(v_r/r^2) = \xi_0 \frac{(\log r)^{2m}}{(2m)!} + \xi_1 \frac{(\log r)^{2m+1}}{(2m+1)!} \\
\cdots &\quad \cdots \quad \cdots \quad \cdots
\end{aligned} \tag{5.79}$$

Expanding (5.76) and using (5.79), we get

$$\begin{aligned}
v(r) &= \xi_0 \left[1 + \frac{(\log r)^2}{2!} + \frac{(\log r)^4}{4!} + \frac{(\log r)^6}{6!} + \cdots\right] \\
&\quad + \xi_1 \left[\log r + \frac{(\log r)^3}{3!} + \frac{(\log r)^5}{5!} + \cdots\right] \\
&= \frac{\xi_0}{2}\left(e^{\log r} + e^{-\log r}\right) + \frac{\xi_1}{2}\left(e^{\log r} + e^{-\log r}\right) \\
&= A_r + \frac{B}{r}
\end{aligned} \tag{5.80}$$

where $A = \frac{1}{2}(\xi_0 + \xi_1)$ and $B = \frac{1}{2}(\xi_0 - \xi_1)$ are two arbitrary constants to be determined by the boundary conditions. Satisfying the boundary conditions (5.71) and (5.72) to (5.80), we have

$$A = \frac{w_2 r_2^2 - w_1 r_1^2}{r_2^2 - r_1^2}$$

$$B = -\frac{r_1^2 r_2^2}{r_2^2 - r_1^2}(w_2 - w_1) \tag{5.81}$$

Putting (5.81) in (5.80), the expression for tangential velocity component is obtained as

$$v(r) = \frac{1}{r_2^2 - r_1^2}\left[\left(w_2 r_2^2 - w_1 r_1^2\right) r - \frac{r_1^2 r_2^2}{r}(w_2 - w_1)\right]$$

5.4.3 Solution by Modified Decomposition

Since the equation (5.70) is an ordinary differential equation with variable coefficients, it is convenient to use the transformation $y = \log r$, which reduces the equation to $\frac{d^2 v}{dy^2} = v$, whose decomposition solution is

$$v(y) = \overline{\xi}_0 + \overline{\xi}_1 y + L_y^{-1} v \tag{5.82}$$

where $L_y^{-1} = \int\int(\cdot)dydy$ is the inverse operator of $L = \frac{d^2}{dy^2}$.

Following the modified decomposition procedure, the solution is written as

$$v(y) = \sum_{m=0}^{\infty} a_m y^m \tag{5.83}$$

Putting (5.83) in the above equation, that is, (5.74), and then performing integrations, we have

$$\sum_{m=0}^{\infty} a_m y^m = \overline{\xi}_0 + \overline{\xi}_1 y + \sum_{m=0}^{\infty} a_m \frac{y^{m+2}}{(m+1)(m+2)}$$

Replacement of m by $(m-2)$ on the right side gives

$$\sum_{m=0}^{\infty} a_m y^m = \overline{\xi}_0 + \overline{\xi}_1 y + \sum_{m=0}^{\infty} a_{m-2} \frac{y^m}{m(m-1)}$$

Comparing the coefficients on both sides, we have

$$a_0 = \overline{\xi}_0, a_1 = \overline{\xi}_1 \tag{5.84}$$

and the recurrence relation for $m \geq 2$

$$a_m = \frac{a_{m-2}}{m(m-1)} \tag{5.85}$$

From the relation (5.85) we obtain

$$\begin{aligned} a_2 &= \frac{a_0}{21}, \qquad a_3 = \frac{a_1}{31} \\ a_4 &= \frac{a_0}{41}, \qquad a_5 = \frac{a_1}{51}, \text{ etc.} \end{aligned} \tag{5.86}$$

Then we use the relations (5.86) in (5.83) to obtain

$$v(r) = Cr + \frac{D}{r} \tag{5.87}$$

where $C = \frac{1}{2}(a_0 + a_1)$ and $D = \frac{1}{2}(a_0 - a_1)$. Satisfying the boundary conditions (5.71) and (5.72) by (5.87), we obtain

$$C = \frac{w_2 r_2^2 - w_1 r_1^2}{r_2^2 - r_1^2}$$

and the expression for $v(r)$ is found to be

$$v(r) = \frac{1}{r_2^2 - r_1^2}\left[\left(w_2 r_2^2 - w_1 r_1^2\right) r - \frac{r_1^2 r_2^2}{r}(w_2 - w_1)\right]$$

5.5 Flow in Convergent and Divergent Channels

In the previous sections, we discussed some physical problems that involve linear differential equations whose solutions are obtained very easily. In the present section, we deal with a problem that also involves a nonlinear differential equation. A simple case, is the flow in convergent and divergent channels. Haldar and Bandyopadhyay [5] studied this problem by the application of such decomposition method, and we also discussed it in Section 5.3.

5.5.1 Equations of Motion

To study the flow field of a steady flow of viscous incompressible fluid between two nonparallel walls, we use the cylindrical polar coordinates (r, ϕ, z) where the z-axis is taken as the line of intersection of the planes and the radial distance r is measured from this line. The rigid walls are the positions of the planes $\phi = \pm\alpha$. where 2α is the angle subtended by the planes at each other Figure 5.3 Case I and Case II.

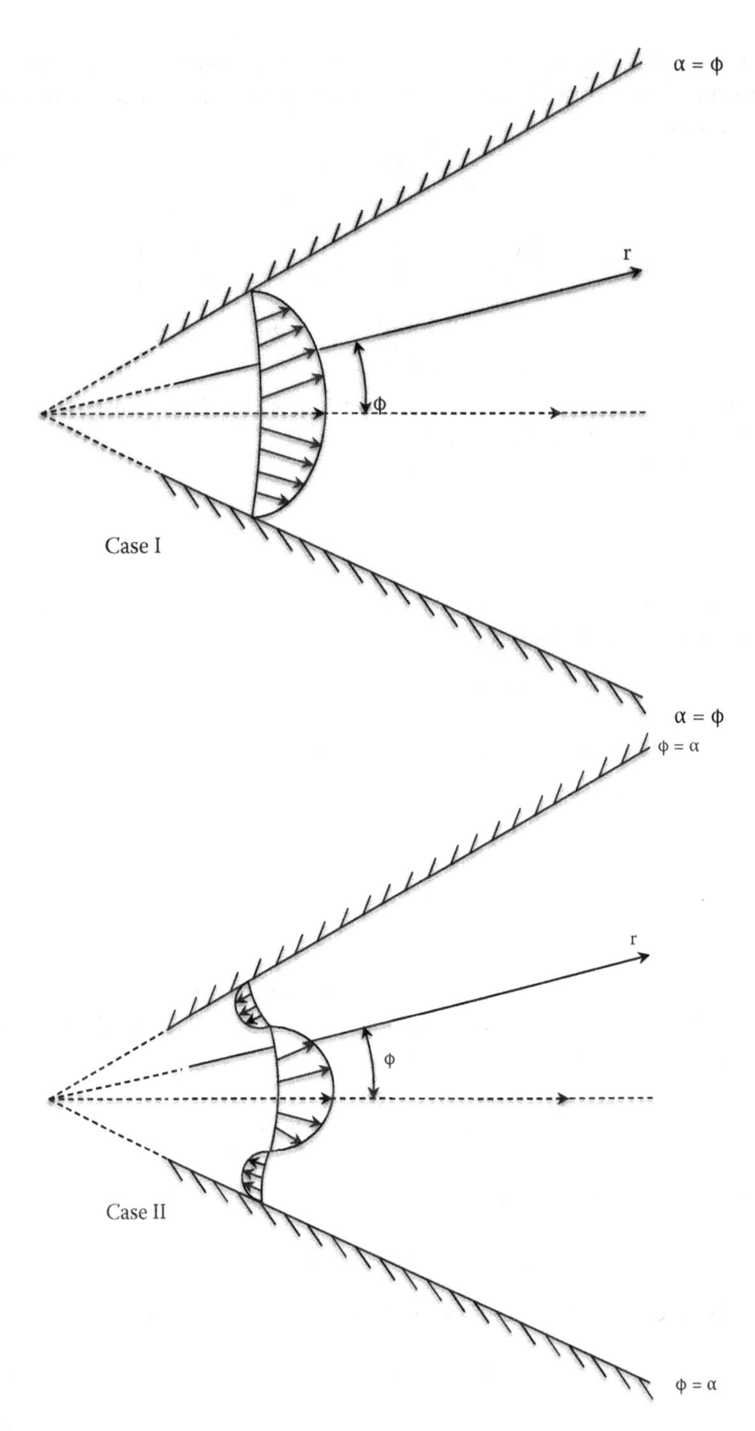

FIGURE 5.3
Flow in a divergent Channel.

Now if the flow is purely radial, then all other components except the radial one are zero. Therefore, the continuity equation and momentum equations in such a case are

$$\frac{\partial}{\partial r}(ru) = 0 \tag{5.88}$$

and

$$u\frac{\partial u}{\partial r} = -\frac{1}{\rho}\frac{\partial p}{\partial r} + \nu\left[\frac{\partial^2 u}{\partial r^2} + \frac{1}{r}\frac{\partial u}{\partial r} + \frac{1}{r^2}\frac{\partial^2 u}{\partial \phi^2} - \frac{u}{r^2}\right] \tag{5.89}$$

$$0 = -\frac{1}{\rho}\frac{1}{r}\frac{\partial p}{\partial \phi} + \nu\frac{2}{r^2}\frac{\partial u}{\partial \phi} \tag{5.90}$$

where u is the radial velocity component, p is the fluid pressure, ρ is the density of fluid, and ν is the kinematic viscosity.

The equation of continuity suggests that

$$u = \frac{F(\phi)}{r} \tag{5.91}$$

where F is an arbitrary function of ϕ to be determined later on. Substituting (5.91) in (5.89) and (5.90) and then eliminating p, we have

$$2FF' + \nu\left(F''' + 4F'\right) = 0 \tag{5.92}$$

where prime denotes differentiation with respect to ϕ.

The boundary conditions on F are

$$F = 0 \quad \text{at} \quad \phi = \pm\,\alpha \tag{5.93}$$

$$F' = 0 \quad \text{at} \quad \phi = 0 \tag{5.94}$$

$$F = F_m \quad \text{at} \quad \phi = 0 \tag{5.95}$$

where F_m is the maximum velocity along the line $\phi = 0$.

Now it is convenient to write the equations from (5.92) to (5.95) with the help of the following transformations:

$$f = \frac{F}{F_m}, \quad \theta = \frac{\phi}{\alpha}, \quad Re = \frac{F_m}{\nu} \tag{5.96}$$

The nondimensional form of (5.92) is

$$f''' + \alpha^2(4f' + 2\,Reff') = 0 \tag{5.97}$$

and the boundary conditions from (5.93) to (5.95) take the forms

$$f = 0 \quad \text{at} \quad \theta = \pm\,1 \tag{5.98}$$

$$f' = 0 \quad \text{at} \quad \theta = 0 \tag{5.99}$$

$$f = 1 \quad \text{at} \quad \theta = 0 \tag{5.100}$$

Now the mathematical model consists of the differential equation (5.97) and the four boundary conditions (5.98), (5.99), and (5.100). Actually, we need only three boundary conditions, as the equation (5.97) is of order three. When the flow is purely convergent or divergent, the function f is symmetrical about $\theta = 0$, and in such a case the value of f may be prescribed at $\theta = 0$. Therefore, out of four boundary conditions, (5.99) and (5.100) are appropriate, and the third condition should be any one of the conditions (5.98).

5.5.2 Solution by Regular Decomposition

Let $L = \frac{d^3}{d\theta^3}$, and the equation (5.97) becomes

$$Lf + \alpha^2(4f' + Nf) = 0 \tag{5.101}$$

where

$$Nf = 2R_e ff' \tag{5.102}$$

Solving (5.101) for Lf, we have

$$Lf = -\alpha^2(4f' + Nf) \tag{5.103}$$

If L^{-1} is the inverse operator of L, then performing the operation with the inverse operator L^{-1} on both sides of the equation (5.103), we get

$$f(\theta) = f_0(\theta) - \alpha^2 L^{-1}(4f' + Nf) \tag{5.104}$$

where

$$f_0(\theta) = \frac{1}{2}A\theta^2 + B\theta + C \tag{5.105}$$

is the solution of $Lf = 0$. Here $L^{-1} = \int\int\int(\cdot)d\theta d\theta d\theta$ and A, B, C are the constants to be evaluated by means of the boundary conditions. Satisfying the boundary conditions (5.98) to (5.100) by (5.105), we see that $A = -2$, $B = 0$, $C = 1$, and the expression for f_0 is found to be

$$f_0(\theta) = 1 - \theta^2 \tag{5.106}$$

Then we write down the decomposition forms of f and the nonlinear term Nf as

$$f(\theta) = \sum_{m=0}^{\infty}\lambda^m f_m \tag{5.107}$$

and

$$Nf = \sum_{m=0}^{\infty}\lambda^m A_m \tag{5.108}$$

where A_m is the Adomian's special polynomials and λ is a counting parameter. The parameterized form of (5.104) is

$$f(\theta) = f_0(\theta) - \lambda\alpha^2 L^{-1}(4f' + Nf) \tag{5.109}$$

If we substitute (5.107) and (5.108) in (5.109) and then equate the coefficients of like-power terms of λ from both sides of the resulting expression, we have the following components of f:

$$\begin{aligned}
f_1(\theta) &= -\alpha^2 L^{-1}\left[4f_0' + A_0\right] \\
f_2(\theta) &= -\alpha^2 L^{-1}\left[4f_1' + A_1\right] \\
f_3(\theta) &= -\alpha^2 L^{-1}\left[4f_2' + A_2\right] \\
\cdots \quad & \cdots \quad \cdots \\
\cdots \quad & \cdots \quad \cdots \\
f_{m+1}(\theta) &= -\alpha^2 L^{-1}\left[4f_m' + A_m\right]
\end{aligned} \tag{5.110}$$

where prime denotes differentiation with respect to θ.

Our next task is to determine the Adomian's special polynomials A_m. To do that, we combine (5.102), (5.107), and (5.108) and then compare the like-power terms of λ on both sides of the resulting expression. The set of the polynomials is given by

$$\begin{aligned}
A_0 &= 2R_e f_0 f_0' \\
A_1 &= 2R_e(f_0 f_1' + f_0' f_1) \\
A_2 &= 2R_e\left(f_0 f_2' + f_1 f_1' + f_2 f_0'\right) \\
\cdots \quad & \cdots \quad \cdots \\
\cdots \quad & \cdots \quad \cdots
\end{aligned} \tag{5.111}$$

Since f_0 is known, it is implied that A_0 is also known because it is a function of f_0 and its derivative. Hence the solution f_1 of (5.110) is computable, and the expression for f_1 is given by

$$f_1(\theta) = 4\alpha^2\left[(R_e+2)\frac{\theta^4}{4!} - 6R_e\frac{\theta^6}{6!}\right] \tag{5.112}$$

Again, as f_0 and f_1 are known, A_1 is also known. The second component function f_2 can also be determined from (5.110), and it is given by

$$f_2(\theta) = -8\alpha^4\left[(R_e+2)^2\frac{\theta^6}{6!} - 36R_e(R_e+2)\frac{\theta^8}{8!} + 336R_e^2\frac{\theta^{10}}{10!}\right] \tag{5.113}$$

Similarly, f_3 and other component functions, such as f_4, f_5, etc., can also be computable, and the expression for f_3 is found to be

$$\begin{aligned}
f_3(\theta) = 16\alpha^6 \Bigg[&(R_e+2)^3\frac{\theta^8}{8!} - 162R_e(R_e+2)^2\frac{\theta^{10}}{10!} \\
&+ 96R_2^2(67R_e+127)\frac{\theta^{12}}{12!} - 336R_e(R_e+11)\frac{\theta^{14}}{14!}\Bigg]
\end{aligned} \tag{5.114}$$

If we consider a four-term approximant of the solution f, then we have

$$f(\theta) = f_0(\theta) + f_1(\theta) + f_2(\theta) + f_3(\theta) \tag{5.115}$$

Substituting (5.106), (5.112), (5.113), and (5.114) in (5.115), we get the expression for f as

$$f(\theta) = 1 - \theta^2 + \alpha_4\theta^4 - \alpha_6\theta^6 + \alpha_8\theta^8 - \alpha_{10}\theta^{10} + \alpha_{12}\theta^{12} - \alpha_{14}\theta^{14} \tag{5.116}$$

where

$$\begin{aligned}
\alpha_4 &= \frac{4\alpha^2}{4!}(R_e + 2) \\
\alpha_6 &= \frac{8\alpha^2}{6!}\left[\alpha^2(R_e+2)^2 + 3R_e\right] \\
\alpha_8 &= \frac{16\alpha^4}{8!}(R_e+2)\left[\alpha^2(R_e+2)^2 + 18R_e\right] \\
\alpha_{10} &= \frac{32\alpha^4}{10!}\left[81\alpha^2(R_e+2)^2 + 84R_e\right] \\
\alpha_{12} &= \frac{16\times 96\,\alpha^6 R_e^2}{12!}[67R_e + 127] \\
\alpha_{14} &= \frac{16\times 336\alpha^6 R_e}{14!}[R_e + 11]
\end{aligned} \tag{5.117}$$

For convergent flow Figure 5.4 the velocity is negative, and the expression (5.91) becomes

$$u = -\frac{F(\phi)}{r} \tag{5.118}$$

If we eliminate p between (5.89) and (5.90) and use (5.117), we get

$$2FF' - \nu(F''' + 4F') = 0 \tag{5.119}$$

whose nondimensional form, by virtue of (5.96), is

$$f''' + \alpha^2(4f' - 2R_e ff') = 0 \tag{5.120}$$

If we compare the two equations (6.97) and (5.120), we see that the equation (5.120) can be obtained replacing R_e by $-R_e$ in the equation (5.97). Hence, the solution of (5.120) can immediately be obtained replacing R_e by $-R_e$ in the solution of (5.97).

5.5.3 Solution by Double Decomposition

Let $L = \frac{d^3}{d\eta^3}$ whose inverse is $L^{-1} = \int\int\int(\cdot)d\eta d\eta d\eta$. Then following the regular decomposition procedure, we get the solution of (5.97) as

$$f(\theta) = f_0 - \alpha^2 L^{-1}[4f' + Nf] \tag{5.121}$$

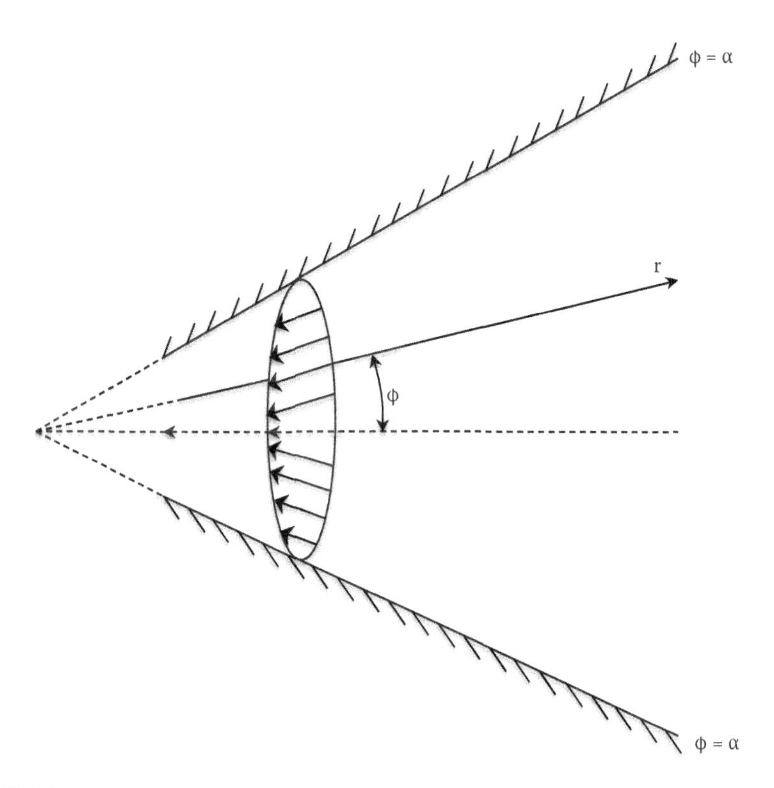

FIGURE 5.4
Flow in a convergent channel.

whose parameterized form is

$$f(\theta) = f_0 - \lambda\alpha^2 L^{-1}[4f' + Nf] \tag{5.122}$$

where λ has its usual meanings. The nonlinear term Nf is defined by (5.108), and f_0 is the solution of $Lf = 0$ given by

$$f_0(\theta) = \alpha_1 + \alpha_2\theta + \frac{1}{2}\alpha_3\theta^2 \tag{5.123}$$

α_1, α_2, and α_3 are the constants of integration.

For the double decomposition procedure, we write the following parameterized forms:

$$\begin{aligned} f(\theta) &= \sum_{m=0}^{\infty} \lambda^m f_m \\ Nf &= \sum_{m=0}^{\infty} \lambda^m A_m \\ f_0(\theta) &= \sum_{m=0}^{\infty} \lambda^m f_{0,m} \end{aligned} \tag{5.124}$$

where the Adomian's special polynomials are given by (5.111). Substituting the above relations in (5.122) and equating the like-power terms of λ from both sides of the resulting expression, we have the following components of f:

$$
\begin{aligned}
f_0(\theta) &= f_{0,0} \\
f_1(\theta) &= f_{0,1} - \alpha^2 L^{-1}\left[4f_0' + A_0\right] \\
f_2(\theta) &= f_{0,2} - \alpha^2 L^{-1}\left[4f_1' + A_1\right] \\
f_3(\theta) &= f_{0,3} - \alpha^2 L^{-1}\left[4f_2' + A_2\right] \\
&\cdots \quad \cdots \quad \cdots \\
&\cdots \quad \cdots \quad \cdots \\
f_{m+1}(\theta) &= f_{0,(m+1)} - \alpha^2 L^{-1}\left[4f_m' + A_m\right]
\end{aligned}
\tag{5.125}
$$

Since the solution (5.123) contains the constants α_1, α_2, and α_3, the parameterized decomposition forms of these constants are given by

$$
\begin{aligned}
\alpha_1 &= \sum_{m=0}^{\infty} \lambda^m \alpha_{1,m} \\
\alpha_2 &= \sum_{m=0}^{\infty} \lambda^m \alpha_{2,m} \\
\alpha_3 &= \sum_{m=0}^{\infty} \lambda^m \alpha_{3,m}
\end{aligned}
\tag{5.126}
$$

Substituting the third relation of (5.124) and (5.126) in (5.123), we have

$$
\begin{aligned}
f_{0,0}(\theta) &= \alpha_{1,0} + \alpha_{2,0}\theta + \frac{1}{2}\alpha_{3,0}\theta^2 \\
f_{0,1}(\theta) &= \alpha_{1,1} + \alpha_{2,1}\theta + \frac{1}{2}\alpha_{3,1}\theta^2 \\
f_{0,2}(\theta) &= \alpha_{1,2} + \alpha_{2,2}\theta + \frac{1}{2}\alpha_{3,2}\theta^2 \\
&\cdots\cdots \quad \cdots \quad \cdots \\
&\cdots\cdots \quad \cdots \quad \cdots \\
f_{0,m}(\theta) &= \alpha_{1,m} + \alpha_{2,m}\theta + \frac{1}{2}\alpha_{3,m}\theta^2 \\
&\cdots\cdots \quad \cdots \quad \cdots
\end{aligned}
\tag{5.127}
$$

Thus the solution is completely known, as the f_0, f_1, f_2, etc., are obtained. But the expression for each component contains some constants, and these constants can be determined by their respective boundary conditions.

To formulate the specific boundary conditions of each component function, we substitute the first relation of (5.124) into the boundary conditions (5.98), (5.99), and (5.100). Then comparing the coefficients of the terms from both sides of the resulting expressions, we have the following sets of boundary conditions for each component function:

$$\left.\begin{aligned} f_0(\theta) &= 0, \quad \text{at} \quad \theta = \pm 1 \\ f_0'(\theta) &= 0, \quad \text{at} \quad \theta = 0 \\ f_0(\theta) &= 1, \quad \text{at} \quad \theta = 1 \end{aligned}\right\} \tag{5.128}$$

and

$$\left.\begin{aligned} f_m(\theta) &= 0, \quad \text{at} \quad \theta = \pm 1 \\ f_m'(\theta) &= 0, \quad \text{at} \quad \theta = 0 \\ f_m(\theta) &= 0, \quad \text{at} \quad \theta = 0 \end{aligned}\right\} \tag{5.129}$$

for all n except zero.

For divergent flow the velocity u, that is, f is positive, and the equation (5.92) is appropriate for this flow. The solution f_0 can be obtained from the first relations of (5.125) and (5.127). But the expression for f_0 contains three unknowns, $\alpha_{1,0}$, $\alpha_{2,0}$, and $\alpha_{3,0}$, which are to be determined by the boundary conditions (5.128). Satisfying these boundary conditions by the solution f_0, we have the constants found to be $\alpha_{1,0} = 1, \alpha_{2,0} = 0, \alpha_{3,0} = -2$, and the resulting expression for f_0 is given by

$$f_0(\theta) = 1 - \theta^2 \tag{5.130}$$

When f_0 is known, then the Adomian's polynomial A_0 is known because A_0 is a function of f_0 and its derivative. Hence, the solution $f_1(\theta)$ of (5.125) is computable because of f_0 and A_0. Satisfying the boundary conditions in (5.129) for $m = 1$ by the function $f_1(\theta)$, we get the constants $\alpha_{1,1}$, $\alpha_{2,1}$, and $\alpha_{3,1}$ as

$$\alpha_{1,1} = \alpha_{2,1} = 0, \frac{1}{2}\alpha_{31} = -\frac{\alpha^2}{15}(2R_e + 5) \tag{5.131}$$

and the expression for $f_1(\theta)$ is given by

$$f_1(\theta) = \frac{2\alpha^2}{15}\left[\beta_1\theta^2 + \beta_2\theta^4 + \beta_3\theta^6\right] \tag{5.132}$$

where

$$\begin{aligned} \beta_1 &= \frac{1}{2!}(2R_e + 5) \\ \beta_2 &= \frac{30}{4!}(R_e + 2) \\ \beta_3 &= -\frac{180}{6!}R_e \end{aligned} \tag{5.133}$$

The component f_2 can also be determined from (5.125) in the same way. The constants involved in it are to be evaluated by satisfying its respective boundary conditions obtained by putting $m = 2$ in (5.129). Thus, the resulting expression for f_2 is given by

$$f_2(\theta) = 4\alpha^4 \left[\gamma_1\theta^2 + \gamma_2\theta^4 + \gamma_3\theta^6 + \gamma_4\theta^8 + \gamma_5\theta^{10}\right] \tag{5.134}$$

where

$$\begin{aligned}
\gamma_1 &= \frac{24}{5}\left(163R_e^2 + 900R_e + 1260\right)\\
\gamma_2 &= \frac{1}{154}\left(2R_e^2 + 9R_e + 10\right)\\
\gamma_3 &= \frac{1}{255}\left(9R_e^2 + 30R_e + 20\right)\\
\gamma_4 &= 9\left(R_e^2 + 2R_e\right)\\
\gamma_5 &= -\frac{336}{5}R_e^2
\end{aligned}$$

Similarly, the component f_3 can also be computable, and its expression is given by

$$f_3(\theta) = 4\alpha^6\left[\overline{L}_1\theta^2 - L_1\right] \tag{5.135}$$

where

$$L_1(\theta) = \delta_1\theta^4 + \delta_2\theta^6 + \delta_3\theta^8 + \delta_4\theta^{10} + \delta_5\theta^{12} + \delta_6\theta^{14} \tag{5.136}$$

$$\begin{aligned}
\delta_1 &= -\frac{24}{13}\left(163R_e^3 + 1226R_e^2 + 3060R_e + 2520\right)\\
\delta_2 &= -\frac{1}{21\times 776}\left(751R_e^3 + 4220R_e^2 + 4830R_e + 4220\right)\\
\delta_3 &= -\frac{2}{29}\left(29R_e^3 + 138R_e^2 + 80R_e + 40\right)\\
\delta_4 &= \frac{4}{85}\left(177R_e^3 + 645R_e^2 + 540R_e\right)\\
\delta_5 &= -4\left(701R_e^3 + 1402R_e^2\right)\\
\delta_6 &= 6633R_e^3
\end{aligned} \tag{5.137}$$

and $\overline{L}_1$ is the value of $L_1(\theta)$ at $\theta = 1$ given by (5.136).

If a four-term approximant of f is the solution of the problem, then we have

$$f(\theta) = f_0(\theta) + f_1(\theta) + f_2(\theta) + f_3(\theta) \tag{5.138}$$

where f_0, f_1, f_2, and f_3 are given by their respective expressions.

In the previous section, we showed that the solution in the convergent channel flow is obtained from the solution of divergent channel flow replacing

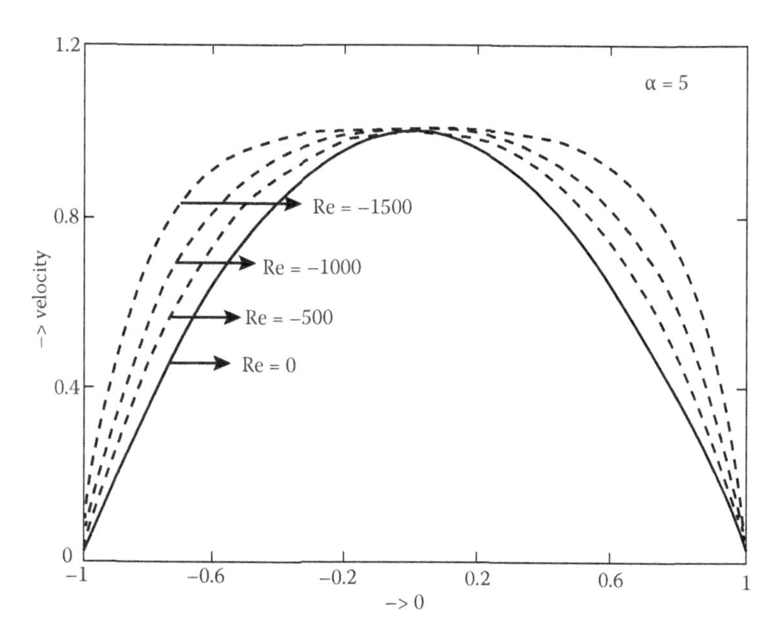

FIGURE 5.5
Velocity distribution for convergent channel flow.

R_e by $-R_e$. Similarly, in the case of the double decomposition procedure, the convergent solution can also be obtained replacing R_e by $-R_e$ in the double decomposition divergent solution.

In the discussion, the character of the numerical solution of the theoretical result is sketched in Figures 5.5, 5.6, and 5.7. The figures describe the families of the axial velocity profiles for convergent and divergent channel flows for different values of Reynolds number and divergent angle. In each case the solutions differ markedly from each other.

In the convergent channel flow (Figure 5.5), the solution is positive over the cross section of the channel. For a higher value of Reynolds number, the solution is nearly constant over the central portion of the channel, and it falls steeply to zero at the wall. This steep fall of the solution indicates the existence of a boundary layer at the wall.

In the divergent channel flow (Figure 5.6), the solutions are markedly affected by the Reynolds numbers and by the angles of divergence also. The solutions are positive in the central region of the channel for small values of Reynolds numbers, but for high Reynolds numbers, negative solutions are observed near the walls indicating the backflow there. The characters of the solutions in the convergent and divergent flows are qualitatively in good agreement with those of Millsaps and Pohlhausen [6].

It is interesting to note that the solution is also affected by the angle of divergence in the divergent channel flow (Figure 5.7). The backflow occurs earlier, that is, at low Reynolds number with larger angles of divergence.

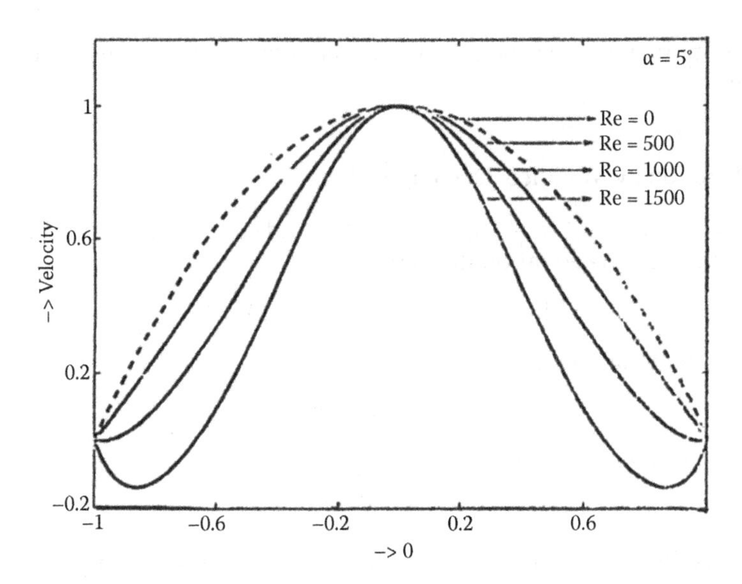

FIGURE 5.6
Velocity distribution for divergent channel flow.

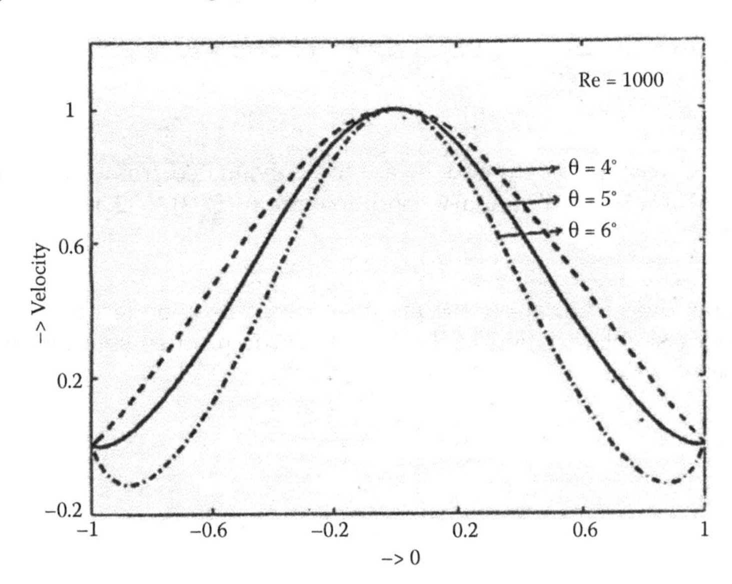

FIGURE 5.7
Velocity distribution in divergent channel flow for different divergent angles.

The numerical solutions for convergent and divergent channel flows are depicted in the figures on the basis of the theoretical result (5.138), which is a four-term approximant of the solution of double decomposition method studied by Haldar and Bandyopadhyay [5]. To improve the solution more, we must include additional terms in the expression of axial velocity.

5.5.4 Solution by Modified Decomposition

Let $L = \frac{d^3}{d\theta^3}$ and $L_1 = \frac{d}{d\theta}$ be two linear third- and first-order differential operators. Then the equation (5.97) becomes

$$Lf + \alpha^2(4L_1 f + 2R_e f L_1 f) = 0$$

which, on solving for Lf, gives

$$Lf = -\alpha^2(4L_1 f + Nf) \tag{5.139}$$

where Nf is the nonlinear term given by

$$Nf = 2R_e f L_1 f \tag{5.140}$$

If L^{-1} is the inverse operator of L, then following the decomposition procedure we have

$$f(\theta) = f_0(\theta) - L^{-1}\left[\alpha^2(4L_1 f + Nf)\right] \tag{5.141}$$

where $f_0(\theta)$ is the solution of $Lf = 0$, and it is found to be

$$f_0(\theta) = B_1 + B_2\theta + \frac{1}{2}B_3\theta^2 \tag{5.142}$$

Here the constants B_1, B_2, and B_3 are to be evaluated from the boundary conditions. Using the boundary conditions from (5.98) to (5.100) in (5.142), we have

$$f_0(\theta) = 1 - \theta^2 \tag{5.143}$$

We now proceed to apply the modified decomposition technique, which produces a slight variation in the regular decomposition solution. For the purpose we write

$$f(\theta) = \sum_{m=0}^{\infty} a_m \theta^m \tag{5.144}$$

$$L_1 f = f'(\theta) = \sum_{m=0}^{\infty} (m+1)a_{m+1}\theta^m \tag{5.145}$$

$$Nf = \sum_{m=0}^{\infty} A_m \theta^m \tag{5.146}$$

where A_m is the special polynomials. Putting the relations from (5.143) to (5.146) in (5.141), we get

$$\sum_{m=0}^{\infty} a_m\theta^m = 1 - \theta^2 - \alpha^2 L^{-1}\left[4\sum_{m=0}^{\infty}(m+1)a_{m+1}\theta^m + \sum_{m=0}^{\infty} A_m\theta^m\right]$$

Carrying out the integrations on the right side, we have

$$\sum_{m=0}^{\infty} a_m\theta^m = 1 - \theta^2$$

$$- \alpha^2 \left[4\sum_{m=0}^{\infty} \frac{(m+1)a_{m+1}\theta^{m+3}}{(m+1)(m+2)(m+3)} + \sum_{m=0}^{\infty} \frac{A_m\theta^{m+3}}{(m+1)(m+2)(m+3)} \right]$$

Then changing m by $(m-3)$ on the right side, we get

$$\sum_{m=0}^{\infty} a_m\theta^m = 1 - \theta^2 - \alpha^2 \sum_{m=3}^{\infty} \left[\frac{4(m-2)a_{m-2} + A_{m-3}}{m(m-1)(m-2)} \right] \theta^m$$

From the above relation, the coefficients are identified as

$$a_0 = 1, \quad a_1 = 0, \quad a_2 = -1 \tag{5.147}$$

and for $m \geq 3$ the recurrence relation

$$a_m = -\alpha^2 \left[\frac{4(m-2)a_{m-2} + A_{m-3}}{m(m-1)(m-2)} \right] \tag{5.148}$$

which gives the other coefficients.

Our next task is to determine A_m. To do that, we substitute (5.144) to (5.146) in (5.140) to get

$$\sum_{m=0}^{\infty} A_m\theta^m = 2R_e \left[\sum_{m=0}^{\infty} a_m\theta^m \right] \left[\sum_{m=0}^{\infty} (m+1)\, a_{m+1}\theta^m \right]$$

Using Cauchy product of two infinite series on the right side of the above relation, we obtain

$$\sum_{m=0}^{\infty} A_m\theta^m = \sum_{m=0}^{\infty} \left[2R_e \sum_{n=0}^{m} a_{m-n}^{m}(n+1)a_{n+1} \right] \theta^m$$

which gives, on comparison of like-power terms of θ,

$$A_m = 2R_e \sum_{n=0}^{m} a_{m-n}^{m}(n+1)a_{n+1}$$

$$A_{m-3} = 2R_e \sum_{n=0}^{m-3} a_{m-n-3}(n+1)a_{n+1} \tag{5.149}$$

Elimination of A_{m-3} between (5.148) and (5.149) gives

$$m(m-1)(m-2)a_m = -\alpha^2 \left[4(m-2)a_{m-2} + 2R_e \sum_{m=0}^{m-3} a_{m-n-3}^{m-3}(n+1)a_{n+1} \right] \tag{5.150}$$

Putting $m = 3, 4, 5$, etc., we get the following set of coefficients:

$$
\begin{aligned}
a_3 &= a_5 = a_7 = \ldots = 0 \text{ (each)} \\
a_4 &= \frac{4\alpha^2(R_e + 2)}{4!} \\
a_6 &= -\frac{8\alpha^2}{6!}\left[\alpha^2 (R_e + 2)^2 + 3R_e\right] \\
a_8 &= \frac{16\alpha^4(R_e + 2)}{8!}\left[\alpha^2 (R_e + 2)^2 + 18R_e\right] \\
a_{10} &= -\frac{32\alpha^4}{10!}\left[\alpha^4 (R_e + 2)^4 + 81R_e\alpha^2 (R_e + 2)^2 + 84R_e^2\right] \\
a_{12} &= \frac{64\alpha^4(R_e + 2)}{12!}\left[\alpha^4 (R_e + 2)^4 - 84R_e\alpha^2 (R_e + 2)^2 + 264R_e^2\right] \\
a_{14} &= -\frac{128\alpha^6}{14!}\left[\alpha^6 (R_e + 2)^6 + 939R_e\alpha^4 (R_e + 2)^4\right. \\
&\quad \left. + 17292R_e^2\alpha^2(R_e + 2)^2 + 9702R_e^3\right]
\end{aligned}
\tag{5.151}
$$

Let the fifteen-term approximant of f be the solution of the problem. Then we have from (5.144)

$$f(\theta) = \sum_{m=0}^{14} a_m\theta^m \tag{5.152}$$

retaining only fifteen terms, and the above analysis is given for divergent channel flow.

For convergent channel flow, the velocity u is negative, and the relation (5.91) takes the form

$$u = -\frac{F(\phi)}{r}$$

Substitution of this relation in (5.89) and (5.90) and then elimination of p give

$$2FF' - \nu(F''' + 4F') = 0$$

whose nondimensional form, by virtue of (5.96), is

$$f''' + \alpha^2(4f' - 2R_e ff') = 0$$

If we compare this dimensionless equation with (5.97), we see that the above equation can easily be obtained replacing R_e by $-R_e$ in (5.97), and the relation (5.152) will immediately be the solution of convergent channel flow if we change R_e by $-R_e$ in all the coefficients given in (5.151).

References

1. Adomian, G.: *Stochastic Systems*, Academic Press (1983).
2. Adomian, G.: *Nonlinear Stochastic Operator Equations*, Academic Press (1986).
3. Adomian, G.: *Nonlinear Stochastic Systems Theory and Applications to Physics*, Kluwer Academic Publishers (1989).
4. Adomian, G.: Solving Frontier Problems of Physics; The Decomposition Method, Kluwer Academic Publishers (1994).
5. Haldar, K., and M. K. Bandyopadhyay: Application of Double Decomposition Method to the Nonlinear Fluid Flow Problems, *Rev. Bull. Cal. Math. Soc.*, 14(2) 109–120 (2006).
6. Millsaps, K., and K. Pohlhausen: Thermal Distribution in Jeffery-Hamel Flows Between Nonparallel Plane Walls, *JAS, 20*, 187–196 (1953).

6

Blood Flow in Artery

6.1 Introduction

The study of blood flow through arteries is very important because of the fact that many cardiovascular diseases are closely related to the nature of blood movement and the dynamic behavior of blood vessels. Cardiovascular diseases—particularly artherosclerosis—are responsible for many clinical deaths. It is a well-known fact that more than 50% of the total number of deaths are due to vascular diseases. Medical surveys show that diseases of blood vessel walls are the causes for about 80% of the total deaths.

Artherosclerosis is the abnormal and unnatural growth in the lumen of an artery. It disturbs the normal blood flow and causes arterial diseases. The important flow characteristics, such as velocity, wall shear stress, pressure, possible separation, and reattachment, which have medical significance, are also disturbed and reveal some alterations in the flow caused by unnatural growth (medically called *stenosis*).

The specific reason for the initiation of abnormal growth in the lumen of an artery is not known, but its effect over the cardiovascular system has been studied by considering blood in its vicinity. The aim of research is to investigate the actual cause of formation of stenosis in the lumen of an artery. Therefore, an adequate knowledge of mechanical properties of blood vessels is absolutely necessary for better understanding and development of treatment against vascular diseases. The relevant information for the treatment of these diseases is very helpful to the bioengineers who are engaged in designing and construction of improved artificial organs.

Many research workers have studied the problem of blood flow through stenosed tube by considering blood as a Newtonian fluid, but blood behaves as a non-Newtonian fluid under certain conditions. Haynes and Burton [15], Merrill et al. [31], Charm and Kurland [6], Hershey et al. [16], Cokelet [11], and Lih [29] have pointed out that blood, being a suspension of corpuseles, behaves like a non-Newtonian fluid at low shear rate. Hershey and Cho [17], Charm and Kurland [6,7], and Huckaba and Hahu [28] have shown that blood flowing through a tube of diameter less than 0.2 mm and at a shear rate less than 20/sec behaves like a power-law fluid. Casson [4], Reiner and Scott Blair [35], Charm and Kurland [5–7], and Merrill et al. [32] have suggested that blood exhibits yield stress and behaves as a Casson model fluid at a shear rate equal to 0.1/sec.

Bugliarello and Sevilla [3] have shown that blood flow through an artery of smaller diameter consists of two parts; one part is the peripheral plasma layer, which is Newtonian in character, and the other part is a central core composed of red blood cells in plasma. Sukla et al. [38] have studied the effect of stenosis on the resistance to blood flow through an artery by considering the behavior of blood as a power-low fluid and a Casson fluid. Sukla et al. [39] have considered a two-layered model in which the peripheral plasma layer and the core are both Newtonian in character. In this analysis they have studied the influence of peripheral layer viscosity on the resistance to blood flow through a stenosed artery. Sukla et al. [40] have again considered a two-layered model in which fluid in both the regions are non-Newtonian in character and have examined the influence of peripheral layer viscosity on the flow resistance. Chaturani and Samy [8] have considered the problem of blood flow through a stenosed artery that contains a two-layered model in which the peripheral plasma layer is a Newtonian fluid, and the core is represented by a Casson fluid. In this analysis they have examined the effects of peripheral layer thickness on the resitance of blood flow and wall shear stress.

Haldar [18,20], Haldar and Andersson [25], and Haldar and Dey [21] have considered the flow of blood through stenosed artery where blood is treated as a non-Newtonian. Haldar [19,22] and Haldar and Ghosh [23,24] have also considered blood flow problems in constricted arteries in which blood is Newtonian in character. Haldar and Andersson [25] have studied two-layered model of blood flow through stenosed artery in which the peripheral plasma layer behaves as Newtonian fluid and the core is represented by non-Newtonian fluid with the supposition that the two fluids are immissible.

Adomian [1] has applied the most powerful decomposition method to the Navier–Stokes equations in Cartesian coordinates to study the flow field theoretically. Haldar [26,27] has also applied this method to the Navier–Stokes equations in cylindrical polar coordinates and studied the steady blood flow through a stenosed artery in the presence of a transverse magnetic field. Mamaloukas et al. [30] have investigated the pulsatile flow of blood through a circular rigid tube provided with a constriction by means of the decomposition method.

In this chapter we consider some nonlinear physical problems of blood flow in a stenosed artery for illustrating the decomposition method clearly in the following sections.

6.2 Steady Flow of Blood through a Constricted Artery

This section deals with the problem of blood flow through an artery provided with cosine-shaped constriction. The decomposition method has been applied to study the flow field in the tube. Theoretical results obtained in this analysis are the expressions for stream function, axial velocity component, and wall shearing stress.

6.2.1 Equations of Motion

The problem of blood flow through a locally constricted tube in the cardiovascular system is formulated with the following assumptions: the stenosis develops in the lumen of an artery in an axially symmetric manner and depends on the axial distance and its growth height; the flow field is steady, laminar, two-dimensional, and fully developed; and the blood flowing through the tube behaves as a Newtonian fluid (Figure 6.1).

The equations that govern the flow field in the tube under the assumed conditions are the continuity equation and the Navier–Stokes equations in cylindrical polar coordinates. It is convenient to write the system of equations in the nondimensional forms with the help of the following transformations:

$$\begin{aligned} u &= \frac{\overline{u}}{u_0}, \quad v = \frac{\overline{v}}{u_0} \\ r &= \frac{\overline{r}}{R_0}, \quad x = \frac{\overline{x}}{R_0} \\ p &= \frac{\overline{p}}{\rho u_0^2} \end{aligned} \tag{6.1}$$

where $(\overline{u}, \overline{v})$ and (u, v) are the dimensional and dimensionless velocity components in the axial and radial directions, respectively, $(\overline{x}, \overline{r})$ and (x, r) are the corresponding coordinate axes, R_0 is the radius of normal tube, u_0 is the characteristic velocity, ρ is the fluid density, and $(\overline{p}, p)$ are the fluid pressures in the dimensional and nondimensional states.

The nondimensional continuity equation of the system is

$$\frac{\partial}{\partial x}(ru) + \frac{\partial}{\partial r}(rv) = 0 \tag{6.2}$$

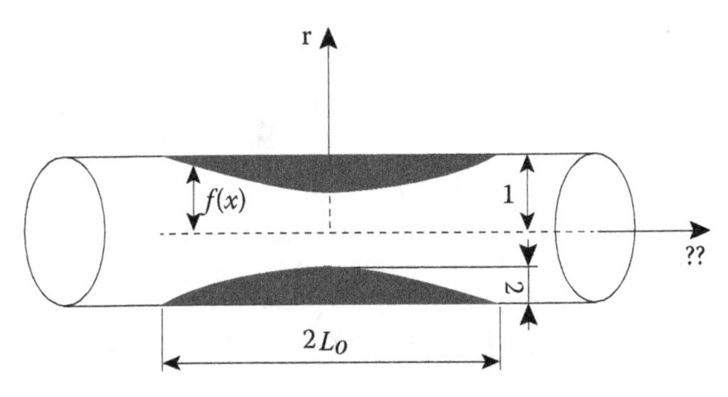

FIGURE 6.1
Geometry of constriction.

The dimensionless Navier–Stokes equations are

$$u\frac{\partial u}{\partial x} + v\frac{\partial u}{\partial r} = -\frac{\partial p}{\partial x} + \frac{1}{R_e}\left[\frac{\partial^2 u}{\partial r^2} + \frac{1}{r}\frac{\partial u}{\partial r} + \frac{\partial^2 u}{\partial x^2}\right] \tag{6.3}$$

in the axial direction and

$$u\frac{\partial v}{\partial x} + v\frac{\partial v}{\partial r} = -\frac{\partial p}{\partial r} + \frac{1}{R_e}\left[\frac{\partial^2 v}{\partial r^2} + \frac{1}{r}\frac{\partial v}{\partial r} - \frac{v}{r^2} + \frac{\partial^2 v}{\partial x^2}\right] \tag{6.4}$$

in the radial direction, where $R_e = u_0 r_0/\nu$ is the Reynolds number and ν is the kinematic viscosity of fluid.

If we introduce nondimensional stream function $\psi(x, r)$ defined by

$$u = -\frac{1}{r}\frac{\partial \psi}{\partial r}, v = \frac{1}{r}\frac{\partial \psi}{\partial x} \tag{6.5}$$

then the continuity equation (6.2) is satisfied identically. Eliminating p between (6.3) and (6.4) and then using (6.5), we have the equation of motion governing the flow field in the stenosed artery in term of the stream function $\psi(x, r)$ as

$$R_e\left[\frac{1}{r}J - \frac{2}{r^2}D^2\psi\frac{\partial \psi}{\partial x}\right] = D^4\psi \tag{6.6}$$

where J is the Jacobian defined by

$$J = \frac{\partial(D^2\psi, \psi)}{\partial(r, x)} = \begin{bmatrix} \frac{\partial}{\partial r}(D^2\psi) & \frac{\partial \psi}{\partial r} \\ \frac{\partial}{\partial x}(D^2\psi) & \frac{\partial \psi}{\partial x} \end{bmatrix} \tag{6.7}$$

and the operator D^2 is given by $D^2 = \frac{\partial^2}{\partial r^2} - \frac{1}{r}\frac{\partial}{\partial r} + \frac{\partial^2}{\partial x^2}$.

The corresponding boundary conditions imposed on $\psi(x, r)$ are

$$\left.\begin{aligned} -\frac{1}{r}\frac{\partial \psi}{\partial r} &= 0 \\ \psi(x, r) &= -\frac{1}{2} \end{aligned}\right\} \quad \text{at} \quad r = f \tag{6.8}$$

$$\left.\begin{aligned} -\frac{\partial}{\partial r}\left(\frac{1}{r}\frac{\partial \psi}{\partial r}\right) &= 0 \\ \psi(x, r) &= 0 \end{aligned}\right\} \quad \text{at} \quad r = 0 \tag{6.9}$$

where f denotes the radius of the stenosed artery and describes the geometry of stenosis. The expression for nondimensional f (Figure 6.1) is given by

$$\begin{aligned} f &= 1 - \frac{\tau}{2}\left(1 + \cos\frac{1}{L_0}\pi x\right), \quad -L_0 \le x \le L_0 \\ &= 0 \quad \text{otherwise} \end{aligned} \tag{6.10}$$

In the relation (6.10), $f = \overline{f}/R_0$ and $\tau = \epsilon/R_0$, where $(\overline{f}, \varepsilon)$ are the dimensional radius and stenosis height of the tube. The equations (6.2)–(6.10) constitute the mathematical model of the present blood flow problem.

6.2.2 Solution by Double Decomposition

Let $L = \frac{\partial^2}{\partial r^2} - \frac{1}{r}\frac{\partial}{\partial r} = r\frac{\partial}{\partial r}\left(\frac{1}{r}\frac{\partial}{\partial r}\right)$ whose inverse operator L^{-1} is defined by (5.23) and L^{-2} is given by (5.24). Then the equation (6.6) becomes

$$L^2\psi = R_e N\psi - \frac{\partial^4\psi}{\partial x^4} - 2\frac{\partial^2}{\partial x^2}(L_\psi) \tag{6.11}$$

where

$$N\psi = \frac{1}{r}J - \frac{2}{r^2}\frac{\partial\psi}{\partial x}D^2\psi \tag{6.12}$$

and the operator D^2 is defined by $D^2 = L + \frac{\partial^2}{\partial x^2}$. Operating with the operator L^{-2} on both sides of the above equation (6.11), we have

$$\psi(x, r) = \overline{\Psi}_r + L^{-2}\left[R_e N\psi - \frac{\partial^4\psi}{\partial x^4} - 2\frac{\partial^2}{\partial x^2}(L\psi)\right] \tag{6.13}$$

where $\overline{\Psi}_r$ is the solution of $L^2\psi = 0$, and it is given by

$$\overline{\psi}_r(x, r) = \frac{1}{16}Br^4 + CL_1^{-1}r\log r + \frac{1}{2}Er^2 + F \tag{6.14}$$

$B, C, E,$ and F being the integration constants.

Let ψ and the nonlinear term $N\psi$ be decomposed into the following forms:

$$\psi(x, r) = \sum_{m=0}^{\infty}\lambda^m\psi_m \tag{6.15}$$

$$N\psi = \sum_{m=0}^{\infty}\lambda^m A_m \tag{6.16}$$

where λ and A_m have their usual meanings. Then the parameterized form of (6.13) is

$$\psi(x, r) = \overline{\Psi}_r(x, r) + \lambda L^{-2}\left[R_e N\psi - \frac{\partial^4\psi}{\partial x^4} - 2\frac{\partial^2}{\partial x^2}(L\psi)\right] \tag{6.17}$$

For double decomposition, we write the parameterized decomposition form of $\overline{\Psi}_r$ as

$$\overline{\Psi}_r(x, r) = \sum_{m=0}^{\infty}\lambda^m\overline{\psi}_{r,m} \tag{6.18}$$

Using (6.15), (6.16), and (6.18) in (6.17) and comparing the terms, we have

$$
\begin{aligned}
\psi_0(x,r) &= \overline{\Psi}_{r,0}(x,r)\\
\psi_1(x,r) &= \overline{\Psi}_{r,1}(x,r) + L^{-2}\left[R_e A_0 - \frac{\partial^4 \psi_0}{\partial x^4} - 2\frac{\partial^2}{\partial x^2}(L\psi_0)\right]\\
\psi_2(x,r) &= \overline{\Psi}_{r,2}(x,r) + L^{-2}\left[R_e A_1 - \frac{\partial^4 \psi_1}{\partial x^4} - 2\frac{\partial^2}{\partial x^2}(L\psi_1)\right]\\
&\cdots \quad \cdots \quad \cdots \quad \cdots\\
&\cdots \quad \cdots \quad \cdots \quad \cdots\\
\psi_{m+1}(x,r) &= \overline{\Psi}_{r(m+1)}(x,r) + L^{-2}\left[R_e A_m - \frac{\partial^4 \psi_m}{\partial x^4} - 2\frac{\partial^2}{\partial x^2}(L\psi_m)\right]\\
&\cdots \quad \cdots \quad \cdots \quad \cdots
\end{aligned}
\tag{6.19}
$$

where A_0, A_1, etc. are special polynomials. To determine these polynomials, we equate (6.12) and (6.16) and then use (6.15) to get

$$
\begin{aligned}
A_0 &= \frac{1}{r}\frac{\partial(D^2\psi_0, \psi_0)}{\partial(r,x)} - \frac{2}{r^2}\cdot\frac{\partial \psi_0}{\partial x}\cdot D^2\psi_0\\
A_1 &= \frac{1}{r}\left[\frac{\partial(D^2\psi_1, \psi_0)}{\partial(r,x)} + \frac{\partial(D^2\psi_0, \psi_1)}{\partial(r,x)}\right]\\
&\quad - \frac{2}{r^2}\left[\frac{\partial \psi_0}{\partial x}\cdot D^2\psi_1 + \frac{\partial \psi_1}{\partial x}\cdot D^2\psi_0\right]\\
&\cdots \quad \cdots \quad \cdots \quad \cdots\\
&\cdots \quad \cdots \quad \cdots \quad \cdots
\end{aligned}
\tag{6.20}
$$

Since the solution $\overline{\Psi}_r(x,r)$ involves four integration constants, B, C, E, and F, following the double decomposition procedure, we write the parameterized decomposition forms of these constants as

$$
\begin{aligned}
B &= \sum_{m=0}^{\infty} \lambda^m B_m\\
C &= \sum_{m=0}^{\infty} \lambda^m C_m\\
E &= \sum_{m=0}^{\infty} \lambda^m E_m\\
F &= \sum_{m=0}^{\infty} \lambda^m F_m
\end{aligned}
\tag{6.21}
$$

Substituting (6.18) and (6.21) in (6.14) and comparing like-power terms of λ, we obtain

$$\begin{aligned}
\overline{\Psi}_{r,0}(x,r) &= \frac{1}{16}B_0r^4 + C_0L_1^{-1}r\log r + \frac{1}{2}E_0r^2 + F_0 \\
\overline{\Psi}_{r,1}(x,r) &= \frac{1}{16}B_1r^4 + C_1L_1^{-1}r\log r + \frac{1}{2}E_1r^2 + F_1 \\
&\cdots \quad \cdots \quad \cdots \quad \cdots \\
&\cdots \quad \cdots \quad \cdots \quad \cdots \\
\overline{\Psi}_{r,m}(x,r) &= \frac{1}{16}B_mr^4 + C_mL_1^{-1}r\log r + \frac{1}{2}E_mr^2 + F_m \\
&\cdots \quad \cdots \quad \cdots \quad \cdots
\end{aligned} \tag{6.22}$$

Again substitution of (6.15) into (6.8) and (6.9) gives the boundary conditions for the component functions ψ_0, ψ_1, ψ_2, etc., as

$$\left.\begin{aligned}
-\frac{1}{r}\frac{\partial\psi_0}{\partial r} = 0, \quad \psi_0 = -\frac{1}{2} \quad \text{at} \quad r = f \\
-\frac{\partial}{\partial r}\left(\frac{1}{r}\frac{\partial\psi_0}{\partial r}\right) = \psi_0 = 0 \quad \text{at} \quad r = 0
\end{aligned}\right\} \tag{6.23}$$

$$\left.\begin{aligned}
-\frac{1}{r}\frac{\partial\psi_m}{\partial r} = \psi_m = 0 \quad \text{at} \quad r = f \\
-\frac{\partial}{\partial r}\left(\frac{1}{r}\frac{\partial\psi_m}{\partial r}\right) = \psi_m = 0 \quad \text{at} \quad r = 0
\end{aligned}\right\} \tag{6.24}$$

for all positive values of m except zero.

Now equating the first relations of (6.19) and (6.22), we have

$$\psi_0(x,r) = \frac{1}{16}B_0r^4 + C_0L_1^{-1}r\log r + \frac{1}{2}E_0r^2 + F_0$$

Then we satisfy the boundary conditions (6.23) by the expression of $\psi_0(x,r)$, we get the following result for $\psi_0(x,r)$:

$$\psi_0(x,r) = \frac{1}{2f^4}(r^4 - 2f^2r^2) \tag{6.25}$$

Again using the second relation of (6.22) and the relation (6.25) in the second relation of (6.19), we get

$$\psi_1(x,r) = l_1r^{10} + l_2r^8 + l_3r^6 + \frac{1}{16}B_1, r^4 + C_1L_1^{-1}r\log r + \frac{1}{2}E_1r^2 + F_1 \tag{6.26}$$

where the constants B_1, C_1, E_1, and F_1 are to be evaluated from the boundary conditions obtained by putting $m = 1$ in (6.24), and these are found to be

$$\begin{aligned}
C_1 &= F_1 = 0 \\
B_1 &= -16f^2(4l_1f^4 + 3l_2f^2 + 2l_3) \\
E_1 &= 2f^4(3l_1f^4 + 2l_2f^2 + l_3)
\end{aligned} \tag{6.27}$$

The expressions for l_1, m_1, and n_1 are given by

$$
\begin{aligned}
l_1 &= \left(R_e/960 f^{11}\right)\left(20 f_1^3 - 13 f f_1 f_2 + f^2 f_3\right) \\
l_2 &= \left(R_e/144 f^9\right)\left(4 f^2 f_3 + 11 f f_1 f_3 - 16 f_1^3\right) \\
&\quad - \left(1/576 f^8\right)\left(15 f^2 f_2^2 + 20 f^2 f_1 f_3 - 180 f f_1^2 f_2 + 210 f_1^4 - f^3 f_4\right) \qquad (6.28) \\
l_3 &= \left(R_e/48 f^7\right)\left(12 f_1^3 - 9 f f_1 f_2 + f^2 f_3 - 8 f_1\right) \\
&\quad - \left(1/96 f^6\right)\left(f^3 f_4 - 12 f^2 f_1 f_3 - 9 f^2 f_2^2 + 72 f f_1^2 f_2 - 60 f_1^4\right. \\
&\quad \left. + \; 80 f_1^2 - 16 f f_2\right)
\end{aligned}
$$

where f_1, f_2, f_3, and f_4 are the derivatives of f with respect to x, indicating the orders according to their suffices, and the ith derivative is given by

$$f_i = (\pi/4)^i \cos\left(i\frac{\pi}{2} + \frac{\pi}{4}x\right) \qquad (6.29)$$

Putting the values of B_1, C_1, E_1, and F_1 from (6.27) in (6.26), we obtain

$$
\begin{aligned}
\psi_1(x, r) &= l_1 r^{10} + l_2 r^8 + l_3 r^6 \\
&\quad - \; f^2(4 l_1 f^4 + 3 l_2 f^2 + 2 l_3) r^4 \\
&\quad + \; f^4(3 l_1 f^4 + 2 l_2 f^2 + l_3) r^2 \qquad (6.30)
\end{aligned}
$$

Similarly, we can compute the other component functions ψ_2, ψ_3, etc., from (6.19).

If we consider a two-term approximant of the solution $\psi(x, r)$, we get from (6.15)

$$\psi(x, r) = \psi_0(x, r) + \psi_1(x, r) \qquad (6.31)$$

where $\psi_0(x, r)$ and $\psi_1(x, r)$ are given by (6.25) and (6.30), remembering that $\lambda = 1$. The expression for the axial velocity component, by means of the first relation of (6.5) and (6.31), is found to be

$$u = -\left[\frac{1}{r}\frac{\partial \psi_0}{\partial r} + \frac{1}{r}\frac{\partial \psi_1}{\partial r}\right] \qquad (6.32)$$

where $\frac{1}{r}\frac{\partial \psi_0}{\partial r}$ and $\frac{1}{r}\frac{\partial \psi_1}{\partial r}$ are given by

$$-\frac{1}{r}\frac{\partial \psi_0}{\partial r} = \frac{2}{f^2}\left(f^2 - r^2\right) \qquad (6.33)$$

and

$$
\begin{aligned}
-\frac{1}{r}\frac{\partial \psi_1}{\partial r} &= 4 f^2\left(4 l_1 f^4 + 3 l_2 f^2 + 2 l_3\right) r^2 \\
&\quad - \left(10 l_1 r^8 + 8 l_2 r^6 + 6 l_3 r^4\right) \\
&\quad - \; 2 f^4(3 l_1 f^4 + 2 l_2 f^2 + l_3) \qquad (6.34)
\end{aligned}
$$

The wall shearing stress is defined by

$$T = -\frac{1}{4}\left(\frac{\partial u}{\partial r}\right)_{r=f}(1 + f_1^2) \tag{6.35}$$

which, by virtue of (6.32), gives

$$T = \frac{1}{4}\left[\frac{\partial}{\partial r}\left(\frac{1}{r}\frac{\partial \psi_0}{\partial r} + \frac{1}{r}\frac{\partial \psi_1}{\partial r}\right)\right]_{r=f}(1 + f_1^2) \tag{6.36}$$

Then using the relations (6.33) and (6.34), we have the resulting expression for T as

$$T = (1/f^3)\left[1 + 2f^6\left(6l_1 f^4 + 3l_2 f^2 + l_3\right)\right](1 + f_1^2) \tag{6.37}$$

6.2.3 Numerical Discussion

The numerical distribution of wall shearing stress along the length of the tube is shown in Figure 6.2 for different values of the Reynolds numbers. The figure does not show any appreciable change of the wall shearing stress over a small range of x starting from the leading edge of the constriction. Beyond this range the solution increases rapidly and attains its maximum value just ahead of the throat of constriction, and its rapid downfall is observed in the diverging section of the tube.

At a high Reynolds number, negative distribution of the wall shearing stress is seen over some length of the tube in the downstream region. This negative behavior of the wall stress indicates the occurrence of separation involving circulation with backflow near the wall. As a result of this backflow, a high velocity core surrounded by the separation region is formed, and a low shear exists at the wall.

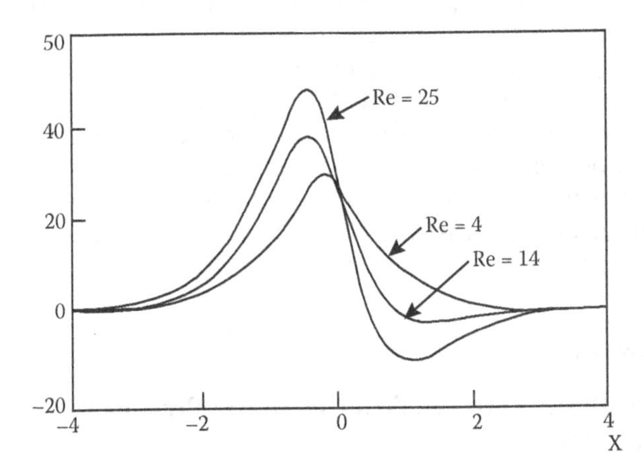

FIGURE 6.2
Distribution of wall shear stress for $\tau = 2/3$.

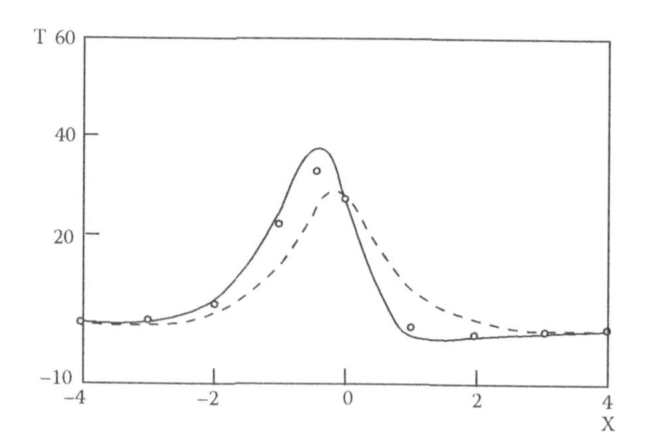

FIGURE 6.3
Distribution of wall shear stress for $Re = 14$ and $\tau = 2/3$. (solid line: decomposition method; dotted line: integral method; broken line: perturbation method).

Since at low Reynolds number, separation is not observed except at higher one, a critical Reynolds number is reached when separation occurs. As the Reynolds number increases beyond this critical Reynolds number, the separation point moves toward the throat of the constriction, and the reattachment point moves downstream with the enlargement of the separated region, which is physiologically unfavorable. The result obtained above is qualitative, and the general pattern of solution is the same as that of Morgan and Young [33].

Figure 6.3 shows the variations of wall shearing stress for Adomian's double decomposition method, the perturbation technique of Haldar [22], and the integral method of Morgan and Young. The decomposition method solution deviates much from the second-order perturbation solution, but it is very close to the integral method solution of Morgan and Young. The decomposition solution can be improved more if the three-term approximant of the solution ψ or more is taken into account.

The advantage of this global methodology is that it leads to an analytical continuous approximate solution that is very rapidly convergent [9,10]. This method does not take any help of linearization or any other simplifications for handling the nonlinear terms. Since the decomposition parameter, in general, is not a perturbation parameter, it follows that nonlinearities in the operator equation can be handled easily, and an accurate approximate solution may be obtained for any physical problem.

6.3 Flow of Blood Through Arteries in the Presence of a Magnetic Field

It has been reported by Barnothy [2] that the biological systems, in general, are affected by the application of an external magnetic field. Gold [13] has obtained an exact solution of one-dimensional steady flow of an electrically conducting fluid through a nonconduction circular tube under the influence of a uniform transverse magnetic field. The corresponding unsteady problem has been studied by Gupta and Bani Singh [14] considering an exponentially decaying pressure gradient. Ramachandra Rao and Deshikachar [36] have studied the physiological type flow in the presence of a transverse magnetic field. Sud and Sekhon [41] have used the finite-element method to study the blood flow through the human arterial system in the presence of a steady magnetic field.

In all the above works, the effects of different types of magnetic fields on flow characteristics in the tubes of uniform circular cross section have been studied, but the corresponding problems in the presence of a constriction are more important from the physiological point of view. Deshikachar and Ramachandra Rao [12] have studied the steady blood flow through a channel of variable cross section in the presence of a transverse magnetic field, and the corresponding unsteady problem has been investigated by Ramachandra Rao and Deshikachar [37]. McMichael and Deutsch [34] have analyzed the steady flow problem in a circular tube of variable cross section under the influence of an axial magnetic field, and the same problem in unsteady case has been investigated by Deshikachar and Ramachandra Rao. Haldar [27] has studied this specific problem of blood flow through a constricted artery with the help of the decomposition method under the influence of an externally applied magnetic field.

6.3.1 Equations of Motion

Consider the steady, laminar, and axially symmetric flow of blood through a locally constricted straight artery of infinite length under the influence of an external transverse magnetic field that is applied uniformly (Figure 6.4). Blood flowing through the tube is supposed to be conducting and Newtonian in character. The assumptions of constant fluid density and viscosity are used here. The appropriate equations governing the flow field in the tube are the

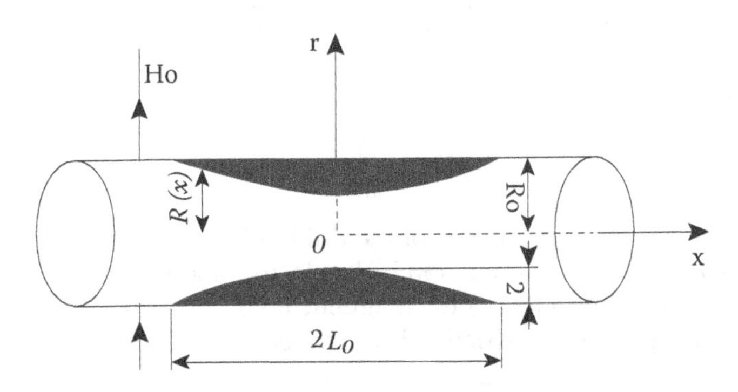

FIGURE 6.4
Geometry of constriction.

momentum equations, and these equations, after introducing the electromagnetic force, in cylindrical polar coordinates $(\bar{x}, \bar{r}, \theta)$ are

$$\vec{U} \cdot \overline{\nabla} \vec{U} = \frac{1}{\rho} \cdot \overline{\nabla} \vec{P} + \nu \overline{\nabla}^2 \vec{U} + \frac{1}{\rho} \left(\vec{I} \times \vec{B} \right) \tag{6.38}$$

where $\vec{U}$ is the velocity vector of the field, P the pressure, ν the kinematic viscosity, ρ the density of the fluid, $\vec{I}$ the current density, and $\vec{B}$ the magnetic field, and the operator $\overline{\nabla}^2$ is given by

$$\overline{\nabla}^2 = \frac{\partial^2}{\partial \bar{r}^2} + \frac{1}{\bar{r}} \frac{\partial}{\partial \bar{r}} + \frac{1}{\bar{r}^2} \frac{\partial^2}{\partial \theta^2} \tag{6.39}$$

The current density and magnetic field are expressed by the Maxwell's equations and Ohm's laws, namely,

$$\vec{I} = \sigma_e \left[\vec{E} + \mu_e \left(\vec{U} \times \vec{B} \right) \right] \tag{6.40}$$

$$\overline{\nabla} \vec{B} = 0 \tag{6.41}$$

$$\overline{\nabla} \vec{E} = 0 \tag{6.42}$$

where $\vec{E}$ is the electric field, γ_e is the conductivity of the fluid, and μ_e is the magnetic permeability.

In the present investigation, it is assumed that the effects of the induced magnetic field and the electric field produced due to the motion of electrically conducting fluid are very small and no external force is applied. With these assumptions and the assumption of the axially symmetric flow of fluid, the governing equations of motion of the fluid are the Navier–Stokes equations in cylindrical polar coordinates

$$\bar{u} \frac{\partial \bar{u}}{\partial \bar{x}} + \bar{v} \frac{\partial \bar{u}}{\partial \bar{r}} = -\frac{1}{\rho} \frac{\partial \bar{p}}{\partial \bar{x}} + \nu \left(\frac{\partial^2 \bar{u}}{\partial \bar{r}^2} + \frac{1}{\bar{r}} \frac{\partial \bar{u}}{\partial \bar{r}} + \frac{\partial^2 \bar{u}}{\partial \bar{x}^2} \right) - \frac{B_0^2}{\rho} \bar{u} \tag{6.43}$$

$$\bar{u} \frac{\partial \bar{v}}{\partial \bar{x}} + \bar{v} \frac{\partial \bar{v}}{\partial \bar{r}} = -\frac{1}{\rho} \frac{\partial \bar{p}}{\partial \bar{r}} + \nu \left(\frac{\partial^2 \bar{v}}{\partial \bar{r}^2} + \frac{1}{\bar{r}} \frac{\partial \bar{v}}{\partial \bar{r}} - \frac{\bar{v}}{r^2} + \frac{\partial^2 \bar{v}}{\partial \bar{x}^2} \right) \tag{6.44}$$

and the continuity equation

$$\frac{\partial}{\partial \bar{x}} (\bar{r}\ \bar{u}) + \frac{\partial}{\partial \bar{r}} (\bar{r}\ \bar{u}) = 0 \tag{6.45}$$

where $(\bar{u}, \bar{v})$ are the components of the fluid velocity in the axial and radial directions, respectively, $B_0 (= \mu_e H_0)$ is the electromagnetic induction, and H_0 is the transverse component of the magnetic field.

The geometry of the constriction is described by

$$\frac{\overline{R}(\bar{x})}{R_0} = 1 - \frac{\bar{\tau}}{R_0} \bar{f}(\bar{x}) \tag{6.46}$$

where R_0 is the radius of the normal tube, $\overline{R}(\overline{x})$ the radius of the tube in the stenotic region, and $\overline{\tau}$ the maximum height of stenosis.

The boundary conditions are

$$\overline{u} = \overline{v} = 0 \quad \text{at} \quad \overline{r} = \overline{R}(\overline{x}) \tag{6.47}$$

$$\frac{\partial \overline{u}}{\partial \overline{r}} = 0 \quad \text{at} \quad \overline{r} = 0 \tag{6.48}$$

$$\int_0^{\overline{R}(x)} \overline{r}\ \overline{u} d\overline{r} = \overline{Q}/2\pi \tag{6.49}$$

where $\overline{Q}$ is the constant volumetric flux across any cross section of the tube.

It is convenient to write system of equations from (6.43) to (6.49) in the nondimensional forms with the help of the following transformations:

$$\begin{aligned} u &= \overline{u}/U_0, v = \overline{v}/U_0 \\ r &= \overline{r}/R_0, x = \overline{x}/R_0 \\ p &= \overline{p}/\rho U_0^2 \end{aligned} \tag{6.50}$$

where (u, v) are the dimensionless velocity components, U_0 is the characteristic velocity, and p is the nondimensional fluid pressure.

The momentum equations (6.43) and (6.44) are

$$u\frac{\partial u}{\partial x} + v\frac{\partial u}{\partial r} = -\frac{\partial p}{\partial x} + \frac{1}{R_e}\left(\frac{\partial^2 u}{\partial r^2} + \frac{1}{r}\frac{\partial u}{\partial r} + \frac{\partial^2 u}{\partial x^2}\right) - M^2 u \tag{6.51}$$

in the axial direction and

$$u\frac{\partial v}{\partial x} + v\frac{\partial u}{\partial r} = -\frac{\partial p}{\partial r} + \frac{1}{R_e}\left(\frac{\partial^2 v}{\partial r^2} + \frac{1}{r}\frac{\partial v}{\partial r} - \frac{v}{r^2} + \frac{\partial^2 v}{\partial x^2}\right) \tag{6.52}$$

in the radial direction where R_e and M are the Reynolds number and Hartmann number defined by

$$\begin{aligned} R_e &= U_0 R_0/\nu \\ M^2 &= B_0^2 \sigma_e R_0^2/\mu \end{aligned} \tag{6.53}$$

Similarly the dimensionless continuity equation (6.45) is

$$\frac{\partial}{\partial x}(r\ u) + \frac{\partial}{\partial r}(r\ v) = 0 \tag{6.54}$$

and the geometry of constriction takes the form (Figure 6.4)

$$\eta(x) = 1 - \tau f(x) \tag{6.55}$$

where

$$\begin{aligned}\eta(x) &= \overline{R}(\overline{x})/R_0 \\ f(x) &= \overline{f}(\overline{x})/R_0 \\ \tau &= \overline{\tau}/R_0\end{aligned} \tag{6.56}$$

The corresponding nondimensional boundary conditions are

$$u = v = 0 \quad \text{at} \quad r = \eta(x) \tag{6.57}$$

$$\frac{\partial u}{\partial r} = 0 \quad \text{at} \quad r = 0 \tag{6.58}$$

$$\int_0^{\eta(x)} r\, u\, d\, r = -\frac{1}{2} \tag{6.59}$$

We next introduce the stream function ψ defined by

$$u = -\frac{1}{r}\frac{\partial \psi}{\partial r}, v = \frac{1}{r}\frac{\partial \psi}{\partial x} \tag{6.60}$$

Then the continuity equation (6.54) is satisfied identically. Using (6.60) elimination of p between (6.51) and (6.52) gives the following governing equation:

$$R_e\left[\frac{1}{r}\cdot J - \frac{2}{r^2}\nabla^2\psi\cdot\frac{\partial \psi}{\partial x}\right] = \nabla^4\psi - M^2 r\left(\frac{1}{r}\frac{\partial \psi}{\partial r}\right) \tag{6.61}$$

where J is the Jacobian defined by

$$J = \frac{\partial(\nabla^2\psi, \psi)}{\partial(r, x)} = \begin{vmatrix} \frac{\partial}{\partial r}(\nabla^2\psi) & \frac{\partial \psi}{\partial r} \\ \frac{\partial}{\partial x}(\nabla^2\psi) & \frac{\partial \psi}{\partial x}\end{vmatrix} \tag{6.62}$$

and the operator ∇^2 is given by

$$\nabla^2 = \frac{\partial^2}{\partial r^2} - \frac{1}{r}\frac{\partial}{\partial r} + \frac{\partial^2}{\partial x^2} \tag{6.63}$$

The boundary conditions in terms of ψ are

$$-\frac{1}{r}\frac{\partial \psi}{\partial r} = 0, \quad \psi = \frac{1}{2} \quad \text{at} \quad r = \eta(x) \tag{6.64}$$

$$-\frac{\partial}{\partial r}\left(\frac{1}{r}\frac{\partial \psi}{\partial r}\right) = \psi = 0 \quad \text{at} \quad r = 0 \tag{6.65}$$

6.3.2 Method of Solution

The equation (6.61) is a nonlinear partial differential equation, and the exact solution of this equation is not always possible. This equation can be solved by using traditional numerical techniques, which result in massive numerical computations. To avoid this difficulty, the decomposition method developed by Adomian has been applied here to obtain analytic approximations to the nonlinear equation (6.61).

Let

$$L = \frac{\partial^2}{\partial r^2} - \frac{1}{r}\frac{\partial}{\partial r}, \tag{6.66}$$

and the equation (6.61) becomes

$$L^2\psi = R_e N\psi - \frac{\partial^4 \psi}{\partial x^4} - 2\frac{\partial^2}{\partial x^2}(L\psi) + M^2(L\psi) \tag{6.67}$$

where

$$N\psi = \frac{1}{r}\cdot J - \frac{2}{r^2}\cdot\frac{\partial \psi}{\partial x}\cdot \Delta^2\psi \tag{6.68}$$

If we operate both sides of (6.61) with the inverse operator L^{-2}, then we get

$$\psi(x,r) = \psi_0(x,r) + L^{-2}\left[R_e N\psi - \frac{\partial^4 \psi}{\partial x^4} - 2\frac{\partial^2}{\partial x^2}(L\psi) + M^2(L\psi)\right] \tag{6.69}$$

Here ψ_0 is the solution of the homogeneous equation

$$L^2\psi_0 = 0 \tag{6.70}$$

and it is given by

$$\psi_0(x,r) = \frac{1}{16}A(x)r^4 + B(x)L_1^{-1}r\log r + \frac{1}{2}C(x)r^2 + F(x) \tag{6.71}$$

where the integration constants $A, B, C,$ and F involved in (6.71) are to be determined from the given boundary conditions (6.64), (6.65), and $L_1^{-1} = \int(\cdot)dr$.

Next, we decompose ψ and $N\psi$ into the following forms:

$$\psi(x,r) = \sum_{m=0}^{\infty}\lambda^m\psi_m \tag{6.72}$$

$$N\psi = \sum_{m=0}^{\infty}\lambda^m P_m \tag{6.73}$$

where P_m and λ have their usual meanings. Then the parameterized form of (6.69) is

$$\psi(x,r) = \psi_0(x,r) + \lambda L^{-2}\left[R_e N\psi - \frac{\partial^4 \psi}{\partial x^4} - 2\frac{\partial^2}{\partial x^2}(L\psi) + M^2(L\psi)\right] \tag{6.74}$$

Substituting (6.72) and (6.73) in (6.74) and then comparing the like-power terms of λ on both sides of it, we get

$$\psi_{m+1}(x,r) = L^{-2}\left[R_e P_m - \frac{\partial^4 \psi_m}{\partial x^4} - 2\frac{\partial^2}{\partial x^2}(L\psi_m) + M^2(L\psi_m)\right] \tag{6.75}$$

where $m = 0, 1, 2$, etc. Once the component $\psi_0(x,r)$ is determined, the other components of $\psi_0(x,r)$, such as $\psi_1(x,r)$, $\psi_2(x,r)$, etc., can be easily obtained from (6.75). The decomposition referred to above is called *regular* or *ordinary decomposition* of ψ.

If we further take the parameterized form of $\psi_0(x,r)$ given by

$$\psi_0(x,r) = \sum_{m=0}^{\infty} \lambda^m \psi_{0m} \tag{6.76}$$

we mean the double decomposition. Substitution of (6.72), (6.73), and (6.76) in (6.74) gives the double decomposition components of $\psi(x,r)$, and these are given by the relation

$$\psi_{m+1}(x,r) = \psi_{0(m+1)}(x,r) + L^{-2}\left[R_e P_m - \frac{\partial^4 \psi_m}{\partial x^4} - 2\frac{\partial^2}{\partial x^2}(L\psi_m) + M^2(L\psi_m)\right] \tag{6.77}$$

n being zero and any positive integer.

Since the expression for $\psi_0(x,r)$ contains the constants A, B, C, and F, the parameterized decomposition forms of all these constants are

$$\begin{aligned} A(x) &= \sum_{m=0}^{\infty} \lambda^m A_m \\ B(x) &= \sum_{m=0}^{\infty} \lambda^m B_m \\ C(x) &= \sum_{m=0}^{\infty} \lambda^m C_m \\ F(x) &= \sum_{m=0}^{\infty} \lambda^m F_m \end{aligned} \tag{6.78}$$

If we substitute (6.76) and (6.78) in (6.71), and then if we compare the terms on both sides of it, we get

$$\psi_{m+1}(x,r) = \frac{1}{16}A_{m+1}r^4 + B_{m+1}L_1^{-1}r\log r + \frac{1}{2}C_{m+1}r^2 + F_{m+1} \tag{6.79}$$

The relations (6.77) and (6.79) together give the components of $\psi(x,r)$. The constants involved in each $\psi_m(x,r)$ will be determined by their respective boundary conditions.

The polynomials $P_0, P_1, \cdots, P_m$ are Adomian's polynomials, and they are defined in such a way that $P_0 \equiv P_m(\psi_0)$, $P_1 \equiv P_1(\psi_0, \psi_1)$, $P_2 \equiv P(\psi_0, \psi_1, \psi_2), \cdots, P_m \equiv P_m(\psi_0, \psi_1, \cdots, \psi_m)$. To determine these polynomials, we use (6.72) and (6.73) in (6.68) and compare the terms on both sides of it to get the following set of Adomian's polynomials:

$$
\begin{aligned}
P_0 &= \frac{1}{r}\frac{\partial(\nabla^2\psi_0, \psi_0)}{\partial(r, x)} - \frac{2}{r^2}\cdot\frac{\partial\psi_0}{\partial x}\cdot\nabla^2\psi_0 \\
P_1 &= \frac{1}{r}\left[\frac{\partial(\nabla^2\psi_0, \psi_1)}{\partial(r, x)} + \frac{\partial(\nabla^2\psi_1, \psi_1)}{\partial(r, x)}\right] \\
&\quad - \frac{2}{r^2}\left[\frac{\partial\psi_0}{\partial x}\cdot\nabla^2\psi_1 + \frac{\partial\psi_1}{\partial x}\cdot\nabla^2\psi_0\right] \\
&\cdots \quad \cdots \quad \cdots \quad \cdots \\
&\cdots \quad \cdots \quad \cdots \quad \cdots
\end{aligned}
\tag{6.80}
$$

Again, substitution of (6.72) in the boundary conditions (6.64) and (6.65) gives the boundary conditions for the respective components ψ_0, ψ_1, etc., as

$$
\left.
\begin{aligned}
-\frac{1}{r}\frac{\partial\psi_0}{\partial r} = 0,\ \psi_0 = -\frac{1}{2} &\qquad \text{at} \quad r = \eta \\
-\frac{\partial}{\partial r}\left(\frac{1}{r}\frac{\partial\psi_0}{\partial r}\right) = \psi_0 = 0 &\qquad \text{at} \quad r = 0
\end{aligned}
\right\}
\tag{6.81}
$$

$$
\left.
\begin{aligned}
-\frac{1}{r}\frac{\partial\psi_m}{\partial r} = \psi_m = 0 &\qquad \text{at} \quad r = \eta \\
-\frac{\partial}{\partial r}\left(\frac{1}{r}\frac{\partial\psi_m}{\partial r}\right) = \psi_m = 0 &\qquad \text{at} \quad r = 0
\end{aligned}
\right\}
\tag{6.82}
$$

for all positive integers of m.

6.3.3 Solution of the Problem

Before proceeding to the solutions, we have to find out the inverse operator L^{-2} and for that we consider the following equation for ψ:

$$L\psi = r\frac{\partial}{\partial r}\left(\frac{1}{r}\frac{\partial\psi}{\partial r}\right) = F \tag{6.83}$$

which, on solving, gives

$$\psi = \left[L_1^{-1}r\left(L_1^{-1}r^{-1}\right)\right]F \tag{6.84}$$

remembering that the boundary condition terms vanish and L_1^{-1} is a onefold indefinite integral. From the relation (6.84), it is obvious that the inverse L^{-1} is identified as

$$L^{-1} = \left[L_1^{-1}r\left(L_1^{-1}r^{-1}\right)\right] \tag{6.85}$$

and hence we get

$$L^{-2} = L_1^{-1}\left[rL_1^{-1}\left\{r^{-1}L_1^{-1}\left(rL_1^{-1}r^{-1}\right)\right\}\right] \quad (6.86)$$

Using the boundary conditions (6.81) in (6.71), we have the expression for ψ_0 as

$$\psi_0(x,r) = \frac{1}{2\eta^4}\left(r^4 - 2\eta^2 r^2\right) \quad (6.87)$$

The expression for ψ_1 can be obtained from (6.77) and (6.79) by putting $m = 0$, and this expression involves the operator L^{-2}. Performing the operation of this inverse operator, we get

$$\psi_1(x,r) = \alpha(x)r^{10} + \beta(x)r^8 + \gamma(x)r^6 + \frac{1}{16}A_1 r^4 + B_1 L_1^{-1} r \log r + \frac{1}{2}C_1 r^2 + F_1 \quad (6.88)$$

where A_1, B_1, C_1, and F_1 are integration constants to be obtained by satisfying the boundary conditions (6.82), putting $m = 1$, and these constants are found to be

$$\begin{aligned} B_1 &= F_1 = 0 \\ A_1 &= -16\eta^2(4\alpha\eta^4 + 3\beta\eta^2 + 2\gamma) \\ C_1 &= 2\eta^4(3\alpha\eta^4 + 2\beta\eta^2 + \alpha) \end{aligned} \quad (6.89)$$

The expressions for α, β, and γ are given by

$$\alpha(x) = (R_e/960\eta^{11})(20\eta_1^3 - 13\eta\eta_1\eta_2 + \eta^2\eta_3) \quad (6.90)$$

$$\begin{aligned} \beta(x) &= (Re/144\eta^9)(4\eta_1 - \eta^2\eta_3 + 11\eta\eta_1\eta_2 - 16\eta_1^3) \\ &- (1/576\eta^8)(15\eta^2\eta_2^2 + 20\eta^2\eta_1\eta_3 - 180\eta\eta_1^2\eta_2 + 210\eta_1^4 - \eta^3\eta_4) \end{aligned} \quad (6.91)$$

$$\begin{aligned} \gamma(x) &= (Re/48\eta^7)(12\eta_1^3 - 9\eta\eta_1\eta_2 + \eta^2\eta_3 - 8\eta_1) \\ &- (1/96\eta^6)(\eta^3\eta_1 - 12\eta^2\eta_1\eta_3 - 9\eta^2\eta_2^2 \\ &+ 72\eta\eta_1^2\eta_2 - 60\eta_1^4 + 80\eta_1^2 - 16\eta\eta_2) \\ &- (M^2/48\eta^4) \end{aligned} \quad (6.92)$$

where η_1, η_2, η_3, and η_4 are the derivatives of η with respect to x indicating the orders according to their suffices. The resulting expression for ψ_1 is found to be

$$\begin{aligned} \psi_1(x,r) &= \alpha r^{10} + \beta r^3 + \gamma r^6 \\ &- \eta^2(4\alpha\eta^4 + 3\beta\eta^2 + 2\gamma)r^4 \\ &+ \eta^4(3\alpha\eta^4 + 2\beta\eta^2 + \gamma)r^2 \end{aligned} \quad (6.93)$$

If we consider a two-term approximant of the solution ψ, we obtain from (6.72)

$$\psi(x,r) = \psi_0(x,r) + \psi_1(x,r) \tag{6.94}$$

where ψ_0 and ψ_1 are given by (6.87) and (6.73) remembering that $\lambda = 1$. The axial velocity component is found to be

$$u(x,r) = -\left[\frac{1}{r}\cdot\frac{\partial\psi_0}{\partial r} + \frac{1}{r}\cdot\frac{\partial\psi_1}{\partial r}\right] \tag{6.95}$$

The wall shearing stress is defined by

$$T = -\frac{1}{4}\left(\frac{\partial u}{\partial r}\right)_{r=\eta}\cdot(1+\eta_1^2) \tag{6.96}$$

which, on substitution of u from (6.95), gives

$$T = (1/\eta^3)\left[1 + 2\eta^6(6\alpha\eta^4 + 3\beta\eta^2 + \gamma)\right](1+\eta_1^2) \tag{6.97}$$

6.3.4 Numerical Discussion

For numerical discussion the function $f(x)$ is described by

$$f(x) = \frac{1}{2}\left(1 + \cos\frac{\pi x}{L_0}\right), \quad -L_0 \le x \le L_0 \tag{6.98}$$

Then the geometry of constriction (6.55) takes the following form

$$\eta(x) = 1 - \frac{1}{2}\tau\left(1 + \cos\frac{\pi x}{L_0}\right), \quad -L_0 \le x \le L_0 \tag{6.99}$$

The variations of the wall shear stress (6.97) along the length of the constricted artery are shown graphically for different values of the Hartmann number. Figure 6.5 shows the variations of the wall shear stress with x for different values of the Hartmann number. It is seen that the negative behavior of the solution observed in the diverging section of the tube decreases with the increasing Hartmann number. As a result the circulation diminishes indicating the favorable physiological condition. Therefore, it can be concluded that the effect of an external transverse magnetic field applied uniformly favors the condition of blood flow.

The theoretical result has been explained numerically for two-term approximation of the solution $\psi(x,r)$. The result can be improved by considering a three-term approximant or more to the solution.

The advantage of the decomposition method is to give an analytic approximate solution of a nonlinear ordinary or partial differential equation that is rapidly convergent [9,10]. The speed of convergency depends on the choice of operator, which may be a highest-ordered differential operator or a combination of linear operators or a multidimensional operator.

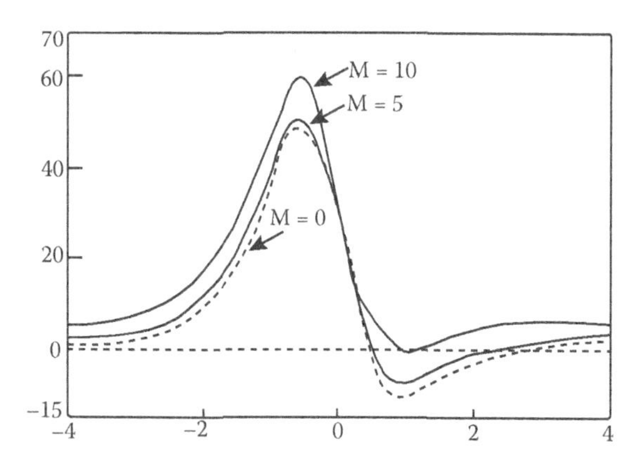

FIGURE 6.5
Distribution of wall shear stress for $Re = 25$ and $\tau = 2/3$.

6.4 Pulsatile Flow of Blood through a Constricted Artery

In the previous section, the flow of blood through a rigid blood vessel with a cosine-shaped constriction was considered. This problem was solved by means of Adomian's double decomposition method. In the present section, Mamaloukas, Haldar, and Mazumdar [30] have studied the pulsatile blood flow in a constricted artery. This constriction is described by a cosine function, and the problem has also been solved with the help of the double decomposition method of Adomian.

6.4.1 Equations of Motion

Let an axially symmetric pulsatile flow of a Newtonian blood in a tube with a constriction be considered. The equations of motion in the tube under the assumed condition are the continuity equation and the Navier–Stokes equations in cylindrical polar coordinates. It is convenient to write the system of equations governing the flow field in the tube in the dimensionless forms by the following transformations:

$$u = \frac{\bar{u}}{u_0}, \quad v = \frac{\bar{v}}{u_0}, \quad t = \omega\bar{t}$$
$$r = \frac{\bar{r}}{r_0}, \quad x = \frac{\bar{x}}{r_0}, \quad p = \frac{\bar{p}}{\nu u_0^2} \tag{6.100}$$

where $(\bar{u}, \bar{v})$ and (u, v) are the dimensional and nondimensional velocity components in the axial and radial directions, respectively, $(\bar{x}, \bar{r})$ and (x, r) are the corresponding coordinate axes, r_0 is the radius of the normal tube, u_0 is the characteristic velocity, ω is the frequency, $(\bar{t}, t)$ are the dimensional and nondimensional times, ρ is the fluid density, and $(\bar{p}, p)$ are the fluid pressures in the dimensional and nondimensional states.

The continuity equation of the system is

$$\frac{\partial}{\partial r}(ru) + \frac{\partial}{\partial x}(rv) = 0 \tag{6.101}$$

and the Navier–Stokes equations are

$$q^2\frac{\partial u}{\partial t} + R_e\left(u\frac{\partial u}{\partial x} + v\frac{\partial u}{\partial r}\right) = -\frac{\partial p}{\partial x} + \left(\frac{\partial^2 u}{\partial r^2} + \frac{1}{r}\frac{\partial u}{\partial r} + \frac{\partial^2 u}{\partial x^2}\right) \tag{6.102}$$

$$q^2\frac{\partial v}{\partial t} + R_e\left(u\frac{\partial v}{\partial x} + v\frac{\partial v}{\partial r}\right) = -\frac{\partial p}{\partial r} + \left(\frac{\partial^2 v}{\partial r^2} + \frac{1}{r}\frac{\partial v}{\partial r} - \frac{v}{r^2} + \frac{\partial^2 v}{\partial x^2}\right) \tag{6.103}$$

where q and R_e are the Wamersley frequency parameter and the Reynolds number defined by

$$\left.\begin{aligned} q^2 &= \omega^2 r_0^2/\nu \\ R_e &= u_0 r_0/\nu \end{aligned}\right\} \tag{6.104}$$

If the nondimensional stream function $\psi(x, r, t)$ defined by

$$\left.\begin{aligned} u &= -\frac{1}{r}\frac{\partial \psi}{\partial r} \\ v &= \frac{1}{r}\frac{\partial \psi}{\partial x} \end{aligned}\right\} \tag{6.105}$$

is introduced, then the equation of continuity (6.101) is identically satisfied, and the governing equation of motion in terms of ψ after eliminating p between (6.102) and (6.103) is

$$q^2 D^2\left(\frac{\partial \psi}{\partial t}\right) + R_e\left[\frac{1}{r}\frac{\partial(D^2\psi, \psi}{\partial(r, x)} - \frac{2}{r^2}\cdot\frac{\partial \psi}{\partial x}\cdot D^2\psi\right] = D^4\psi \tag{6.106}$$

where the operator D^2 is given by

$$D^2 = r\frac{\partial}{\partial r}\left(\frac{1}{r}\frac{\partial}{\partial r}\right) + \frac{\partial^2}{\partial x^2} \tag{6.107}$$

The boundary conditions imposed on ψ are

$$-\frac{1}{r}\frac{\partial \psi}{\partial r} = 0,\ \psi = -\frac{1}{2}\cos t \quad \text{at} \quad r = f \tag{6.108}$$

$$-\frac{\partial}{\partial r}\left(\frac{1}{r}\frac{\partial \psi}{\partial r}\right) = \psi = 0 \quad \text{at} \quad r = 0 \tag{6.109}$$

where f denotes the radius of the stenosed artery and describes the geometry of stenosis whose nondimensional form (Figure 6.1) is given by

$$\begin{aligned} f(x) &= 1 - \frac{\tau}{2}\left(1 + \cos\frac{\pi}{L_0}x\right), -L_0 \leq x \leq L_0 \\ &= 0 \text{ otherwise} \end{aligned} \tag{6.110}$$

In the relation (6.110), $f = \bar{f}/R_0$ and $\eta = \varepsilon/R_0$, where $(\bar{f}, \varepsilon)$ are the dimensional radius and stenosis height in the tube. The equation (6.106), together with the boundary conditions (6.108) and (6.109), constitutes the mathematical model of the blood flow problem in the artery.

The equation (6.106) is a nonlinear partial differential equation, and it is not always possible to get the analytic solution of this equation. The traditional numerical techniques that result in massive numerical computations can give only an exact solution of the equation (6.106). The modern powerful decomposition method developed by Adomian is now being applied here to obtain analytic approximations to this nonlinear equation.

6.4.2 Solution by Double Decomposition

Let $L = r\frac{\partial}{\partial r}\left(\frac{1}{r}\frac{\partial}{\partial r}\right)$, and the operator (6.107) takes the form $D^2 = L + \frac{\partial^2}{\partial x^2}$. The equation (6.106) becomes

$$L^2\psi = R_e N\psi - \frac{\partial^4\psi}{\partial x^4} - 2\frac{\partial^2}{\partial x^2}(L\psi) + q^2 D^2\left(\frac{\partial\psi}{\partial t}\right) \tag{6.111}$$

where

$$N\psi = \frac{1}{r}\frac{\partial(D^2\psi, \psi)}{\partial(r, x)} - \frac{2}{r^2}\frac{\partial\psi}{\partial t}\cdot D^2\psi \tag{6.112}$$

Operating with the inverse operator L^{-2} defined by (5.24) on both sides of (6.111), we have

$$\psi(x, r, t) = \psi_0(x, r, t) + L^{-2}\left[R_e N\psi - \frac{\partial^4\psi}{\partial x^4} - 2\frac{\partial^2}{\partial x^2}(L\psi) + q^2 D^2\left(\frac{\partial\psi}{\partial t}\right)\right] \tag{6.113}$$

where $\psi_0(x, r, t)$ is the solution of the homogeneous equation $L^2\psi = 0$ and is given by

$$\psi_0(x, r, t) = \frac{1}{16}r^4\alpha_1(x, t) + \alpha_2(x, t)L_1^{-1}r\log r + \frac{1}{2}\alpha_3(x, t)r^2 + \alpha^4(x, t) \tag{6.114}$$

Here $L_1^{-1} = \int(\cdot)dr$ and $\alpha_1(x, t), \alpha_2(x, t), \alpha_3(x, t), \alpha_4(x, t)$ are the integration constants to be evaluated by the given boundary conditions.

We now decompose ψ and the nonlinear term $N\psi$ into the following parameterized forms:

$$\psi(x, r, t) = \sum_{m=0}^{\infty}\lambda^m\psi_m \tag{6.115}$$

$$N\psi(x, r, t) = \sum_{m=0}^{\infty}\lambda^m A_m \tag{6.116}$$

where λ and A_m have their usual meanings. Then we write the parameterized form of (6.113) as

$$\psi(x, r, t) = \psi_0(x, r, t) + \lambda L^{-2}\left[R_e N\psi - \frac{\partial^4 \psi}{\partial x^4} - 2\frac{\partial^2}{\partial x^2}(L\psi) + q^2 D^2\left(\frac{\partial \psi}{\partial t}\right)\right] \tag{6.117}$$

If we again take parameterized form of $\psi_0(x, r, t)$ given by

$$\psi_0(x, r, t) = \sum_{m=0}^{\infty} \lambda^m \psi_{0m} \tag{6.118}$$

Substituting (6.115), (6.116), and (6.118) in (6.117), and then equating the like-power terms of λ on both sides, we get

$$\psi_0(x, r, t) = \psi_{00}(x, r, t)$$

$$\psi(x, r, t) = \psi_{01}(x, r, t) + L^{-2}\left[R_e A_0 - \frac{\partial^4 \psi_0}{\partial x^4} - 2\frac{\partial^2}{\partial x^2}(L\psi_0) + q^2 D^2\left(\frac{\partial \psi_0}{\partial t}\right)\right]$$

$$\psi(x, r, t) = \psi_{02}(x, r, t) + L^{-2}\left[R_e A_1 - \frac{\partial^4 \psi_1}{\partial x^4} - 2\frac{\partial^2}{\partial x^2}(L\psi_1) + q^2 D^2\left(\frac{\partial \psi_1}{\partial t}\right)\right]$$

$$\cdots \quad \cdots \quad \cdots \quad \cdots \tag{6.119}$$

$$\cdots \quad \cdots \quad \cdots \quad \cdots$$

$$\psi_{m+1}(x, r, t) = \psi_{02}(x, r, t) + L^{-2}\left[R_e A_m - \frac{\partial^4 \psi_1}{\partial x^4} - 2\frac{\partial^2}{\partial x^2}(L\psi_m) + q^2 D^2\left(\frac{\partial \psi_m}{\partial t}\right)\right]$$

We next proceed to determine Adomian's special polynomials and the components of ψ_0. To evaluate each A_m, we combine (6.112), (6.115), and (6.117), and then equating the terms of like-powers of λ, we get the following expressions for A_m:

$$A_0 = \frac{1}{r}\frac{\partial(D^2\psi_0, \psi_0)}{\partial(r, x)} - \frac{2}{r^2}\cdot\frac{\partial \psi_0}{\partial x}\cdot D^2\psi_0$$

$$A_1 = \frac{1}{r}\left[\frac{\partial(D^2\psi_1, \psi_0)}{\partial(r, x)} + \frac{\partial(D^2\psi_0, \psi_1)}{\partial(r, x)}\right] - \frac{2}{r^2}\left[\frac{\partial \psi_0}{\partial x}\cdot D^2\psi_1 + \frac{\partial \psi_1}{\partial x}\cdot D^2\psi_0\right] \tag{6.120}$$

$$\cdots\cdots \quad \cdots \quad \cdots \quad \cdots$$

$$\cdots\cdots \quad \cdots \quad \cdots \quad \cdots$$

Since the integration constants involved in the solution of $\psi(x, r, t)$ are $\alpha_1(x, t), \alpha_2(x, t), \alpha_3(x, t)$, and $\alpha_4(x, t)$, parameterized decompositions of these constants are

$$\begin{aligned}
\alpha_1(x, t) &= \sum_{m=0}^{\infty} \lambda^m \alpha_{1m} \\
\alpha_2(x, t) &= \sum_{m=0}^{\infty} \lambda^m \alpha_{2m} \\
\alpha_3(x, t) &= \sum_{m=0}^{\infty} \lambda^m \alpha_{3m} \\
\alpha_4(x, t) &= \sum_{m=0}^{\infty} \lambda^m \alpha_{4m}
\end{aligned} \tag{6.121}$$

Substitution of (6.118) and (6.121) in (6.114), and then comparison of the coefficients of like power-terms of λ give

$$\begin{aligned}
\psi_{00} &= \frac{r^4}{16}\alpha_{10} + L_1^{-1} r \log r \alpha_{20} + \frac{r^2}{2}\alpha_{30} + \alpha_{40} \\
\psi_{01} &= \frac{r^4}{16}\alpha_{11} + L_1^{-1} r \log r \alpha_{21} + \frac{r^2}{2}\alpha_{31} + \alpha_{41} \\
&\cdots\cdots \qquad \cdots \qquad \cdots \\
&\cdots\cdots \qquad \cdots \qquad \cdots \\
\psi_{0m} &= \frac{r^4}{16}\alpha_{1m} + L^{-1} r \log r \alpha_{2m} + \frac{r^2}{2}\alpha_{3m} + \alpha_{4m}
\end{aligned} \tag{6.122}$$

In this analysis, the solution is obtained as the components of ψ are determined. But each component contains some integration constants that are to be evaluated by the respective specific conditions. Substituting ψ from (6.115) in (6.108) and (6.109), and then comparing the terms on both sides, we have the following boundary conditions for $\psi_0, \psi_1, \cdots$, etc. For ψ_0

$$-\frac{1}{r}\frac{\partial \psi_0}{\partial r} = 0, \ \psi = -\frac{1}{2}\cos t \quad \text{at} \quad r = f \tag{6.123}$$

$$-\frac{\partial}{\partial r}\left(\frac{1}{r}\frac{\partial \psi_0}{\partial r}\right) = 0, \ \psi_0 = 0 \quad \text{at} \quad r = 0 \tag{6.124}$$

for ψ_1

$$-\frac{1}{r}\frac{\partial \psi_1}{\partial r} = \psi_1 = 0, \quad \text{at} \quad r = f \tag{6.125}$$

$$-\frac{\partial}{\partial r}\left(\frac{1}{r}\frac{\partial \psi_1}{\partial r}\right) = \psi_1 = 0, \quad \text{at} \quad r = 0 \tag{6.126}$$

and so on.

Satisfying the boundary conditions (6.123) and (6.124) by the first relation of (6.119), the integration constants $\alpha_{10}(x,t)$, $\alpha_{20}(x,t)$, $\alpha_{30}(x,t)$, and $\alpha_{40}(x,t)$ are found to be

$$\alpha_{20}(x,t) = \alpha_{40}(x,t) = 0$$

$$\alpha_{10}(x,t) = \frac{8}{f^4}\cos t \tag{6.127}$$

$$\alpha_{30}(x,t) = -\frac{2}{f^2}\cos t$$

and the resulting expression for $\psi_0(x,r,t)$ is

$$\psi_0(x,r,t) = \frac{1}{2f^4}\left(r^4 - 2f^2r^2\right)\cos t \tag{6.128}$$

Since ψ_0 is known, A_0 is also known, as it is a function of ψ_0 and its derivatives. Hence, the right-hand side of the second relation of (6.119) is known because of ψ_0 and A_0. Satisfying the boundary conditions (6.125) and (6.126) by the function ψ_1 in (6.119), the integration constants $\alpha_{11}(x,t)$, $\alpha_{21}(x,t)$, $\alpha_{31}(x,t)$, and $\alpha_{41}(x,t)$ are obtained as

$$\begin{aligned}
\alpha_{11}(x,t) = {} & q^2\left[\frac{3Sf^4}{72} + \frac{N_1f^2}{6}\right]\cos^2 t \\
& + \left[\frac{N_2f^4}{24} + \frac{N_3f^2}{6}\right]\cos t \\
& - R_e\left[\frac{M_1f^6}{60} + \frac{M_2f^4}{24} + \frac{M_3f^4}{96}\right]\cos^2 t
\end{aligned} \tag{6.129}$$

$$\begin{aligned}
\alpha_{31}(x,t) = {} & -q^2\left[\frac{Sf^6}{228} + \frac{N_1f^4}{96}\right]\sin t \\
& - \left[\frac{N_2f^6}{288} + \frac{N_3f^4}{96}\right]\cos t \\
& + R_e\left[\frac{M_1f^8}{640} + \frac{M_2f^6}{288} + \frac{M_3f^4}{6}\right]\cos^2 t
\end{aligned}$$

$$\alpha_{21}(x,t) = \alpha_{41}(x,t) = 0$$

where

$$S(x) = \frac{2}{f^6}\left(5f_1^2 - ff_2\right)$$

$$N_1(x) = \frac{2}{f^4}\left(2 + ff_2 - 3f_1^2\right)$$

$$N_2(x) = \frac{2}{f^8}\left(15f^2f_2^2 + 20f^2f_1f_3 - 180ff_1^2f_2 + 210f_1^4 - f^3f_4\right)$$

$$N_3(x) = H(x) + 16S(x) \tag{6.130}$$

$$H(x) = \frac{2}{f^6}\left(f^3f_4 - 9f^2f_2^2 - 12f^2f_1f_3 + 72ff_1^2f_2 - 60f_1^4\right)$$

$$M_1(x) = \frac{4}{f^{11}}\left(20f_1^3 - 13ff_1f_2 + f^2f_3\right)$$

$$M_2(x) = -\frac{4}{f^9}\left(32f_1^3 - 22ff_1f_2 + 2f^2f_3 - 8f_1\right)$$

$$M_3(x) = \frac{4}{f^7}\left(12f_1^3 - 9ff_1f_2 + f^2f_3 - 8f_1\right)$$

The quantities f_1, f_2, f_3, and f_4 are the derivatives of f with respect to x indicating the orders according to their suffices, and the ith derivative is given by

$$f_i = \frac{\tau}{2}\left(\frac{\pi}{4}\right)^i \cos\left(i\frac{\pi}{2} + \frac{\pi}{4}x\right) \tag{6.131}$$

and the expression for ψ_1 is

$$\begin{aligned}\psi_1(x, r, t) = R_e &\left[\frac{M_1 f^{10}}{3840} + \frac{M_2 f^8}{1152} + \frac{M_3 f^6}{192}\right]\cos^2 t \\ &- q^2\left[\frac{Sf^8}{1152} + \frac{N_1 f^6}{192}\right]\sin t \\ &- \left[\frac{N_2 f^8}{1152} + \frac{N_3 f^6}{192}\right]\cos t \\ &+ \frac{1}{16}\alpha_{11}(x, t)f^4 + \frac{1}{2}\alpha_{31}(x, t)f^2\end{aligned} \tag{6.132}$$

If a two-term approximant of the solution ψ is considered to be sufficiently accurate, then we have

$$\psi(x, r, t) = \psi_0(x, r, t) + \psi_1(x, r, t) \tag{6.133}$$

where ψ_0 and ψ_1 are given by (6.128) and (6.132).

The axial velocity component, by means of (6.105) and (6.133), is found to be

$$u(x, r, t) = -\left[\frac{1}{r}\frac{\partial \psi_0}{\partial r} + \frac{1}{r}\frac{\partial \psi_1}{\partial r}\right] \tag{6.134}$$

where $\frac{1}{r}\frac{\partial \psi_0}{\partial r}$ and $\frac{1}{r}\frac{\partial \psi_1}{\partial r}$ are given by

$$\frac{1}{r}\frac{\partial \psi_0}{\partial r} = \frac{2}{f^4}\left(r^2 - f^2\right)\cos t \tag{6.135}$$

and

$$\begin{aligned}\frac{1}{r}\frac{\partial \psi_1}{\partial r} = R_e &\left[\frac{M_1 r^8}{384} + \frac{M_2 r^6}{144} + \frac{M_3 r^4}{32}\right]\cos^2 t \\ &- q^2\left[\frac{S r^6}{144} + \frac{N_1 r^4}{32}\right]\sin t \\ &- \left[\frac{N_2 r^6}{144} + \frac{N_3 r^4}{32}\right]\cos t \\ &+ \frac{1}{4}\alpha_{11}(x, t) r^2 + \alpha_{31}(x, t)\end{aligned} \tag{6.136}$$

The wall shearing stress is defined by

$$T = -\frac{1}{4}\left(\frac{\partial u}{\partial r}\right)_{r=f} \cdot (1 + f_1^2) \tag{6.137}$$

which gives, using (6.134),

$$T = -\frac{1}{4}\left[\frac{\partial}{\partial r}\left\{\frac{1}{r}\frac{\partial \psi_0}{\partial r} + \frac{1}{r}\frac{\partial \psi_1}{\partial r}\right\}\right]_{r=f} \cdot (1 + f_1^2) \tag{6.138}$$

Then the use of the relations (6.135) and (6.136) leads (6.137) to the following form

$$T = -\frac{1}{4}\left[\frac{4}{f^3} + R_e \overline{\alpha}\cos^2 t - \overline{\beta}\cos t - q^2 \overline{\gamma}\sin t\right] \cdot (1 + f_1^2) \tag{6.139}$$

where

$$\overline{\alpha} = \frac{1}{60f^4}\left(20f_1^2 - 19ff_1f_2 + 3f^2f_3 - 40f_1\right)$$
$$\overline{\beta} = \frac{1}{24f^3}(f^3f_4 - 36ff_1^2f_2 + 160f_1^2 - 3f^2f_2^2 - 4f^2f_1f_3 + 90f_1^4 - 32ff_2 - 96) \tag{6.140}$$
$$\overline{\gamma} = \frac{1}{24f}(4 - f_1^2 + ff_2)$$

6.4.3 Numerical Discussions

In this analysis, the distribution of axial velocity and the wall shear stress was determined for the pulsatile flow of blood through a constricted rigid blood vessel following Adomian's decomposition method. Values of the axial velocity component $u(x, r, t)$ at different time points during the flow have been determined from the equation (6.134) and plotted against r (Figures 6.6–6.8). Figures 6.6, 6.7, and 6.8 correspond, respectively, to the values of the Wonersley parameter $M = 1, 4,$ and 3. It is observed that the axial velocity profiles change sharply with time t. In Figure 6.6 the peak velocity reduces from $t = 0$ until $t = 1.5$, while, as calculated here, the flow reversals occur as flow proceeds from $t = 2$ and $t = 2.5$.

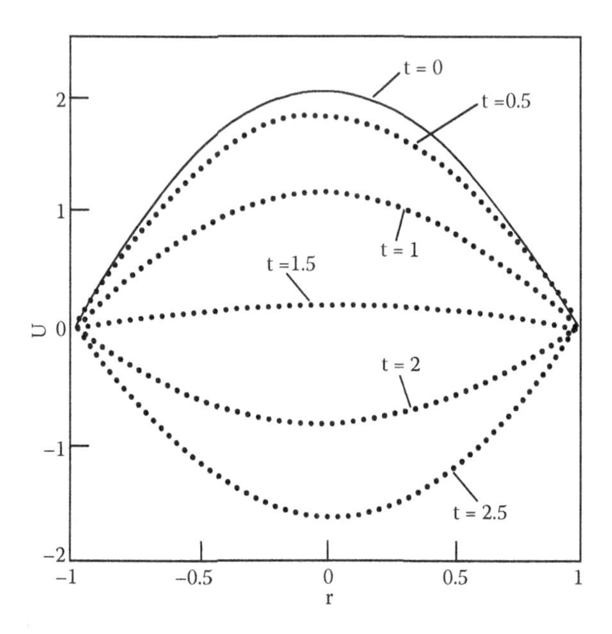

FIGURE 6.6
Axial velocity distribution at $x = 0$ cross section for $q = 1$ and $t = 0.2$.

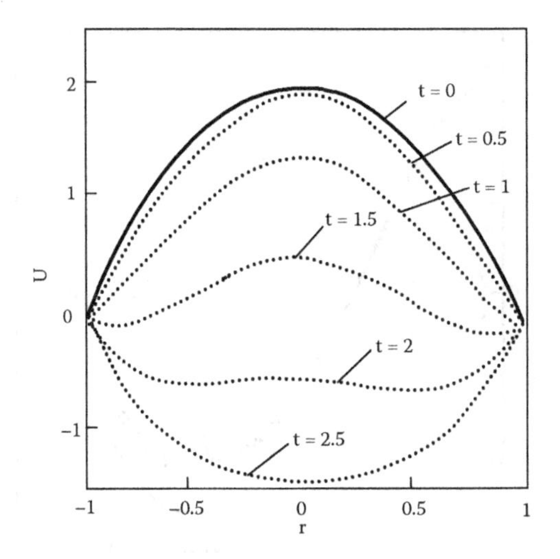

FIGURE 6.7
Axial velocity distribution at $x = 0$ cross section for $q = 3$ and $\tau = 0.2$.

In Figure 6.7, it is observed that the peak velocity decreases from $t = 0$ until $t = 1.5$. Backflow is found to occur at $t = 1.5$ near the wall region, whereas the flow near the tube axis is in the forward direction. This tendency of forward flow near the axis of the tube, however, reduces with increasing time until it is approximately uniform at $t = 2.5$.

In Figure 6.8, the peak velocity increases from $t = 0$ until $t = 0.5$, and then reduction in the velocity profile starts. Other features are more or less similar

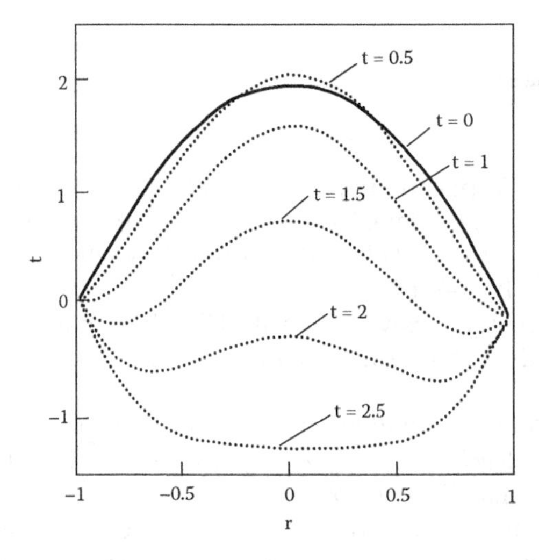

FIGURE 6.8
Axial velocity distribution at $x = 0$ cross section for $q = 4$ and $\tau = 0.2$.

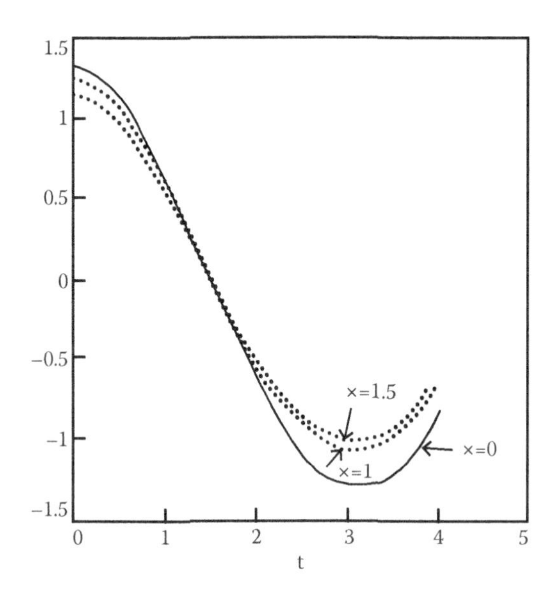

FIGURE 6.9
Axial velocity distribution at $x = 0$ cross section for $q = 1$ and $\tau = 0.2$.

to those in Figure 6.7. Variation of stress distribution with time at different stations, $x = 0$, $x = 1$, and $x = 1.5$, is shown in Figure 6.9. It is clear that shear stress distribution is negative for some time before it tends toward positive direction again. The period of negative stress corresponds to flow separation.

References

1. Adomian, G.: Application of the Decomposition Method to the Navier–Stokes Equations, *Jr. Math. Anal. Appl.*, *119*, 340–360 (1986).
2. Barnothy, M. F. (Ed.): *Biological Effects of Magnetic Fields*, Vols. 1 and 2, Penum Press (1964–1969).
3. Bugliarello, G. and J. Sevill: Velocity Distribution and Other Characteristics of Steady and Pulsatile Blood in Finite Glass Tubes, *Biorheology 7*, 85–107 (1970).
4. Casson, N.: A Flow Equation for Pigment Oil Suspensions of the Printing Ink Type. In *Rheology of Disperse Systems*, C. C. Mill (Ed.), pp. 84–102, London Pergaman Press (1959).
5. Charm, S. F., and G. S. Kurland: Tube Flow Behaviour and Shear Stress, Shear Rate and Characteristics of Canine Blood, *Am. J. Physist.*, *203*, 417 (1962).

6. Charm, S. F., and G. S. Kurland: Viscometry of Human Blood for Shear Rate of 0–100.00 Sec.$^{-1}$, *Nature, 206*, 617–618 (1965).
7. Charm, S. F., and G. S. Kurland: *Blood and Microcirculation*, Wiley (1975).
8. Chaturani, P., and S. P. Samy: A Two-Layered Model for Blood Flow Through Stenosed Arteries, *Proc. 11th National Congress on Fluid Mechanics and Fluid Power*, Vol. 1. pp. CF 16 (1982).
9. Cherruault, Y.: Convergence of Adomian's Method, *Kybernetes, 18*(2), 31–39 (1989).
10. Cherruault, Y.: New Results for the Convergences of Adomian's Method Applied to Integral Equations, *Math. Comp. Modelling, 16*(2), 85–93 (1992).
11. Cokelet, G. R: The Rheology of Human Blood. In Biomechanics, Y. C. Fung et al. (eds.), p. 63, Prentice Hall (1972).
12. Deshikachar, K. S., and A. Ramachandra Rao: Effects of a Magnetic Field on the Flow and Blood Oxygeneration in Channels of Variable Cross-Section, *Int. J. Engng. Sci., 23*, 1121–1133 (1985).
13. Gold, R. R.: Magnetohydrodynamics Pipe Flow, Part 1, *J. Fluid Mech., 13*, 505–512 (1962).
14. Gupta, S. C., and Bani Singh: Unsteady Magnetohydrodynamics Flow in a Circular Pipe Under a Transverse Magnetic Field, *Phys. Fluids, 13*, 346–352 (1970).
15. Haynes, R. H., and A. C. Burton: Role of Non-Newtonian Behaviour of Blood in Hydrodynamics. *Am. J. Physiol. 197*, 943 (1959).
16. Hershey, D., R. E. Byrnes, R. L. Deddens, and A. M. Rao: *Blood Rheology Temperature Dependence of the Power Law Model*, Paper presented at A. I. Ch. L. meeting, Boston (December) (1964).
17. Hershey, D., and S. J. Cho: Blood Flow in Rigid Tubes: Thickness and Slip Velocity of Plasma Film at the Wall, *J. Appl. Physiol., 21*, 27 (1966).
18. Haldar, K.: Effects of the Shape of Stenosis on the Resistance to Blood Flow Through an Artery, *Bull. Math. Biol., 47*(4), 545–550 (1985).
19. Haldar, K: Oscillatory Flow of Blood in a Stenosed Artery, *Bull. Math. Biol., 49*(3), 279–287 (1987).
20. Haldar, K.: A Note on the Periodic Motion of a Visco-Elastic Fluid in a Radially Non-Symmetric Constricted Tube, *Rheol Acta, 27*, 434-436 (1988).
21. Haldar, K., and K. N. Dey: Effects of Erythrocytes on the Flow Characteristics of Blood in an Indented Tube, *Arch. Mech., 42*(1), 109–114, Warszawa (1990).
22. Haldar, K.: Analysis of Separation of Blood Flow in Constricted Arteries, *Arch. Mech., 43*(3), 103–109, Warszawa (1991).
23. Haldar, K., and S. N. Ghosh: Effects of Body Force on the Pulsating Blood Flow in Arteries, *Engng. Trans., 41*(41), 157–166 (1993).
24. Haldar, K., and S. N. Ghosh: Effects of a Magnetic Field on Blood Flow Through an Indented Tube in the Presence of Erythrocytes, *Ind. Jr. Pure and Appld. Maths., 25*(3), 345–352 (1994).
25. Haldar, K., and H. L. Andersson: Two-Layered Model of Blood Flow Through Stenosed Arteries, *Acta Mechanica, 117*, 221–228 (1996).

26. Haldar, K.: Application of Adomian's Approximations to the Navier–Stokes Equations in Cylindrical Coordinates, *Appl. Math. Lett., 9*(4), 109–113 (1996).
27. Haldar, K.: Application of Adomian's Approximation to Blood Flow Through Arteries in the Presence of a Magnetic Field, *Int. Jr. Appl. Math. Comp., 12*(1–2), 267–279 (2003).
28. Huckaba, C. E., and A. W. Hahu: A Generalised Approach to the Modelling of Arterial Blood Flow, *Bull. Math. Biophys., 30*, 645–662 (1968).
29. Lih, M. M.: *Transport Phenomena in Medicine and Biology*, Wiley (1975).
30. Mamaloukas, C., K. Haldar, and H. P. Mazumdar: Application of Double Decomposition to Pulsatile Flow, *Int. Jr. Appl. Math. Comp., 10*(1–2), 193–207 (2002).
31. Merrill, E. W., G. R. Cokelet, A. Britten, and R. E. Wells: Non-Newtonian Rheology of Human Blood Effect of Fibrinogen Deduced by Subtraction, *Circulat. Res., 13*(48) (1963).
32. Merrill, E. W., G. R. Cokelet, and A. Britten: Rheology of Human Blood and the Red Cell Plasma Membrance, *Biblply Anat., 4*, 51 (1964).
33. Morgan, B. E., and D. F. Young: An Integral Method for the Analysis of Flow in Arterial Stenoses, *Jr. Math. Biol, 36*, 39–53 (1974).
34. McMichael, J. M., and S. Deutsch: Magnetolydrodynamics of Laminar Flow in Slowly Varying Tubes in an Axial Magnetic Field, *Phys. Fluids, 27*, 110–118 (1984).
35. Reiner, M., and G. W. Scott Blair: The Flow of Blood Through Narrow Tubes, *Nature, 184*, 354 (1959).
36. Ramchandra Rao, A. and K. S. Deshikachar,: Physiological Type Flow in a Circular Pipe in the Presence of a Transverse Magnetic Field. *Indian Inst. Sci., 68*, 247–260 (1988).
37. Ramchandra Rao, A. and K. S. Deshikachar,: MHD Oscillatory Flow of Blood Through Channels of Variable Cross-Section, *Int. J. Engng. Sci., 24*, 1615–1628 (1986).
38. Shukla, J. B., Parihar, R. S. and Rao, B. R. P.,: Effects of Stenosis on Non-Newtonian Flow of Blood in an Artery, *Bull. Math. Biol., 42*, 283–294 (1980a).
39. Shukla, J. B., Parihar, R. S. and Rao, B. R. P.,: Effects of Peripheral Layer Viscosity on Blood Flow Through Artery With Mild Stenosis, *Bull. Math. Biol., 42*, 797–805 (1980b).
40. Shukla, J. B., Parihar, R. S. and Rao, B. R. P.,: Biorheological Aspects of Blood Flow Through Artery With Mild Stenosis, *Biorheology, 17*, 403–410 (1980c).
41. Sud, V. K. and Sekhon, G. S.,: Blood Flow Through the Human Arterial System in the Presence of a Steady Magnetic Field, *Phys. Med. Biol., 34*, 795–805 (1989).

7

Steady Subsonic Flow

7.1 Introduction

In this chapter, we discuss the steady subsonic flows of inviscid gasdynamics by means of the decomposition method, which is divided into the four classes mentioned earlier. Out of these four methods, the regular and modified decomposition methods have been used here for homogeneous partial differential equations, and a few problems of inviscid gasdynamics have been considered for clear illustration of the theories.

7.2 Equations of Motion

The problems of inviscid gasdynamics for steady plane or axisymmetric flows are described by partial differential equations subject to certain boundary conditions. If we consider the plane and axisymmetric subsonic flows, then the two-dimensional linearized gasdynamic equations [6] for these flows are respectively given by

$$\left(1 - M_\infty^2\right)\varphi_{xx} + \varphi_{yy} = 0 \tag{7.1}$$

and

$$\left(1 - M_\infty^2\right)\varphi_{xx} + \varphi_{yy} + \frac{1}{y}\varphi_y = 0 \tag{7.2}$$

Here $\varphi(x, y)$ is the perturbation velocity potential defined by

$$\frac{\partial\varphi}{\partial x} = u - u_\infty, \ \frac{\partial\varphi}{\partial y} = v \tag{7.3}$$

where the x-axis is taken in the flow direction and the y-axis is perpendicular to it, u is the velocity component in the x direction, v is the velocity component in the y-direction, u_∞ is the free stream velocity, and $M_\infty = \frac{u_\infty}{c}$ is the Mach number, c being the velocity of sound

The linearized partial differential equations [6] for plane and axisymmetric transonic flows are

$$\left(1 - M_\infty^2\right)\varphi_{xx} + \varphi_{yy} = M_\infty^2(\gamma + 1)\varphi_x\varphi_{xx} \tag{7.4}$$

and

$$\left(1 - M_\infty^2\right)\varphi_{xx} + \varphi_{yy} + \frac{1}{y}\varphi_y = M_\infty^2(\gamma + 1)\varphi_x\varphi_{xx} \tag{7.5}$$

where $\gamma = \frac{c_p}{c_v} = 1.4$, c_p and c_v being the specific heats at constant pressure and constant volume.

The equation (7.1) or (7.2) is a linear second-order partial differential equation. This equation is of the elliptic type for a purely subsonic flow $(0.3 < M_\infty < 1)$ and the hyperbolic type for a purely supersonic flow $(1 < M_\infty < 3)$. The equations (7.4) or (7.5) are a second-order quasi-linear partial differential equation of mixed elliptic-hyperbolic type and is valid for transonic flow $(0.8 < M_\infty < 1.2)$.

7.3 Application of Regular Decomposition to a Linearized Gas Dynamic Equation for Plane Flow

The decomposition method prepares a single method that is used for multidimensional linear and nonlinear problems. This method has also been applied to different frontier problems in other disciplines. The method is used in the development of numerical techniques in order to get the solution of nonlinear partial differential equations. For detailed analysis of the theory, we consider the homogeneous partial differential equation related to the problems of subsonic flow in the gasdynamics.

The x-Partial Solution

For the application of the regular decomposition method, we begin with the linear second-order partial differential equation for subsonic flow, that is,

$$(1 - M_\infty^2)\varphi_{xx} + \varphi_{yy} = 0 \tag{7.6}$$

whose operator form is

$$\beta^2 L_x\varphi + L_y\varphi = 0 \tag{7.7}$$

where $L_x = \frac{\partial^2}{\partial x^2}$ and $L_y = \frac{\partial^2}{\partial y^2}$ are two second-order linear operators and $\beta^2 = 1 - M_\infty^2$.

Solving (7.7) for $L_x\varphi$ and then applying the regular decomposition technique, we have the x-partial solution as

$$\varphi(x, y) = \varphi_0(x, y) - L_x^{-1}\left[\beta^{-2}L_y\varphi\right] \tag{7.8}$$

where

$$L_x^{-1} = \iint (\cdot)\, dx\, dx \tag{7.9}$$

and

$$\varphi_0(x, y) = \xi_0(y) + \xi_1(y)x \tag{7.10}$$

is the solution of $L_x\varphi = 0$. Here the constants $\xi_0(y)$ and $\xi_1(y)$ are to be evaluated by the prescribed boundary conditions.

We now decompose $\varphi(x, y)$ into the following form

$$\varphi(x, y) = \sum_{m=0}^{\infty} \lambda^m \varphi_m \tag{7.11}$$

Then we write the parameterized form [2] of (7.8) as

$$\varphi(x, y) = \varphi_0(x, y) - \lambda L_x^{-1}\left[\beta^{-2} L_y \varphi\right] \tag{7.12}$$

Putting (7.11) into (7.12) and then equating the coefficients of like-power terms of λ from both sides of the resulting expression, we get

$$\begin{aligned}
\varphi_1(x, y) &= -L_x^{-1}\left[\beta^{-2} L_y \varphi_0\right] \\
\varphi_2(x, y) &= -L_x^{-1}\left[\beta^{-2} L_y \varphi_1\right] \\
\cdots \quad & \cdots \quad \cdots \\
\cdots \quad & \cdots \quad \cdots \\
\varphi_{m+1}(x, y) &= -L_x^{-1}\left[\beta^{-2} L_y \varphi_m\right] \\
\cdots \quad & \cdots \quad \cdots
\end{aligned} \tag{7.13}$$

Thus, we see that the components of φ are determined, and the final solution (7.11) is, therefore, computable, remembering that $\lambda = 1$.

The y-Partial Solution

For the y-partial solution, we write the equation (7.7) in the following form:

$$L_y \varphi = -\beta^2 L_x \varphi \tag{7.14}$$

Then operating with L_y^{-1} defined by $L_y^{-1} = \int\int(\cdot)\,dy\,dy$, we get

$$\varphi(x, y) = \overline{\varphi}_0(x, y) - L_y^{-1}\left[\beta^2 L_x \varphi\right] \tag{7.15}$$

whose parameterized form [2] is

$$\varphi(x, y) = \overline{\varphi}_0(x, y) - \lambda L_y^{-1}\left[\beta^2 L_x \varphi\right] \tag{7.16}$$

when we write $\varphi(x, y)$ into the following decomposed form:

$$\varphi(x, y) = \sum_{m=0}^{\infty} \lambda^m \varphi_m \tag{7.17}$$

where

$$\overline{\varphi}_0(x, y) = \eta_0(x) + \eta_1(x)y \tag{7.18}$$

$\eta_0(x)$ and $\eta_1(x)$ having their usual meanings.

If we put (7.17) into (7.16), we get on comparison of terms

$$\begin{aligned}
\varphi_0(x, y) &= -\overline{\varphi}_0(x, y) \\
\varphi_1(x, y) &= -L_y^{-1}\left[\beta^2 L_x \varphi_0\right] \\
\varphi_2(x, y) &= -L_y^{-1}\left[\beta^2 L_x \varphi_1\right] \\
\cdots \quad & \cdots \quad \cdots \\
\cdots \quad & \cdots \quad \cdots \\
\varphi_{m+1}(x, y) &= -L_y^{-1}\left[\beta^2 L_x \varphi_m\right] \\
\cdots \quad & \cdots \quad \cdots
\end{aligned} \tag{7.19}$$

Thus, the components of $\varphi(x, y)$ are determined and the final solution (7.17) is computable.

7.4 Application of Modified Decomposition to a Linearized Gas Dynamic Equation for Plane Flow

The method that produces a slight variation in the solution of the regular decomposition method [3] is called the *modified decomposition method* [3]. This method can be applied to the linearized gasdynamic equations of physical problems. We discuss this method in details for the linear second-order partial differential equation, which is relevant to the real problems of gasdynamics.

The x-Partial Solution

For the x-partial solution of the modified decomposition technique, we consider the same equation (7.7), that is,

$$\beta^2 L_x \varphi + L_y \varphi = 0$$

whose regular decomposition solution is

$$\varphi(x, y) = \varphi_0(x, y) - L_x^{-1}\left[\beta^{-2} L_y \varphi\right] \tag{7.20}$$

where $\varphi_0(x, y)$ is given by

$$\varphi_0(x, y) = \xi_0(y) + \xi_1(y)x \tag{7.21}$$

Then we follow the modified decomposition procedure and set for the purpose

$$\varphi(x, y) = \sum_{m=0}^{\infty} a_m(y) x^m \tag{7.22}$$

Using (7.21) and (7.22) in (7.20), we obtain

$$\sum_{m=0}^{\infty} a_m(y)x^m = \xi_0(y) + \xi_1(y)x - L_x^{-1}\left[\beta^{-2}\frac{\partial^2}{\partial y^2}\sum_{m=0}^{\infty} a_m(y)x^m\right]$$

$$= \xi_0(y) + \xi_1(y)x - \sum_{m=0}^{\infty} \beta^2 \frac{\partial^2}{\partial y^2} a_m(y)\frac{x^{m+2}}{(m+1)(m+2)}$$

If we replace m by $(m-2)$ on the right side of the above relation, we get

$$\sum_{m=0}^{\infty} a_m(y)x^m = \xi_0(y) + \xi_1(y)x - \sum_{m=2}^{\infty} \beta^{-2}\frac{\partial^2}{\partial y^2} a_{m-2}(y)\frac{x^m}{m(m-1)} \tag{7.23}$$

The coefficients are identified by

$$\left.\begin{aligned} \xi_0(y) &= a_0(y) \\ \xi_1(y) &= a_1(y) \end{aligned}\right\} \tag{7.24}$$

and for $m \geq 2$ the recurrence relation

$$a_m(y) = \frac{-\beta^{-2}\frac{\partial^2}{\partial y^2} a_{m-2}(y)}{m(m-1)} \tag{7.25}$$

The relation (7.22) together with the relation (7.25) give the complete solution of the problem.

The y-Partial Solution

For the y-partial solution of the equation (7.7) by the application of the modified decomposition method, we again consider its regular decomposition solution (7.15), that is,

$$\varphi(x, y) = \overline{\varphi}_0(x, y) - L_y^{-1}\left[\beta^2 L_x \varphi\right] \tag{7.26}$$

where

$$\overline{\varphi}_0(x, y) = \eta_0(x) + \eta_1(x)y$$

We now follow the modified decomposition procedure and assume that

$$\varphi(x, y) = \sum_{m=0}^{\infty} b_m(x)y^m \tag{7.27}$$

Putting (7.27) into (7.26) and then integrating, we get after somewhat straightforward calculations

$$\sum_{m=0}^{\infty} b_m(x)y^m = \eta_0(x) + \eta_1(x)y - \sum_{m=2}^{\infty} \beta^2 \frac{\frac{\partial^2}{\partial x^2} b_{m-2}(x)}{m(m-1)} y^m \tag{7.28}$$

Then we equate the coefficients of like-power terms of y from both sides of (7.28) and obtain

$$\left.\begin{aligned}\eta_0(x) &= b_0(x)\\ \eta_1(x) &= b_1(x)\end{aligned}\right\} \tag{7.29}$$

and the recurrence relation for $m \geq 2$

$$b_m(x) = -\frac{\beta^2 \frac{\partial^2}{\partial x^2} b_{m-2}(x)}{m(m-1)} \tag{7.30}$$

Thus, the relation (7.27) together with the relations (7.29) and (7.30) constitute the solution of the problem.

7.5 Flow Past a Wavy Wall

This section deals with the problems of the linearized plane subsonic flow past a wavy wall for a clear illustration of the decomposition method. Ackeret [1] obtained the exact solution for a linearized subsonic flow past a wave-shaped wall. We discuss the same problem (1) in unlimited fluid, (2) in an open throat wind tunnel, and (3) in a closed throat wind tunnel by the application of the decomposition technique. In this section, we have applied the modified decomposition technique to investigate the subsonic flow past a wave-shaped wall in each of the above-mentioned three cases. We can also use the regular or ordinary decomposition method for the same problem, and this is left to the readers.

7.5.1 Steady Plane Subsonic Flow Past a Wave-Shaped Wall in an Unlimited Fluid

Consider two-dimensional subsonic flow past a wavy wall whose geometry (Figure 7.1) is described by

$$h(x) = \tau \sin nx \tag{7.31}$$

where $\tau << 1$ is the amplitude of the wave and $l = 2\pi/n$ is the wave length.

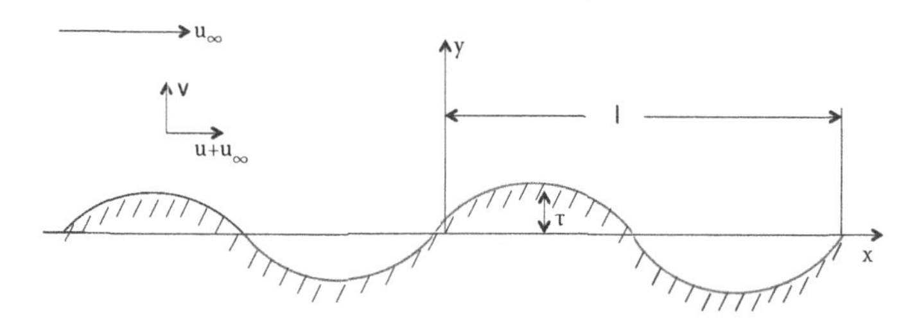

FIGURE 7.1
Flow past a wave-shaped wall.

If φ is the perturbation velocity potential defined by

$$u - u_\infty = \frac{\partial \varphi}{\partial x}$$
$$v = \frac{\partial \varphi}{\partial y} \tag{7.32}$$

where (u, v) are the velocity components in the x and y directions, respectively, then the inearized gasdynamic equation governing the flow field over the wall is

$$\beta^2 \frac{\partial^2 \varphi}{\partial x^2} + \frac{\partial^2 \varphi}{\partial y^2} = 0 \tag{7.33}$$

where $\beta^2 = 1 - M_\infty^2$. For solving the equation (7.33), it is convenient to use the operator form of it, and we write

$$\beta^2 L_x \varphi + L_y \varphi = 0 \tag{7.34}$$

where $L_x = \frac{\partial^2}{\partial x^2}$ and $L_y = \frac{\partial^2}{\partial y^2}$ are the two linear operators, and their inverse operators are defined by

$$L_x^{-1} = \iint (\cdot)\, dx\, dx \tag{7.35}$$

and

$$L_y^{-1} = \iint (\cdot)\, dy\, dy \tag{7.36}$$

The boundary conditions imposed on φ are

$$\varphi_y(x, y) = \frac{dh}{dx} = \tau n \cos nx \quad \text{at} \quad y = 0 \tag{7.37}$$

and

$$\varphi_y = 0 \quad \text{at} \quad y = \infty \tag{7.38}$$

The equation (7.34) together with the boundary conditions (7.37) and (7.38) constitute the mathematical model of the problem.

We can have two types of solutions of the equation (7.34), namely, the x-partial solution and y-partial solution. Out of these two solutions, the y-partial solution is appropriate here according to the boundary conditions, and we proceed to get that solution.

For the y-partial solution of the problem, we solve for $L_y\varphi$ and obtain from (7.34)

$$L_y \varphi = -\beta^2 L_x \varphi \tag{7.39}$$

Then we perform the operation of the operator (7.36) on (7.39) in order to get

$$\varphi(x, y) = a_0(x) + a_1(x) y - L^{-1}\left(\beta^2 L_x \varphi\right) \tag{7.40}$$

$a_0(x)$ and $a_1(x)$ being the constants of integration to be evaluated from the boundary conditions.

We now proceed for the modified decomposition procedure and write $\varphi(x, y)$ in the following form:

$$\varphi(x, y) = \sum_{m=0}^{\infty} a_m(x) y^m \tag{7.41}$$

Substituting (7.41) in (7.40), we have

$$\begin{aligned}\sum_{m=0}^{\infty} a_m(x) y^m &= a_0(x) + a_1(x) y \\ &\quad - \iint \left[\beta^2 \frac{\partial^2}{\partial x^2} \sum_{m=0}^{\infty} a_m(x) y^m \right] dy\, dy \\ &= a_0(x) + a_1(x) y \\ &\quad - \sum_{m=0}^{\infty} \frac{\partial^2}{\partial x^2} a_m(x) \frac{\beta^2}{(m+1)(m+2)} y^{m+2} \\ &= a_0(x) + a_1(x) y \\ &\quad - \sum_{m=2}^{\infty} \frac{\partial^2}{\partial x^2} a_{m-2}(x) \frac{\beta^2}{m(m-1)} y^m \end{aligned} \tag{7.42}$$

which gives the recurrence relation for $m \geq 2$

$$a_m(x) = -\frac{\partial^2}{\partial x^2} a_{m-2}(x) \frac{\beta^2}{m(m-1)} \tag{7.43}$$

Then we satisfy the boundary condition (7.37) by the expanded form of (7.41) and get

$$a_1(x) = \tau n \cos nx \tag{7.44}$$

Putting $m = 2, 3, 4$, etc., in (7.43) and then using (7.44), we have

$$\begin{aligned} a_2(x) &= -a_{0,2}(x) \frac{\beta^2}{2!}, \quad a_3(x) = \frac{\tau}{\beta} \frac{(n\beta)^3}{3!} \cos nx \\ a_4(x) &= a_{0,4}(x) \frac{\beta^4}{4!}, \quad a_5(x) = \frac{\tau}{\beta} \frac{(n\beta)^5}{5!} \cos nx \\ a_6(x) &= -a_{0,6}(x) \frac{\beta^6}{6}, \quad \text{etc.} \end{aligned} \tag{7.45}$$

where $a_{0,n}$ is the nth derivative of a_0 with respect to x for any positive integer of n. By virtue of (7.44) and (7.45), the solution (7.41) takes the form

$$\varphi(x, y) = \left[a_0 - a_{0,2}(x)\frac{(\beta y)^2}{2!} + a_{0,4}(x)\frac{(\beta y)^4}{4!} - \cdots\right]$$

$$+ \frac{\tau}{2\beta}\left(e^{n\beta y} - e^{-n\beta y}\right)\cos n x \tag{7.46}$$

We now proceed to find out the unknown function $a_0(x)$ in order to complete the solution of the problem. For this purpose, we set the function $a_0(x)$ in the form given by

$$a_0(x) = k_1 \cos n x + k_2 \sin n x \tag{7.47}$$

where k_1 and k_2 are to be determined from the boundary condition (7.38).

Using (7.47) in (7.46), we obtain

$$\varphi(x, y) = \frac{1}{2}\left[\left(k_1 + \frac{\tau}{\beta}\right)e^{n\beta y} + \left(k_1 - \frac{\tau}{\beta}\right)e^{-n\beta y}\right]\cos n x$$

$$+ \frac{k_2}{2}\left(e^{n\beta y} + e^{-n\beta y}\right)\sin n x \tag{7.48}$$

which gives, on differentiation with respect to y,

$$\varphi_y(x, y) = \frac{n\beta}{2}\left[\left(k_1 + \frac{\tau}{\beta}\right)e^{n\beta y} - \left(k_1 - \frac{\tau}{\beta}\right)e^{-n\beta y}\right]\cos n x$$
$$+ \frac{k_2 n\beta}{2}\left(e^{n\beta y} - e^{-n\beta y}\right)\sin n x \tag{7.49}$$

Then the boundary condition (7.38) is to be satisfied by (7.49). The condition is satisfied if and only if we put

$$k_1 = -\frac{\tau}{\beta} \text{ and } k_2 = 0 \tag{7.50}$$

and the expression for $a_0(x)$ is

$$a_0(x) = -\frac{\tau}{\beta}\cos n x \tag{7.51}$$

Therefore, the required velocity potential (7.48), by virtue of (7.50), becomes

$$\varphi(x, y) = -\frac{\tau}{\beta}e^{-n\beta y}\cos n x \tag{7.52}$$

which is the exact solution of the problem.

7.5.2 Steady Plane Subsonic Flow Past a Wave-Shaped Wall in an Open-Throat Wind Tunnel

The geometry of the wavy wall is given by

$$h(x) = \tau \sin n x \tag{7.53}$$

where τ is the maximum height of the wave and $2\pi/n$ is the wave length.

The partial differential equation governing the steady subsonic flow over the wavy wall in an open throat wind tunnel is

$$\beta^2 \varphi_{xx} + \varphi_{yy} = 0 \tag{7.54}$$

where $\beta^2 = 1 - M_\infty^2$. The corresponding boundary conditions are

$$\varphi_y(x, y) = \frac{dh}{dx} = \tau n \cos nx \quad \text{at} \quad y = 0 \tag{7.55}$$

and

$$\varphi_x(x, y) = 0 \quad \text{at} \quad y = H \tag{7.56}$$

Here H is the distance between the wave-shaped wall and the wall of the wind tunnel.

Let $L_x = \frac{\partial^2}{\partial x^2}$ and $L_y = \frac{\partial^2}{\partial y^2}$. Then the equation (7.54) becomes

$$\beta^2 L_x + L_y \varphi = 0 \tag{7.57}$$

which can be written for solving $L_y\varphi$ as

$$L_y\varphi = -\beta^2 L_x \varphi \tag{7.58}$$

Then we operate on both sides of (7.58) with the inverse operator L_y^{-1} defined by (7.36) and get

$$\varphi(x, y) = \xi_0(x) + \xi_1(x)y - L_y^{-1}(\beta^2 L_x \varphi) \tag{7.59}$$

which is a regular or ordinary decomposition solution, $\xi_0(x)$ and $\xi_1(x)$ being the constants to be determined from the boundary conditions.

We now apply the modified decomposition procedure to (7.59) and write for this purpose

$$\varphi(x, y) = \sum_{m=0}^{\infty} a_m(x) y^m \tag{7.60}$$

Putting (7.60) in (7.59), we obtain

$$\sum_{m=0}^{\infty} a_m(x) y^m = \xi_0(x) + \xi_1(x)y - \sum_{m=0}^{\infty} \frac{\partial^2}{\partial x^2} a_{m-2}(x) \frac{\beta^2}{m(m-1)} y^m \tag{7.61}$$

Comparison of like-power terms of y on both sides of the above relation gives

$$\xi_0(x) = a_0(x)$$
$$\xi_1(x) = a_1(x) \tag{7.62}$$

and for $m \geq 2$ the recurrence relation

$$a_m(x) = -\frac{\partial^2}{\partial x^2} a_{m-2}(x) \frac{\beta^2}{m(m-1)} \tag{7.63}$$

The relation (7.63) determines the other coefficients in terms of $a_0(x)$ and $a_1(x)$. Putting $m = 2, 3, 4$, etc., in (7.63), we have

$$a_2(x) = -\frac{\beta^2}{2!} a_{0,2}(x)$$
$$a_3(x) = -\frac{\beta^2}{3!} a_{1,2}(x) \tag{7.64}$$
$$a_4(x) = \frac{\beta^4}{4!} a_{0,4}(x)$$
$$a_5(x) = \frac{\beta^4}{5!} a_{1,4}(x), \text{ etc.}$$

where $a_{1,n}$ is the nth derivative of a_1 with respect to x for any positive integer of n. Substituting the coefficients (7.64) in the expansion of (7.60), we have the velocity potential $\varphi(x, y)$ as

$$\varphi(x, y) = \left[a_0(x) - a_{0,2}(x)\frac{(\beta y)^2}{2!} + a_{0,4}(x)\frac{(\beta y)^4}{4!} - \cdots\right] + \frac{1}{\beta}\left[a_1(x)\frac{(\beta y)}{1!} - a_{1,2}(x)\frac{(\beta y)^3}{3!} + a_{1,4}(x)\frac{(\beta y)^5}{5!} - \cdots\right] \tag{7.65}$$

Now we satisfy the boundary condition (7.55) by (7.65) to get

$$a_1(x) = \tau\, n \cos n x \tag{7.66}$$

and the expression (7.65) takes the form

$$\varphi(x, y) = \left[a_0(x) - a_{0,2}(x)\frac{(\beta y)^2}{2!} + a_{0,4}(x)\frac{(\beta y)^4}{4!} - \cdots\right] + \frac{\tau}{2\beta}\left[e^{n\beta y} - e^{-n\beta y}\right] \cos n x \tag{7.67}$$

Again, we satisfy the boundary condition (7.56) by (7.67) and obtain

$$0 = \left[a_{0,1}(x) - a_{0,3}(x)\frac{(\beta H)^2}{2!} + a_{0,5}(x)\frac{(\beta H)42}{4!} \cdots\cdots\right] - \frac{\tau n}{2\beta}\left[e^{n\beta H} - e^{-n\beta H}\right] \sin nx \tag{7.68}$$

Let us assume the function $a_0(x)$ in the form given by

$$a_0(x) = C_1 \cos nx + C_2 \sin nx \tag{7.69}$$

Putting (7.69) in (7.68), we get

$$0 = C_2 \left[e^{n\beta H} + e^{-n\beta H}\right] \cos nx - \left[\frac{\tau}{\beta}\left(e^{n\beta H} - e^{-n\beta H}\right) + C_1\left(e^{n\beta H} + e^{-n\beta H}\right)\right] \sin nx$$

The above relation holds good if and only if we put

$$C_1 = -\frac{\tau}{\beta} \cdot \frac{e^{n\beta H} - e^{-n\beta H}}{e^{n\beta H} + e^{-n\beta H}}, \; C_2 = 0 \tag{7.70}$$

and the unknown function $a_0(x)$ is given by

$$a_0(x) = -\frac{\tau}{\beta}\frac{e^{n\beta H} - e^{-n\beta H}}{e^{n\beta H} + e^{-n\beta H}}, \cos nx \tag{7.71}$$

Therefore, using (7.71) in (7.67), we have the required expression for velocity potential as

$$\varphi(x, y) = \frac{\tau}{\beta} \cdot \frac{\sin h\left[n\sqrt{1 - M_\infty^2}(y - H)\right]}{\cos h\left[n\sqrt{1 - M_\infty^2} H\right]} \cos nx \tag{7.72}$$

which represents the closed-form solution of the problem.

7.5.3 Steady Plane Subsonic Flow Past a Wave-Shaped Wall in a Closed-Throat Wind Tunnel

The mathematical model for steady plane subsonic flow past a wavy wall in a closed-throat wind tunnel is described by the partial differential equation

$$\beta^2 \varphi_{xx} + \varphi_{yy} = 0 \tag{7.73}$$

subject to the boundary conditions

$$\varphi_y(x, y) = \frac{dh}{dx} = \tau n \cos nx \quad \text{at} \quad y = 0 \tag{7.74}$$

and

$$\varphi_y(x, y) = 0 \quad \text{at} \quad y = H \tag{7.75}$$

where the wavy wall is given by

$$h(x) = \tau \text{ Sin } nx \tag{7.76}$$

The notations used in this model have their usual meanings, just like those used in the previous section.

Proceeding exactly in the same way as in Section 5.2, we write the solution of (7.73) as

$$\varphi(x, y) = \sum_{m=u}^{\infty} a_m(x) y^m \tag{7.77}$$

where

$$a_m(x) = -\frac{\partial^2}{\partial x^2} a_{m-2}(x) \cdot \frac{\beta^2}{m(m-1)} \tag{7.78}$$

for $m \geq 2$. Using (7.78) in (7.77), we have

$$\begin{aligned} \varphi(x, y) = &\left[a_0(x) - a_{0,2}(x)\frac{(\beta y)^2}{2!} + a_{0,4}(x)\frac{\beta y)^4}{4!} - \cdots \right] \\ &+ \frac{1}{\beta}\left[a_1(x)\frac{(\beta y)}{1!} - a_{1,2}(x)\frac{(\beta y)^3}{3!} + a_{1,4}(x)\frac{(\beta y)^5}{5!} - \cdots \right] \end{aligned} \tag{7.79}$$

Now satisfying the boundary condition (7.74) by (7.79), we have

$$a_1(x) = \tau n \cos nx \tag{7.80}$$

and the solution (7.79) takes the form

$$\begin{aligned} \varphi(x, y) = &\left[a_0(x) - a_{0,2}(x)\frac{(\beta y)^2}{2!} + a_{0,4}(x)\frac{(\beta y)^4}{4!} - \cdots \right] \\ &+ \frac{\tau}{2\beta}\left[e^{n\beta y} - e^{-n\beta y} \right] \cos nx \end{aligned} \tag{7.81}$$

Again, we satisfy the boundary condition (7.75) by (7.81) to obtain

$$\begin{aligned} 0 = &\left[-a_{0,2}(x)\frac{(\beta H)}{1!} + a_{0,4}(x)\frac{(\beta H)}{3!} - \cdots \right] \\ &+ \frac{\tau n}{2\beta}\left[e^{n\beta H} + e^{-n\beta H} \right] \cos nx \end{aligned} \tag{7.82}$$

Now we have to find out the function $a_0(x)$, and for this purpose we assume $a_0(x)$ in the following form

$$a_0(x) = l_1 \cos nx + l_2 \sin nx \tag{7.83}$$

Using (7.83) in (7.82), we have

$$\begin{aligned} 0 = &\left[\left(e^{n\beta H} - e^{-n\beta H} \right) l_1 + \frac{\tau}{\beta}\left(e^{n\beta H} + e^{-n\beta H} \right) \right] \cos nx \\ &+ l_2 \left[e^{n\beta H} + e^{-n\beta H} \right] \sin nx \end{aligned} \tag{7.84}$$

The relation (7.84) is satisfied only if we put

$$l_1 = -\frac{\tau}{\beta}\frac{e^{n\beta H} + e^{-n\beta H}}{e^{n\beta H} - e^{-n\beta H}}, l_2 = 0 \tag{7.85}$$

Therefore, the form of the unknown function $a_0(x)$ is given by

$$a_0(x) = -\frac{\tau}{\beta}\frac{e^{n\beta H} + e^{-n\beta H}}{e^{n\beta H} - e^{-n\beta H}} \cos nx \tag{7.86}$$

and the expression for $\varphi(x, y)$ is found to be

$$\varphi(x, y) = -\frac{\tau}{\beta} \cdot \frac{\cos h\left[n\sqrt{1 - M_\infty^2}(y - H)\right]}{\sin h\left[n\sqrt{1 - M_\infty^2}H\right]} \tag{7.87}$$

which is a closed solution of the problem.

7.6 Application of Regular Decomposition to a Linearized Gasdynamic Equation for Axisymmetric Flow

In this section, we discuss the application of the regular decomposition method to the partial differential equation that is related to the problem of steady axisymmetric subsonic flow in gasdynamics.

We begin with the partial differential equation

$$\beta^2\frac{\partial^2\varphi}{\partial x^2} + \frac{1}{y} \cdot \frac{\partial\varphi}{\partial y} + \frac{\partial^2\varphi}{\partial y^2} = 0 \tag{7.88}$$

which can be written as

$$\beta^2\frac{\partial^2\varphi}{\partial x^2} + \frac{1}{y}\frac{\partial}{\partial y}\left(y\frac{\partial\varphi}{\partial y}\right) = 0 \tag{7.89}$$

where $\beta^2 = 1 - M_\infty^2$.

Let $L_x = \frac{\partial^2}{\partial x^2}$ and $L_y = \frac{1}{y}\frac{\partial}{\partial y}(\frac{1}{y}\frac{\partial}{\partial y})$ be the differential operators. Then the equation (7.89) becomes

$$\beta^2 L_x\varphi + L_y\varphi = 0 \tag{7.90}$$

The inverse of L_x is defined by (7.35), whereas the inverse of L_y [2] is determined in the following way:

Consider the equation

$$L_y\varphi = F \tag{7.91}$$

which can be written as

$$\frac{1}{y}\frac{\partial}{\partial y}\left(y\frac{\partial\varphi}{\partial y}\right) = F \tag{7.92}$$

Solving this equation for φ, we have

$$\varphi = \left[L_1^{-1}\left\{\frac{1}{y}\left(L_1^{-1}y\right)\right\}\right] F \tag{7.93}$$

remembering that the boundary condition terms vanish, and the L_1^{-1} is a onefold indefinite integral, that is, $L_1^{-1} = \int(\cdot)dy$. The expression (7.93) shows that the inverse L_y^{-1} is identified as

$$L_y^{-1} = L_1^{-1}\left\{\frac{1}{y}\left(L_1^{-1}y\right)\right\} \tag{7.94}$$

The x-Partial Solution

For the x-partial solution, we solve (7.90) for $L_x\varphi$ and get

$$L_x\varphi = -\beta^{-1}L_y\varphi \tag{7.95}$$

Then applying the inverse operator L_x^{-1} on (7.95), we have

$$\varphi(x, y) = \Phi(x, y) - \beta^{-2}L_x^{-1}(L_y\varphi) \tag{7.96}$$

where

$$\Phi(x, y) = \eta_0(y) + \eta_1(y)x \tag{7.97}$$

$\eta_0 y$ and $\eta_1 y$ being the constants to be determined by the boundary conditions.

We now decompose $\varphi(x, y)$ into the following form

$$\varphi(x, y) = \sum_{m=0}^{\infty}\lambda^m\varphi_m \tag{7.98}$$

where λ has its usual meaning and the parameterized form [2] of (7.96) is

$$\varphi(x, y) = \Phi(x, y) - \lambda\beta^{-2}L_x^{-1}(L_y\varphi) \tag{7.99}$$

Putting (7.98) into (7.99) and then equating the coefficients of like-power terms of λ from both sides of it, we obtain

$$\begin{aligned}
\varphi_0(x, y) &= \Phi(x, y)\\
\varphi_1(x, y) &= -\beta^{-2}L_x^{-1}(L_y\varphi_0)\\
\varphi_2(x, y) &= -\beta^{-2}L_x^{-1}(L_y\varphi_1)\\
&\cdots \quad \cdots \quad \cdots\\
&\cdots \quad \cdots \quad \cdots\\
\varphi_{m+1}(x, y) &= -\frac{1}{a}L_x^{-1}(L_y\varphi_m)\\
&\cdots \quad \cdots \quad \cdots
\end{aligned} \tag{7.100}$$

Thus we see that the components of $\varphi(x, y)$ are determined, and the final solution (7.98) is, therefore, computable remembering that $\lambda = 1$.

The y-Partial Solution

For the y-partial solution, we write the equation (7.90) in the form

$$L_y\varphi = -\beta^2 L_x\varphi \tag{7.101}$$

Operating with L_y^{-1} defined by (8.36), we get

$$\varphi(x, y) = \overline{\Phi}(x, y) - L_y^{-1}\left(\beta^2 L_x\varphi\right) \tag{7.102}$$

where $\bar{\Phi}(x, y)$ is the solution of

$$L_y\varphi = 0 \tag{7.103}$$

Then we decompose $\varphi(x, y)$ into

$$\varphi(x, y) = \sum_{m=0}^{\infty}\lambda^m\varphi_m \tag{7.104}$$

and write the parameterized form [2] of (7.102) as

$$\varphi(x, y) = \overline{\Phi}(x, y) - \lambda L_y^{-1}(\beta^2 L_x\varphi) \tag{7.105}$$

Substitution of (7.104) in (7.105) and comparison of like-power terms of λ give

$$\begin{aligned}
\varphi_0(x, y) &= \overline{\Phi}(x, y)\\
\varphi_1(x, y) &= -\beta^2 L_y^{-1}(L_x\varphi_0)\\
\varphi_2(x, y) &= -\beta^2 L_y^{-1}(L_x\varphi_1)\\
\dots \quad & \dots \quad \dots\\
\dots \quad & \dots \quad \dots\\
\varphi_{m+1}(x, y) &= -\beta^2 L_y^{-1}(L_x\varphi_m)\\
\dots \quad & \dots \quad \dots
\end{aligned} \tag{7.106}$$

Thus the components of $\varphi(x, y)$ are determined, and the final solution is computable.

7.7 Flow Past a Corrugated Circular Cylinder

In this section we discuss some problems of gasdynamics on the basis of the regular decomposition method. This method is used to study steady axisymmetric subsonic flow past a corrugated circular cylinder to make the theory clear as far as possible. Reissner [7] has investigated the linearized axisymmetric subsonic flow past a corrugated circular cylinder in an unlimited fluid. Haldar [4,5] has also studied the same problem in an open throat wind tunnel and a closed throat wind tunnel, respectively. In the following sections, we discuss these problems by the application of the ordinary decomposition technique.

7.7.1 Steady Axisymmetric Subsonic Flow Past a Corrugated Circular Cylinder in an Unlimited Fluid [7]

Let $u + U_\infty$ and v be the axial and radial velocity components, respectively, of fluid flowing over an infinitely long corrugated circular cylinder, where U_∞ is a free stream velocity. Let φ be the perturbation velocity potential defined by

$$\left.\begin{aligned} u - U_\infty &= \frac{\partial\varphi}{\partial x} \\ v &= \frac{\partial\varphi}{\partial r} \end{aligned}\right\} \tag{7.107}$$

The linearized partial differential gasdynamic equation for $\varphi(x, r)$ governing the steady axisymmetric subsonic flow over the corrugated cylinder is

$$\beta^2\varphi_{xx} + \frac{1}{r}\cdot\varphi_r + \varphi_{rr} = 0 \tag{7.108}$$

where (x, r) are the axial and radial directions, respectively, $\beta^2 = 1 - M_\infty^2$, and M_∞ is the free stream Mach number.

Let

$$R(x) = R_0 + \tau \sin nx \tag{7.109}$$

be the equation describing the geometry of the corrugated circular cylinder where R_0 is the radius of the cylinder, τ is the roughness parameter, and $2\pi/n$ is the wave length.

The boundary condition at the surface of the cylinder (7.109) is

$$\varphi_r(x, r) = U_\infty \frac{dR}{dx} = U_\infty \tau n \cos nx \quad \text{at} \quad r = R_0 \tag{7.110}$$

and the condition imposed on φ at infinity is

$$\varphi(x, r) = \text{ finite at } r = \infty \tag{7.111}$$

Two partial solutions, such as the x-partial solution and y-partial solution, for the equation (7.108) are possible, and out of these partial solutions, the r-partial solution is appropriate here. To solve this equation for the r-partial solution, we write it in the form

$$\frac{1}{r}\frac{\partial}{\partial r}(r\varphi_r) = k^2\varphi_{xx} \tag{7.112}$$

where $k^2 = (i\beta)^2$.

Let $L_r = \frac{1}{r}\frac{\partial}{\partial r}(r\frac{\partial}{\partial r})$ and $L_x = \frac{\partial^2}{\partial x^2}$. Then the inverse of the operator L_r is $L_r^{-1} = L_1^{-1}[\frac{1}{r}(L_1^{-1}r)]$, where $L_1^{-1} = \int(\cdot)dr$. Now operating on both sides of (7.112) with the inverse operator L_r^{-1}, we have

$$\varphi(x, r) = \varphi_0(x, r) + L_r^{-1}\left[k^2 L_x\varphi\right] \tag{7.113}$$

where

$$\varphi_0(x,r) = a_0(x) + a_1(x)\log r \tag{7.114}$$

Here the constants $a_0(x)$ and $a_1(x)$ are to be determined by the boundary conditions.

Then we decompose $\varphi(x,r)$ into

$$\varphi(x,r) = \sum_{m=0}^{\infty} \lambda^m \varphi_m \tag{7.115}$$

and the parameterized form [2] of (7.113) is

$$\varphi(x,r) = \varphi_0(x,r) + \lambda L_r^{-1}\left[k^2 L_x \varphi\right] \tag{7.116}$$

Substituting (7.115) in (7.116) and then equating like-power terms of λ from both sides of the resulting expression, we obtain

$$\varphi_{m+1}(x,r) = L_r^{-1}\left[k^2 L_x \varphi_m\right] \tag{7.117}$$

where $m = 0, 1, 2$, etc. Putting $m = 0, 1, 2$, etc., in (7.117) successively, we have the following set of expressions for the components of velocity potential:

$$\begin{aligned}
\varphi_1(x,r) &= k^2 L_r^{-1}\left(\frac{\partial^2 \varphi_0}{\partial x^2}\right)\\
\varphi_2(x,r) &= k^2 L_r^{-1}\left(\frac{\partial^2 \varphi_1}{\partial x^2}\right)\\
\cdots \quad & \cdots \quad \cdots\\
\cdots \quad & \cdots \quad \cdots\\
\varphi_{m+1}(x,r) &= k^2 L_r^{-1}\left(\frac{\partial^2 \varphi_m}{\partial x^2}\right)\\
\cdots \quad & \cdots \quad \cdots
\end{aligned} \tag{7.118}$$

Using (7.114) in (7.118), we get

$$\begin{aligned}
\varphi_1(x,r) &= a_{0,2}(x)\frac{(kx)^2}{2^2} + a_{1,2}(x)\left[\log r\frac{(kr)^2}{2^2} - \frac{(kr)^2}{2^2}(1)\right]\\
\varphi_2(x,r) &= a_{0,4}(x)\frac{(kr)^4}{2^2\cdot 4^2} + a_{1,4}(x)\left[\log r\frac{(kr)^4}{2^2\cdot 4^2} - \frac{(kr)^4}{2^2\cdot 4^2}\left(1+\frac{1}{2}\right)\right]\\
\cdots \quad & \cdots \quad \cdots \quad \cdots\\
\cdots \quad & \cdots \quad \cdots \quad \cdots\\
\varphi_m(x,r) &= a_{0,2m}(x)\frac{(kr)^{2m}}{2^2\cdot 4^2\cdots(2m)^2}\\
&\quad + a_{1,2m}(x)\left[\log r\frac{(kr)^{2m}}{2^2\cdot 4^2\cdots(2m)^2} - \frac{(kr)^{2m}}{2^2\cdot 4^2\ldots(2m)^2}\left(1+\frac{1}{2}+\cdots+\frac{1}{m}\right)\right]\\
\cdots \quad & \cdots \quad \cdots \quad \cdots
\end{aligned} \tag{7.119}$$

We now add the components $\varphi_0(x, r), \varphi_1(x, r)$, etc., given in (7.114) and (7.119) and then we use the expanded form of (7.115) remembering that $\lambda = 1$. Finally, we get

$$\varphi(x,r) = \left[a_0(x) + a_{0,2}(x)\frac{(kr)^2}{2^2} + a_{0,4}(x)\frac{(kr)^4}{2^2 \cdot 4^2} + a_{0,6}(x)\frac{(kr)^6}{2^2 \cdot 4^2 \cdot 6^2} + \cdots\right]$$

$$+ \log r\left[a_1(x) + a_{1,2}(x)\frac{(kr)^2}{2^2} + a_{1,4}(x)\frac{(kr)^4}{2^2 \cdot 4^2} + a_{1,6}(x)\frac{(kr)^6}{2^2 \cdot 4^2 \cdot 6^2} + \cdots\right]$$

$$- \left[a_{1,2}(x)\frac{(kr)^2}{2^2}(1) + a_{1,4}(x)\frac{(kr)^4}{2^2 \cdot 4^2}\left(1 + \frac{1}{2}\right)\right.$$

$$\left. + a_{1,6}(x)\frac{(kr)^6}{2^2 \cdot 4^2 \ldots 6^2}\left(1 + \frac{1}{2} + \frac{1}{3}\right) + \cdots\right] \tag{7.120}$$

The solution (7.120) contains two unknown functions $a_0(x)$ and $a_1(x)$, which are to be determined now. For this purpose we set the functions in the following forms looking at the boundary conditions:

$$\left.\begin{aligned} a_0(x) &= k_1 \cos nx \\ a_1(x) &= k_2 \cos nx \end{aligned}\right\} \tag{7.121}$$

Then we substitute (7.121) in (7.120) and write the expression for $\varphi(x, r)$ as

$$\varphi(x,r) = k_1\left[1 - \frac{(nkr)^2}{2^2} + \frac{(nkr)^4}{2^2 \cdot 4^2} - \frac{(nkr)^6}{2^2 \cdot 4^2 \cdot 6^2} + \cdots\right]\cos nx$$

$$+ k_2\left[\log r\left\{1 - \frac{(nkr)^2}{2^2} + \frac{(nkr)^4}{2^2 \cdot 4^2} - \frac{(nkr)^6}{2^2 \cdot 4^2 \cdot 6^2} + \cdots\right\}\right.$$

$$+ \left\{\frac{(nkr)^2}{2^2}(1) - \frac{(nkr)^4}{2^2 \cdot 4^2}\left(1 + \frac{1}{2}\right)\right.$$

$$\left.\left. + \frac{(nkr)^6}{2^2 \cdot 4^2 \cdot 6^2}\left(1 + \frac{1}{2} + \frac{1}{3}\right) - \cdots\right\}\right]\cos nx$$

$$= k_1\, J_0(i\, n\, \beta\, r)\, \cos\, n\, x + k_2\left[\log\, r\, .\, J_0(i\, n\, \beta\, r)\right.$$

$$+ \left\{\frac{(in\beta r)^2}{2^2}(1) - \frac{(in\beta r)^4}{2^2 \ldots 4^2}\left(1 + \frac{1}{2}\right)\right.$$

$$\left.\left. + \frac{(in\beta r)^6}{2^2 \ldots 4^2 \cdot 6^2}\left(1 + \frac{1}{2} + \frac{1}{3}\right) - \cdots\right\}\right]\cos nx$$

$$= k_1\, J_0(i\, n\, \beta\, r)\, \cos\, n\, x + k_2\left[\log\, r\, .\, J_0(i\, n\, \beta\, r)\right.$$

$$- \sum_{p=1}^{\infty} (-1)^p \frac{1}{p^2} \cdot \left(\frac{in\beta r}{2} \right)^{2p} \left(1 + \frac{1}{2} + \frac{1}{3} + \cdots + \frac{1}{p} \right) \Bigg] \cos nx$$

$$= [k_1 J_0(i\,n\,\beta\,r) + k_2 Y_0(i\,n\,\beta\,r)] \cos nx$$

$$= [k_1 I_0(n\,\beta\,r) + k_2 K_0(n\,\beta\,r)] \cos nx \tag{7.122}$$

where I_0 and K_0 are modified Bessel's functions of order zero, and k_1, k_2 are arbitrary constants to be determined from the boundary conditions.

The boundary condition (7.111) shows that the velocity potential $\varphi(x, r)$ is finite at infinity. This means that the asymptotic behavior of the function I_0 requires that k_1 should be zero, and the relation (7.122) reduces to

$$\varphi(x, r) = k_2 K_0(n\,\beta\,r) \cos n\,x \tag{7.123}$$

Then satisfying the boundary condition (7.110) by (7.123), we get

$$k_2 = -\frac{\tau U_\infty}{\beta K_1(n\beta\, R_0)} \tag{7.124}$$

Putting (7.124) in (7.123), we have the required expression for $\varphi(x, r)$ as

$$\varphi(x, r) = -\frac{\tau U_\infty}{\beta} \frac{K_0(n\,\beta\,r)}{K_1(n\,\beta\,R_0)} \cos n,\, x \tag{7.125}$$

which is a closed-form solution of the problem.

7.7.2 Steady Axisymmetric Subsonic Flow Past a Corrugated Circular Cylinder in an Open-Throat Wind Tunnel [4]

Let the boundary of a corrugated circular cylinder be

$$R(x) = R_0 + \tau \sin n\,x \tag{7.126}$$

where R_0, τ, and $2\pi/n$ have their usual meanings. The linearized gasdynamic equation that governs the steady axisymmetric subsonic flow over the circular cylinder (7.126) in an open-throat wind tunnel is

$$\beta^2 \varphi_{xx} + \frac{1}{r} \varphi_r + \varphi_{rr} = 0 \tag{7.127}$$

where $\varphi(x, r)$ is the perturloation velocity potential and β is defined by $\beta^2 = 1 - M_\infty^2$.

If H is the distance of the wind tunnel wall from the axis of the cylinder, then the boundary conditions at the surface of the cylinder and at the surface of the tunnel wall are, respectively,

$$\varphi_r(x, r) = U_\infty \frac{dR}{dx} = U_\infty \tau n \cos nx \quad \text{at} \quad r = R_0 \tag{7.128}$$

and

$$\varphi_x(x, r) = 0 \quad \text{at} \quad r = H \tag{7.129}$$

Now we have to solve the equation (7.127) under the boundary conditions (7.128) and (7.129). To do that, we use the regular decomposition procedure for the r-partial solution, which is appropriate here. Proceeding exactly in the same way as in Section 7.1, we consider (7.122) as the general solution of (7.127), that is,

$$\varphi(x, r) = [C_1 I_0(n\beta r) + C_2 K_0(n\beta r)] \cos nx \tag{7.130}$$

where C_1 and C_2 are arbitrary constants to be determined by the boundary conditions.

Satisfying the boundary conditions (7.128) and (7.129) by (7.130), we get

$$C_1 I_1(n\beta R_0) + C_2 K_1(n\beta R_0) = \frac{U_\infty \tau}{\beta} \tag{7.131}$$

and

$$C_1 I_0(n\beta H) + C_2 K_0(n\beta H) = 0 \tag{7.132}$$

which, on solving for C_1 and C_2, give

$$\left.\begin{aligned} C_1 &= \frac{U_\infty \tau}{\beta} \frac{K_0(n\beta H)}{I_0(n\beta H) K_1(n\beta R_0) + I_1(n\beta R_0) K_0(n\beta H)} \\ C_2 &= -\frac{U_\infty \tau}{\beta} \frac{I_0(n\beta H)}{I_0(n\beta H) K_1(n\beta R_0) + I_1(n\beta R_0) K_0(n\beta H)} \end{aligned}\right\} \tag{7.133}$$

Using (7.133) in (7.130) we get the required expression for $\varphi(x, r)$ as

$$\varphi(x, r) = \frac{U_\infty \tau}{\beta} \frac{K_0(n\beta H) I_o(n\beta r) - I_0(n\beta H) K_0(n\beta r)}{I_0(n\beta H) K_1(n\beta R_0) + I_1(n\beta R_0) K_0(n\beta H)} \cos nx \tag{7.134}$$

which is the exact solution of the problem.

7.7.3 Steady Axisymmetic Subsonic Flow Past a Corrugated Circular Cylinder in a Closed-Throat Wind Tunnel [5]

The mathematical model for axisymmetric subsonic flow past a corrugated circular cylinder in a closed-throat wind tunnel is described by the partial differential equation

$$\beta^2 \varphi_{xx} + \frac{1}{r} \varphi_r + \varphi_{rr} = 0 \tag{7.135}$$

subject to the boundary conditions

$$\varphi_r(x, r) = U_\infty \frac{dR}{dx} = U_\infty \tau n \cos nx \quad \text{at} \quad r = R_0 \tag{7.136}$$

and

$$\varphi_r(x, y) = 0 \quad \text{at} \quad r = H \tag{7.137}$$

where the cylinder is given by the equation

$$R(x) = R_0 + \tau \sin nx \tag{7.138}$$

The symbols used here have their usual meanings.

To solve the equation (7.135), we take the help of the analysis of Section 7.1 and write again the solution as

$$\varphi(x, r) = \mu_1 I_0(n\beta r) + \mu_2 K_0(n\beta r) \tag{7.139}$$

where μ_1 and μ_2 are arbitrary constants to be evaluated with the help of the boundary conditions.

Using the boundary condition (7.136) and (7.137), we obtain

$$\mu_1 I_1(n\beta R_0) - \mu_2 K_1(n\beta R_0) = \frac{\tau}{\beta} \tag{7.140}$$

and

$$\mu_1 I_1(n\beta H) - \mu_2 K_1(n\beta H) = 0 \tag{7.141}$$

Then we solve (7.140) and (7.141) for μ_1 and μ_2 and get

$$\mu_1 = \frac{\tau}{\beta} \frac{K_1(n\beta H)}{I_1(n\beta R_0)K_1(n\beta H) - I_1(n\beta H)K_1(n\beta R_0)} \tag{7.142}$$

and

$$\mu_2 = \frac{\tau}{\beta} \cdot \frac{I_1(n\beta H)}{I_1(n\beta R_0)K_1(n\beta H) - I_1(n\beta H)K_1(n\beta R_0)} \tag{7.143}$$

Substituting (7.142) and (7.143) in (7.139), we write the expression for $\varphi(x, r)$ as

$$\varphi(x, r) = \frac{\tau}{\beta} \cdot \frac{K_1(n\beta H)I_0(n\beta r) + I_1(n\beta H)K_0(n\beta r)}{I_1(n\beta R_0)K_1(n\beta H) - I_1(n\beta H)K_1(n\beta R_0)} \cos nx \tag{7.144}$$

which is a closed-form solution of the problem.

References

1. Ackeret, J.: Über Luftkräfte bei Sehr grossen Geschwindigkeiten insbesondere bei ebenen strömungen. *Helvetica Physica Acta, 1*(5), 301–322 (1928).
2. Adomian, G.: *Nonlinear Stochastic System Theory and Application to Physics*, Kluwer Academic Publishers (1989).

3. Adomian, G.: *Solving Frontier Problems of Physics: The Decomposition Method*, Kluwer Academic Publishers (1994).
4. Haldar, K.: Second Order Subsonic Solution for a Corrugated Circular Cylinder in an Open Throat Wind Tunnel, *Bull. Cal. Math. Soc., 62*, 115–121, (1970).
5. Haldar, K.: Second Order Subsonic Solution for a Corrugated Circular Cylinder in a Closed Throat Wind Tunnel, *Bull. Cal. Math. Soc., 67*, 37–46, (1975).
6. Niyogi, P.: *Inviscid Gasdynamics*, The Macmillan Company of India Limited (1971).
7. Reissner, E.: *On Compressibility Corrections for Subsonic Flow Over Bodiees of Revolution*, NACA, Tech. Rept. No. 1816 (1949).

8

Steady Transonic Flow

8.1 Introduction

The basic nonlinear gasdynamic equation is linearized for a purely subsonic or supersonic flow under the assumption of small perturbation. But in the transonic range, this basic equation still remains nonlinear in character under the small perturbation theory. Kaplan [3] has studied the transonic flow past a wave-shaped wall of small amplitude extending to infinity with the help of the Prandtl-Busemann small perturbation method. Haldar [2] has also investigated the same problem by means of the decomposition method, which avoids simplifications. We discuss this problem in detail in this chapter.

8.2 Transonic Solution by Regular Decomposition

We begin with the two-dimensional transonic equation (7.4), that is,

$$\beta^2 \varphi_{xx} + \varphi_{yy} = M_\infty^2(\gamma + 1)\varphi_x \varphi_{xx} \tag{8.1}$$

where $\beta^2 = 1 - M_\infty^2$, M_∞ is the free stream Mach number defined by $M_\infty = u_\infty / c$, u_∞ and c being the free stream velocity and the speed of sound, respectively, $\gamma (= 1.4)$ is the ratio of specific heats, and φ is the perturbation velocity potential.

Let $L_x = \frac{\partial^2}{\partial x^2}$ and $L_y = \frac{\partial^2}{\partial y^2}$. Then the transonic equation (8.1) converted to the operator form is

$$\beta^2 L_x \varphi + L_y \varphi = K^2 N\varphi \tag{8.2}$$

where $K^2 = M_\infty^2(\gamma + 1)$ and the nonlinear term $N\varphi$ is defined by

$$N\varphi = \varphi_x \varphi_{xx} \tag{8.3}$$

Solving for $L_x \varphi$ we have from (8.2)

$$L_x \varphi = \beta^{-2} \left[K^{-2} N\varphi - L_y \varphi \right] \tag{8.4}$$

and then the application of the inverse operator $L_x^{-1} = \iint(\cdot)dxdx$ on both sides of (8.4) gives

$$\varphi(x, y) = \varphi_0(x, y) + L_x^{-1}\beta^{-2}\left[K^{-2}N\varphi - L_y\varphi\right] \tag{8.5}$$

where $\varphi_0(x, y)$ is the solution of $L_x\varphi = 0$. It is given by

$$\varphi_0(x, y) = \bar{C}_1(y) + \bar{C}_2(y)x \tag{8.6}$$

$C_1(y)$ and $C_2(y)$ being the integration constants to be evaluated by the boundary conditions.

We now decompose $\varphi(x, y)$ and $N\varphi$ into the following forms:

$$\varphi(x, y) = \sum_{m=0}^{\infty}\lambda^m\varphi_m \tag{8.7}$$

$$N\varphi = \sum_{m=0}^{\infty}\lambda^m A_m \tag{8.8}$$

Then the parameterized form of (8.5) is

$$\varphi(x, y) = \varphi_0(x, y) + \lambda L_x^{-1}\beta^{-2}\left[K^2N\varphi - L_y\varphi\right] \tag{8.9}$$

Putting (8.7), (8.8) into (8.9) and comparing the terms, we get

$$\begin{aligned}
\varphi_1(x, y) &= \beta^{-2}L_x^{-1}\left[K^2A_0 - L_y\varphi_0\right]\\
\varphi_2(x, y) &= \beta^{-2}L_x^{-1}\left[K^2A_1 - L_y\varphi_1\right]\\
\cdots \quad &\cdots \quad \cdots\\
\cdots \quad &\cdots \quad \cdots\\
\varphi_{m+1}(x, y) &= \beta^{-2}L_x^{-1}\left[K^2A_m - L_y\varphi_m\right]\\
\cdots \quad &\cdots \quad \cdots
\end{aligned} \tag{8.10}$$

Now we have to find out the Adomian's polynomials A_m to complete the problem. For the purpose we put (8.7), (8.8) into (8.3), and then we compare the terms to get the following set of polynomials:

$$\begin{aligned}
A_0(x, y) &= \frac{1}{2}\frac{\partial}{\partial x}\varphi_{0,x}^2\\
A_1(x, y) &= \frac{\partial}{\partial x}\left(\varphi_{0,x}\varphi_{1,x}\right)\\
A_2(x, y) &= \frac{\partial}{\partial x}\left(\varphi_{0,x}\varphi_{2,x} + \frac{1}{2}\varphi_{1,x}^2\right)\\
\cdots \quad &\cdots \quad \cdots\\
\cdots \quad &\cdots \quad \cdots
\end{aligned} \tag{8.11}$$

The relations (8.7), (8.10), and (8.11) together complete the solution of the problem, and this solution is called the x-partial solution.

Similarly, we can find out the y-partial solution by writing (8.2) in the form

$$\varphi(x, y) = \overline{\varphi}_0(x, y) + L_y^{-1}\left[K^2 N\varphi - \beta^2 \varphi_{xx}\right] \tag{8.12}$$

where $\overline{\varphi}(x, y) = \overline{C}_1(x) + \overline{C}_2(x)y$ and $L_y^{-1} = \int\int(\cdot)\,dy\,dy$.

Proceeding exactly in the same way as for the x-partial solution, we have the following set of components $\varphi_0(x, y)$, $\varphi_1(x, y)$, $\varphi_2(x, y)$, etc., of the solution $\varphi(x, y)$:

$$\begin{aligned}
\varphi_0(x, y) &= \overline{\varphi}_0(x, y) = \overline{C}_1(x) + \overline{C}_2(x)y \\
\varphi_1(x, y) &= L_y^{-1}\left[K^2 A_0 - \beta^2 \varphi_{0,xx}\right] \\
\varphi_2(x, y) &= L_y^{-1}\left[K^2 A_1 - \beta^2 \varphi_{1,xx}\right] \\
&\cdots \quad \cdots \quad \cdots \\
&\cdots \quad \cdots \quad \cdots \\
\varphi_{m+1}(x, y) &= L_y^{-1}\left[K^2 A_m - \beta^2 \varphi_{m,xx}\right] \\
&\cdots \quad \cdots \quad \cdots \\
&\cdots \quad \cdots \quad \cdots
\end{aligned} \tag{8.13}$$

where the polynomials A_0, A_1, A_2, etc., are given by (8.11). The relations (8.7) and (8.13) constitute the complete solution of the problem.

8.3 Transonic Solution by Modified Decomposition

Consider again a steady two-dimensional transonic equation in the operator form

$$\beta^2 L_x \varphi + l_y \varphi = K^2 N\varphi \tag{8.14}$$

where $\beta^2 = 1 - M_\infty^2$, $L_x = \partial^2/\partial x^2$, $L_y = \partial^2/\partial y^2$, $K^2 = M_\infty^2(\gamma + 1)$, and the nonlinear term $N\varphi$ is defined by (8.3), where M_∞ and γ have their usual meanings.

Then we follow the ordinary decomposition technique and write the x-partial solution as

$$\varphi(x, y) = k_1(y) + xk_2(y) + \beta^{-2} L_x^{-1}\left[K^2 N\varphi - L_y \varphi\right] \tag{8.15}$$

where $L_x^{-1} = \int\int(\cdot)dx\,dx$ and the integration constants $k_1(y)$, $k_2(y)$ can be found out by the boundary conditions. We now apply the modified decomposition technique [1], which has been discussed for the ordinary differential equation. Thus for the x-partial solution, we let

$$\varphi(x, y) = \sum_{m=0}^{\infty} a_m(y) x^m \tag{8.16}$$

$$N\varphi(x, y) = \sum_{m=0}^{\infty} A_m(y)x^m \tag{8.17}$$

Here A_m are the Adomian's polynomials. Putting (8.16) and (8.17) in (8.15), we get

$$\sum_{m=0}^{\infty} a_m(y)x^m = k_1(y) + xk_2(y) + \beta^{-2}L_x^{-1}\left[K^2\sum_{m=0}^{\infty} A_m(y)x^m - L_y\sum_{m=0}^{\infty} a_m(y)x^m\right]$$

Carrying out integrations on the right side and then replacing m by $(m-2)$, we obtain

$$\begin{aligned}
\sum_{m=0}^{\infty} a_m(y)x^m &= k_1(y) + xk_2(y) \\
&\quad + \beta^{-2}\left[K^2\sum_{m=2}^{\infty}\frac{A_{m-2}(y)}{m(m-1)} - \sum_{m=2}^{\infty}\frac{(\partial^2 \mid \partial y^2)a_{m-2}(y)}{m(m-1)}\right]x^m \\
&= k_1(y) + xk_2(y) \\
&\quad + \sum_{m=2}^{\infty}\beta^{-2}\left[\frac{A_{m-2}(y)K^2 - (\partial^2 \mid \partial y^2)a_{m-2}(y)}{m(m-1)}\right]x^m
\end{aligned}$$

The coefficients are identified by

$$\begin{aligned}
k_1(y) &= a_0(y) \\
k_2(y) &= a_1(y)
\end{aligned} \tag{8.18}$$

and the recurrence relation for $m \geq 2$

$$a_m(y) = \beta^{-2}\left[A_{m-2}(y)K^2 - (\partial^2/\partial y^2)a_{m-2}(y)\right]/m(m-1) \tag{8.19}$$

To complete the solution of the problem, we have to find out the Adomian's polynomials. Differentiating (8.16) with respect to x, we have

$$\varphi_x(x, y) = \sum_{m=0}^{\infty}(m+1)a_{m+1}(y)x^m \tag{8.20}$$

and

$$\varphi_{xx}(x, y) = \sum_{m=0}^{\infty}(m+1)(m+2)a_{m+2}(y)x^m \tag{8.21}$$

Using (8.16), (8.17), (8.20), and (8.21) in (8.3), we get

$$\sum_{m=0}^{\infty} A_m(y)x^m = \left[\sum_{m=0}^{\infty}(m+1)a_{m+1}(y)x^m\right]\left[\sum_{m=0}^{\infty}(m+1)(m+2)a_{m+2}(y)x^m\right]$$

Then carrying out the Cauchy product [1] of infinite series, we get

$$\sum_{m=0}^{\infty} A_m(y)x^m = \sum_{m=0}^{\infty}\left[\sum_{n=0}^{m}(m-n+1)a_{m-n+1}(y)(n+1)(n+2)a_{n+2}(y)\right]x^m$$

Comparing the terms on both sides, we have

$$A_m(y) = \sum_{n=0}^{m}(m-n+1)a_{m-n+1}(y)(n+1)(n+2)a_{n+2}(y) \tag{8.22}$$

The relation (8.16) together with the relations (8.19) and (8.22) complete the solution of the problem.

Similarly, the y-partial solution can be obtained letting φ and $N\varphi$ in the following forms:

$$\varphi(x, y) = \sum_{m=0}^{\infty} b_m(x)y^m$$

$$N\varphi = \sum_{m=0}^{\infty} \lambda^m P_m \tag{8.23}$$

where

$$b_m(x) = \left[P_{m-2}(x) - \beta^2(\partial^2/\partial x^2)b_{m-2}(x)\right]/m(m-1) \tag{8.24}$$

and

$$P_m(x) = \sum_{n=0}^{m}(m-n+1)b_{m-n+1}(x)(n+1)(n+2)b_{n+2}(x) \tag{8.25}$$

8.4 Transonic Solution by Multidimensional Operator

We begin with the transonic equation (8.2), that is,

$$\beta^2 L_x\varphi + L_y\varphi = K^2 N\varphi \tag{8.26}$$

where β^2, K^2 are defined earlier, and the nonlinear term $N\varphi$ is defined by (8.3), that is,

$$N\varphi = \varphi_x\varphi_{xx} \tag{8.27}$$

Let $L = \beta^2 L_x + L_y$. Then the equation (8.26) becomes

$$L\varphi = K^2 N\varphi \tag{8.28}$$

Now we decompose φ and $N\varphi$ into the following forms:

$$\varphi(x, y) = \sum_{m=0}^{\infty} \lambda^m \varphi_m \tag{8.29}$$

$$N\varphi = \sum_{m=0}^{\infty} \lambda^m A_m \tag{8.30}$$

Then we write the parameterized form of (8.28) as

$$L\varphi = \lambda K^2 N\varphi \tag{8.31}$$

Putting (8.29), (8.30) into (8.31) and comparing the terms, we obtain a set of differential equations for the components $\varphi_0(x, y)$, $\varphi_1(x, y)$, $\varphi_2(x, y)$, etc., as

$$\begin{aligned} L\varphi_0 &= 0 \\ L\varphi_1 &= K^2 A_0 \\ L\varphi_2 &= K^2 A_1 \\ \cdots \quad &\cdots \quad \cdots \\ L\varphi_{m+1} &= K^2 A_m \\ \cdots \quad &\cdots \quad \cdots \end{aligned} \tag{8.32}$$

where $A_0(x, y)$, $A_1(x, y)$, $A_2(x, y)$, etc., are the Adomian's polynomials to be determined by using (8.29) and (8.30) in (8.27). Thus, we have

$$\begin{aligned} A_0(x, y) &= \frac{1}{2}\frac{\partial}{\partial x}(\varphi_{0,x})^2 \\ A_1(x, y) &= \frac{\partial}{\partial x}(\varphi_{0,x}\varphi_{1,x}) \\ \cdots \quad &\cdots \quad \cdots \\ \cdots \quad &\cdots \quad \cdots \end{aligned} \tag{8.33}$$

Now each of the equations in (8.32) can be easily solved under their respective boundary conditions. Once the components $\varphi_0(x, y)$, $\varphi_1(x, y)$, etc., are computable, the complete solution $\varphi(x, y)$ is known. For clear illustration of this procedure, we consider a concrete example in the next section.

8.5 Transonic Flow Past a Wavy Wall

In this section we discuss the solution of a quasi-linear partial differential equation of mixed elliptic-hyperbolic type for two-dimensional transonic flow past an infinitely long wavy wall, which has already been studied by Kaplan [3].

8.5.1 Equations of Motion

Consider the steady two-dimensional transonic flow of an incompressible fluid past an infinitely long wave-shaped wall whose geometry is given by

$$y = \varepsilon \operatorname{Sin} x \tag{8.34}$$

where ε is the maximum height of the wall and $l = 2\pi$ is its wave length.

The partial differential equation that governs the transonic flow field over the wavy wall is

$$(1 - M_\infty^2)\varphi_{xx} + \varphi_{yy} = M_\infty^2(\gamma + 1)\varphi_x\varphi_{xx} \tag{8.35}$$

where M_∞, γ have their usual meanings and $\varphi(x, y)$ is the perturbation velocity potential difined by

$$\varphi_x = \frac{u - u_\infty}{u_\infty}, \quad \varphi_y = \frac{v}{u_\infty} \tag{8.36}$$

Here u_∞ is the free stream velocity and (u, v) are the velocity components in the x-direction parallel to the flow of fluid and y-direction perpendicular to the flow direction, respectively.

The corresponding boundary conditions imposed on φ are

$$\varphi_y(x, y) = \frac{dy}{dx} = \varepsilon \cos x \quad \text{at} \quad y = 0 \tag{8.37}$$

and

$$\varphi_y(x, y) = 0 \qquad \text{at} \quad y = \infty \tag{8.38}$$

It is now convenient to use the following transformations:

$$\begin{aligned} \xi &= x \\ \eta &= (1 - M_\infty^2)^{\frac{1}{2}} y \\ \phi(\xi, \eta) &= \frac{M_\infty^2(\gamma + 1)}{1 - M_\infty^2}\varphi(x, y) \end{aligned} \tag{8.39}$$

By virtue of (8.39) the gasdynamic equation (8.35) takes the form

$$\phi_{\xi\xi} + \phi_{\eta\eta} = \phi_\xi \phi_{\xi\xi} \tag{8.40}$$

The boundary conditions (8.37) and (8.38) become

$$\phi_\eta(\xi, \eta) = T \cos \xi \quad \text{at} \quad \eta = 0 \tag{8.41}$$

and

$$\phi_\eta(\xi, \eta) = 0 \qquad \text{at} \quad \eta = \infty \tag{8.42}$$

where T is the reduced thickness ratio given by

$$T = 1/k^{3/2} \tag{8.43}$$

and k is the transonic similarity parameter defined by

$$k = \frac{1 - M_\infty^2}{\left[M_\infty^2(\gamma + 1) \in\right]^{2/3}} \tag{8.44}$$

The equation (8.40) together with the boundary conditions (8.41) and (8.42) constitute the mathematical model of the problem.

8.5.2 Method of Solution

Let

$$L = \frac{\partial^2}{\partial \xi^2} + \frac{\partial^2}{\partial \eta^2} \tag{8.45}$$

be the multidimensional operator. Using (8.45), the equation (8.40) takes the form

$$L\phi = N\phi \tag{8.46}$$

where $N\phi$ is the nonlinear term defined by

$$N\phi = \phi_\xi \phi_{\xi\xi} \tag{8.47}$$

We now decompose ϕ and $N\phi$ into

$$\phi(\xi, \eta) = \sum_{m=0}^{\infty} \lambda^m \phi_m \tag{8.48}$$

$$N\phi = \sum_{m=0}^{\infty} \lambda^m A_m \tag{8.49}$$

and then write the parameterixed form of (8.46) as

$$L\phi = \lambda N\phi \tag{8.50}$$

where λ and A_m have their usual meanings. Putting (8.48) and (8.49) into (8.50) we have, on comparison of terms, the following set of partial differential equations for the components $\phi_0(\xi, \eta)$, $\phi_1(\xi, \eta)$, $\phi_2(\xi, \eta)$, etc.:

$$L\phi_m = A_{m-1} \tag{8.51}$$

for all positive integers remembering that $A - 1 = 0$.

Next we proceed to determine A_0, A_1, A_2, etc., which are defined in such a way that each A_m depends on ϕ_0, ϕ_1, etc., that is, $A_0 = A_0(\phi_0)$, $A_1 = A_1(\phi_0, \phi_1)$, $A_2 = A_2(\phi_0, \phi_1, \phi_2)$, etc. To do that, we substitute (8.48) and (8.49) in (8.47), and then compare the terms on both sides of the resulting equation to get the following relation:

$$A_m = \sum_{n=0}^{m} \phi_{\eta,\xi} \phi_{m-n}, \xi\xi \tag{8.52}$$

which gives the Adomian's special polynomials for $m = 0, 1, 2$, etc.

Again, substitution of (8.48) in (8.41) and (8.42) gives the boundary conditions for the components ϕ_0, ϕ_1, ϕ_2, etc., at the body and at infinity as

$$\phi_{m,\eta}(\xi, \eta) = K_m T \cos_\xi \quad \text{at} \quad \eta = 0 \tag{8.53}$$

where

$$K_m = 1 \quad \text{when } m = 0 \qquad (8.54)$$
$$= 0 \quad \text{when } m \geq 1$$

and

$$\phi_{m,\eta}(\xi, \eta) = 0 \quad \text{at} \quad \eta = \infty \qquad (8.55)$$

for all positive integers of m.

8.5.3 Solution of the Problem

Model M_0

Putting $m = 0$ in (8.51), (8.53), and (8.55), the mathematical model M_0 is described by Laplace's equation

$$L_{\phi_0} = 0 \qquad (8.56)$$

subject to the boundary conditions

$$\phi_{0,\eta}(\xi, \eta) = T \cos \xi \quad \text{at} \quad \eta = 0 \qquad (8.57)$$

and

$$\phi_{0,\eta} = 0 \quad \text{at} \quad \eta = \infty \qquad (8.58)$$

To solve this model, we can use the powerful standard method known as the *separation of variables* or *decomposition method* (regular or modified). However, the solution of the model M_0 is given by

$$\phi_0(\xi, \eta) = -Te^{-\eta} \cos \xi \qquad (8.59)$$

Model M_1

The model M_1 can be obtained by putting $m = 1$ in (8.51), (8.52), (8.53), and (8.55). It is given by the partial differential equation

$$L_{\phi_1} = A_0 = \phi_{0,\xi}\phi_{0,\xi\xi} \qquad (8.60)$$

and the boundary conditions

$$\left.\begin{aligned} \phi_{1,\eta}(\xi, \eta) &= 0 \quad \text{at} \quad \eta = 0 \\ \phi_{1,\eta}(\xi, \eta) &= 0 \quad \text{at} \quad \eta = \infty \end{aligned}\right\} \qquad (8.61)$$

Since $\phi_0(\xi, \eta)$ is known, the right-hand side of the equation (8.60) is also known, and using (8.59), we write the equation (8.60) as

$$\frac{\partial^2 \phi_1}{\partial \xi^2} + \frac{\partial^2 \phi_1}{\partial \eta^2} = \frac{1}{2} T^2 e^{-2\eta} \operatorname{Sin} 2\xi \qquad (8.62)$$

To solve the equation (8.62), we assume its solution in the form

$$\phi_1(\xi, \eta) = f_1(\eta) \operatorname{Sin} 2\xi \tag{8.63}$$

Substituting (8.63) in (8.62), we have an ordinary differential equation for $f_1(\eta)$ given below:

$$\frac{d^2 f_1}{d\eta^2} - 4f_1 = \frac{1}{2}T^2 e^{-2\eta} \tag{8.64}$$

The corresponding boundary conditions (8.61), by virtue of (8.63), become

$$\frac{df_1}{d\eta} = 0 \quad \text{at} \quad \eta = 0 \;\&\; \eta = \infty \tag{8.65}$$

The solution of the equation (8.64) subject to the boundary condition (8.65) is given by

$$f_1(\eta) = -\frac{T^2}{16}(1 + 2\eta)e^{-2\eta} \tag{8.66}$$

Therefore, the required solution of the model M_1 is

$$\phi_1(\xi, \eta) = -\frac{T^2}{16}(1 + 2\eta)e^{-2\eta} \sin 2\xi \tag{8.67}$$

Model M_2

Similarly, the model M_2 is described by the differential equation

$$L_{\phi_2} = A_1 \tag{8.68}$$

and the boundary conditions

$$\phi_{2,\eta} = 0 \quad \text{at} \quad \eta = 0 \quad \& \; \eta = \infty \tag{8.69}$$

Since $\phi_0(\xi, \eta)$ and $\phi_1(\xi, \eta)$ are known, the right-hand side of the equation (8.68) is known, and this equation, by virtue of (8.59) and (8.67), takes the form

$$\phi_{2,\xi\xi} + \phi_{2,\eta\eta} = -\frac{T^3}{16}e^{-3\eta}(1 + 2\eta)(3\cos 3\xi - \cos \xi) \tag{8.70}$$

For solving this equation, we assume its solution in the following form:

$$\phi_2(\xi, \eta) = (-T^3/16)[f_2(\eta)\cos 3\xi - \bar{f}_2(\eta)\cos \xi] \tag{8.71}$$

where $f_2(\eta)$ and $\bar{f}_2(\eta)$ are functions of η only. Differentiating (8.71) twice separately with respect ξ and η and then adding, we get

$$\begin{aligned}\phi_{2,\xi\xi} + \phi_{2,\eta\eta} = (-T^3/16) &\left[\left(\frac{d^2 f_2}{d\eta^2} - 9f_2\right)\cos 3\xi \right.\\ &\left. + \left(\frac{d^2 \bar{f}_2}{d\eta^2} - \bar{f}_2\right)\cos \xi\right]\end{aligned} \tag{8.72}$$

Comparing (8.70) and (8.72), we have

$$\frac{d^2 f_2}{d\eta^2} - 9f_2 = 3e^{-3\eta}(1+2\eta) \tag{8.73}$$

and

$$\frac{d^2 \bar{f}_2}{d\eta^2} - \bar{f}_2 = e^{-3\eta}(1+2\eta) \tag{8.74}$$

Using (8.71), the boundary conditions (8.69) are found to be

$$\frac{df_2}{d\eta} = 0 \quad \text{at} \quad \eta = 0 \quad \& \; \eta = \infty \tag{8.75}$$

and

$$\frac{d\bar{f}_2}{d\eta} = 0 \quad \text{at} \quad \eta = 0 \quad \& \; \eta = \infty \tag{8.76}$$

The solution of the equation (8.73) subject to the boundary conditions (8.75) is

$$f_2(\eta) = -\frac{1}{16} e^{-3\eta}(9\eta^2 + 12\eta + 4) \tag{8.77}$$

and that of the equation (8.74) under the boundary conditions (8.76) is

$$\bar{f}_2(\eta) = \frac{1}{18}\left[e^{-3\eta}(4\eta + 5) - 11e^{-\eta}\right] \tag{8.78}$$

Substituting (8.77) and (8.78) in (8.71), we have the required solution of the equation (8.75) as

$$\begin{aligned} \phi_2(\xi, \eta) = \frac{T^3}{32 \times 72} & \left[8e^{-3\eta}(9\eta^2 + 12\eta + 4)\cos 3\xi \right. \\ & \left. - 9\left\{e^{-3\eta}(4\eta + 5) - 11e^{-\eta}\right\}\cos\eta\right] \end{aligned} \tag{8.79}$$

Proceeding in the same way, we can also compute other components of $\phi(\xi, \eta)$, such as $\phi_3(\xi, \eta)$, $\phi_4(\xi, \eta)$, etc. If we consider a four-term approximant of the solution, then we have

$$\phi(\xi, \eta) = \phi_0(\xi, \eta) + \phi_1(\xi, \eta) + \phi_2(\xi, \eta) + \phi_3(\xi, \eta) \tag{8.80}$$

where $\phi_0(\xi, \eta)$, $\phi_1(\xi, \eta)$, $\phi_2(\xi, \eta)$ are given by (8.59), (8.67), (8.79), and the expression for $\phi_3(\xi, \eta)$ is

$$\begin{aligned} \phi_3(\xi, \eta) = \frac{T^4}{8.24.32} & \left[\left\{2e^{-4\eta}(24\eta^2 + 68\eta + 57) + e^{-2\eta}(66\eta - 127)\right\}\sin\xi \right. \\ & + 4e^{-4\eta}(16\eta^3 + 36\eta^2 + 28\eta + 7)\sin 4\xi \end{aligned} \tag{8.81}$$

The expression (8.80) is the solution of the problem for a four-term approximant of $\phi(\xi, \eta)$ and can also be improved more by adding additional component functions of $\phi(\xi, \eta)$ to it.

References

1. Adomian, G.: *Solving Frontier Problems of Physics: The Decomposition Method*, Kluwer Academic Publishers (1994).
2. Haldar, K.: Nonperturbative Analytic Solution of Transonic Flow Past a Wavy Wall, *Bull. Cal. Math. Soc.* (in press).
3. Kaplan, C.: On a solution of the Nonlinear Differential Equation for Transonic Flow Past a Wave-shaped wall, NACA Report. 1069 (1952).

9

Laplace's Equation

9.1 Introduction

In applied mathematics, the most important partial differential equation is Laplace's equation. The partial differential equations that describe the heat flow in two- and three-dimensional spaces are given by

$$\frac{\partial T}{\partial t} = c^2 \left(\frac{\partial^2 T}{\partial x^2} + \frac{\partial^2 T}{\partial y^2} \right) \tag{9.1}$$

and

$$\frac{\partial T}{\partial t} = c^2 \left(\frac{\partial^2 T}{\partial x^2} + \frac{\partial^2 T}{\partial y^2} + \frac{\partial^2 T}{\partial z^2} \right) \tag{9.2}$$

respectively. Here T is the temperature, and c^2 is defined by

$$c^2 = \kappa/\sigma\rho \tag{9.3}$$

where κ is the thermal conductivity, σ is the specific heat, and ρ is the density of the material.

If the temperature T reaches a steady state, that is, independent of time t, then the time derivative of T vanishes, that is, $\frac{\partial T}{\partial t} = 0$, and the equations (9.1) and (9.2) reduce to Laplace's equations as

$$\frac{\partial^2 T}{\partial x^2} + \frac{\partial^2 T}{\partial y^2} = 0 \tag{9.4}$$

and

$$\frac{\partial^2 T}{\partial x^2} + \frac{\partial^2 T}{\partial y^2} + \frac{\partial^2 T}{\partial z^2} = 0 \tag{9.5}$$

Laplace's equation is applied (1) to describe the temperature in the steady state heat flow, (2) to describe the gravitational potential in the regions where attracting matter is absent, (3) to describe the electric potential in the theory of steady electric current flow in solid conductors, (4) to describe the magnetic potential in the theory of magnetostatics, and (5) to describe the velocity potential of moving homogeneous liquid irrotationally in the hydrodynamical problems. Laplace's equation is sometimes called a *potential equation*.

In this chapter, we study Laplace's equation in two-dimensional polar coordinates for a circular disc and a circular annulus. Haldar [1] has investigated the same problem by applying the modified decomposition method.

Now if (x, y) and (r, θ) are the Cartesian and polar coordinates of any point in space, then we have

$$x = r\cos\theta, \quad y = r\sin\theta \tag{9.6}$$

Using (9.6) in (9.4), we have Laplace's equation in two-dimensional polar form as

$$\frac{\partial^2 T}{\partial r^2} + \frac{1}{r}\frac{\partial T}{\partial r} + \frac{1}{r^2}\frac{\partial^2 T}{\partial \theta^2} = 0 \tag{9.7}$$

9.2 Solution of Laplace's Equation by Regular Decomposition

We begin with Laplace's equation in polar coordinates (9.7), that is,

$$\frac{\partial^2 T}{\partial r^2} + \frac{1}{r}\frac{\partial T}{\partial r} + \frac{1}{r^2}\frac{\partial^2 T}{\partial \theta^2} = 0$$

which, by virtue of the transformation $z = \log r$, takes the following form

$$\frac{\partial^2 T}{\partial z^2} + \frac{\partial^2 T}{\partial \theta^2} = 0 \tag{9.8}$$

Let $L_z = \frac{\partial^2}{\partial z^2}$ and $L_\theta = \frac{\partial^2}{\partial \theta^2}$. Then the equation (9.8) becomes

$$L_z T + L_\theta T = 0 \tag{9.9}$$

Now we solve (9.9) for $L_z T$ and then integrate twice with respect to z to write

$$T(z, \theta) = T_0(z, \theta) - L_z^{-1}(L_\theta T) \tag{9.10}$$

where $L_z^{-1} = \int\int(\cdot)dzdz$ and $T_0(z, \theta)$ is given by

$$T_0(z, \theta) = C_0(\theta) + zC_1(\theta) \tag{9.11}$$

$C_0(\theta)$ and $C_1(\theta)$ being integration constants to be determined from the boundary conditions of the problem.

Decomposing T into

$$T(z, \theta) = \sum_{m=0}^{\infty} \lambda^m T_m \tag{9.12}$$

and then using this relation in the parameterized form of (9.10) as

$$T(z, \theta) = T_0(z, \theta) - \lambda L_z^{-1}(L_\theta T) \tag{9.13}$$

we get, after comparing the like-power terms of λ, the following set of components T_1, T_2, T_3, etc.:

$$
\begin{aligned}
T_1(z,\theta) &= -L_z^{-1}(L_\theta T_0)\\
T_2(z,\theta) &= -L_z^{-1}(L_\theta T_1)\\
&\cdots \quad \cdots \quad \cdots\\
&\cdots \quad \cdots \quad \cdots\\
T_{m+1}(z,\theta) &= -L_z^{-1}(L_\theta T_m)\\
&\cdots \quad \cdots \quad \cdots\\
&\cdots \quad \cdots \quad \cdots
\end{aligned}
\tag{9.14}
$$

Summing up all the components given in (9.11) and (9.14) and then using this sum in the expansion of (9.12), we obtain the complete solution of the problem for $\lambda = 1$.

9.3 Solution of Laplace's Equation by Modified Decomposition

Consider again Laplace's equation in polar coordinates

$$\frac{\partial^2 T}{\partial r^2} + \frac{1}{r}\frac{\partial T}{\partial r} + \frac{1}{r^2}\frac{\partial^2 T}{\partial \theta^2} \tag{9.15}$$

If we use the transformation $z = \log r$, the equation (9.15) becomes

$$\frac{\partial^2 T}{\partial z^2} + \frac{\partial^2 T}{\partial \theta^2} = 0 \tag{9.16}$$

which takes the operator form as

$$L_z T + L_\theta T = 0 \tag{9.17}$$

where $L_z = \frac{\partial^2}{\partial z^2}$ and $L_\theta = \frac{\partial^2}{\partial \theta^2}$. Using the regular decomposition method, we write the z-partial solution of equation (9.17) in the form

$$T(z,\theta) = T_0(z,\theta) - L_z^{-1}(L_\theta T) \tag{9.18}$$

where

$$T(z,\theta) = B_0(\theta) + zB_1(\theta) \tag{9.19}$$

The constants involved in (9.19) are to be evaluated by the specified boundary conditions, and the operator L_z^{-1} is defined as the twofold indefinite integral $L_z^{-1} = \int\int(\cdot)dzdz$.

Now we apply the modified decomposition method to the regular decomposition solution (9.18) in order to make a variation in it. Thus, we let

$$T(z,\theta) = \sum_{m=0}^{\infty} a_m(\theta) z^m \tag{9.20}$$

Substituting (9.20) in (9.18), we have

$$\begin{aligned}
\sum_{m=0}^{\infty} a_m(\theta) &= B_0(\theta) + zB_1(\theta) - \iint (\partial^2/\partial\theta^2) \sum_{m=0}^{\infty} a_m(\theta) z^m dz dz \\
&= B_0(\theta) + zB_1(\theta) - \sum_{m=0}^{\infty} \frac{z^{m+2}}{(m+1)(m+2)} \frac{\partial^2}{\partial\theta^2} a_m(\theta) \\
&= B_0(\theta) + zB_1(\theta) - \sum_{m=2}^{\infty} \frac{(\partial^2/\partial\theta^2)\, a_{m-2}(\theta)}{m(m-1)} z^m
\end{aligned}$$

The coefficients are identified by

$$\left.\begin{aligned} a_0(\theta) &= B_0(\theta) \\ a_1(\theta) &= B_1(\theta) \end{aligned}\right\} \tag{9.21}$$

and for $m \geq 2$ the recurrence relation

$$a_m(\theta) = -\frac{\partial^2}{\partial\theta^2} a_{m-2}(\theta) \frac{1}{m(m-1)} \tag{9.22}$$

The relation (9.20) together with the relations (9.21) and (9.22) give the series solution of the problem.

The modified decomposition method discussed above is illustrated by considering some physical problems for a circular disc and also for a circular annulus quoted from Wazwaz [2].

9.4 Laplace's Equation for a Circular Disc

In this section, we study the Laplace's equation in polar coordinates (r, θ) for a circular disc of radius a. The upper and lower surfaces of the disc are insulated, and the boundary condition at the circumference of the disc is prescribed. When the steady state of heat flow inside the disc is reached, then the temperature distribution is governed by Laplace's equation in polar coordinates, and the phenomenon is expressed by the boundary value problem

$$\frac{\partial^2 T}{\partial r^2} + \frac{1}{r}\frac{\partial T}{\partial r} + \frac{1}{r^2}\frac{\partial^2 T}{\partial\theta^2} = 0 \quad 0 < r < a, 0 \leq \theta \leq 2\pi \tag{9.23}$$

$$T(a, \theta) = f(\theta) \tag{9.24}$$

Since the equation (9.23) is a second-order partial differential equation, its solution involves two arbitrary constants. To determine these constants, we need two conditions of which one is the boundary condition given by (9.24) and the other is the condition that states that the temperature T should be finite at the center of the disc.

9.4.1 Problems of Circular Disc

Problem 1. Use the modified decomposition method to solve the Dirichlet problem for a disc formulated by the mathematical model

$$\frac{\partial^2 T}{\partial r^2} + \frac{1}{r}\frac{\partial T}{\partial r} + \frac{1}{r^2}\frac{\partial^2 T}{\partial \theta^2} = 0, \quad 0 < r < 1, 0 \leq \theta \leq 2\pi \tag{9.25}$$

$$T(1, \theta) = \cos^2 \theta \tag{9.26}$$

Solution. To solve the problem, we use the transformation $z = \log r$. The partial differential equation (9.25) and the boundary condition (9.26) become

$$\frac{\partial^2 T}{\partial z^2} + \frac{\partial^2 T}{\partial \theta^2} = 0 \tag{9.27}$$

$$T(0, \theta) = \cos^2 \theta \tag{9.28}$$

The solution of the equation (9.27), as discussed in the Section 3, is

$$T(z, \theta) = T_0(z, \theta) + \sum_{m=2}^{\infty} a_m(\theta) z^m \tag{9.29}$$

where

$$T_0(z, \theta) = a_0(\theta) + z a_1(\theta) \tag{9.30}$$

and

$$a_m(\theta) = -\frac{\partial^2}{\partial \theta^2} a_{m-2}(\theta) \frac{1}{m(m-1)} \tag{9.31}$$

Satisfying the boundary condition (9.28) by the solution (9.29), we have

$$a_0(\theta) = \cos^2 \theta \tag{9.32}$$

Putting $m = 2, 3, 4$, etc., into the relation (9.31), we have

$$\begin{aligned} a_2 &= \frac{2\cos 2\theta}{2!}, a_3 = -\frac{a_1''}{3!} \\ a_4 &= \frac{8\cos 2\theta}{4!}, a_5 = \frac{a_1^{iv}}{5!} \\ a_6 &= \frac{32\cos 2\theta}{6!}, \text{ etc.} \end{aligned} \tag{9.33}$$

Combining (9.29), (9.30), and (9.33), we get

$$T(z,\theta) = \frac{1}{2} + \frac{\cos 2\theta}{2}\left[1 + \frac{(2z)^2}{2!} + \frac{(2z)^4}{4!} + \frac{(2z)^6}{6!} + \cdots\right] + \left[\frac{a_1}{2}\frac{(2z)}{1!} - \frac{a_1^{ii}}{8}\frac{(2z)^3}{3!} + \frac{a_1^{iv}}{32}\frac{(2z)^5}{5!} - \cdots\right]$$

or in polar coordinates

$$T(r,\theta) = \frac{1}{2} + \frac{\cos 2\theta}{2}\left[1 + \frac{(2\log r)^2}{2!} + \frac{(2\log r)^4}{4!} + \frac{(2\log r)^6}{6!} + \cdots\right] + \left[\frac{a_1}{2}\frac{(2\log r)}{1!} - \frac{a_1^{ii}}{8}\frac{(2\log r)^3}{3!} + \frac{a_1^{iv}}{32}\frac{(2\log r)^5}{5!} - \cdots\right] \tag{9.34}$$

The solution $T(r,\theta)$ should be bounded at $r = 0$. Therefore, the expression for $T(r,\theta)$ should not contain any term that involves the factor $\frac{1}{r}$ of any positive power. Looking at this view, we have to choose the function $a_1(\theta)$ properly so that the terms involving the factor $\frac{1}{r}$ should be vanished. Thus, we let

$$a_1(\theta) = k_1 \cos 2\theta + k_2 \tag{9.35}$$

where k_1 and k_2 are constants to be determined, and we have

$$-\frac{a_1^{ii}}{4} = \frac{a_1^{iv}}{8} = -\ldots = k_1 \cos 2\theta \tag{9.36}$$

Using the relations (9.35) and (9.36) in (9.34), we have

$$T(r,\theta) = \frac{1}{2} + \frac{\cos 2\theta}{4}\left(r^2 + \frac{1}{r^2}\right) + \frac{k_1 \cos 2\theta}{4}\left(r^2 - \frac{1}{r^2}\right) + k_2 \log r \tag{9.37}$$

But $T(r,\theta)$ should be finite at $r = 0$; for this reason we put $k_1 = 1$, $k_2 = 0$, and the solution (9.37) becomes

$$T(r,\theta) = \frac{1}{2} + \frac{r^2}{2}\cos 2\theta \tag{9.38}$$

which is a closed-form solution.

Problem 2. Use the modified decomposition method for the following Dirichlet problem for a circular disc:

$$\frac{\partial^2 T}{\partial r^2} + \frac{1}{r}\frac{\partial T}{\partial r} + \frac{1}{r^2}\frac{\partial^2 T}{\partial \theta^2} = 0, \quad 0 < r < 1, 0 \le \theta \le 2\pi \tag{9.39}$$

$$T(1,\theta) = 2\sin^2\theta + \sin\theta \tag{9.40}$$

Solution. Proceeding exactly in the same way as in Problem 1, we write the solution of the equation (9.39) as

$$T(z,\theta) = a_0(\theta) + za_1(\theta) + \sum_{m=2}^{\infty} a_m(\theta) z^m \tag{9.41}$$

where

$$a_m(\theta) = -\frac{\partial^2}{\partial\theta^2}\, a_{m-2}(\theta) \frac{1}{m(m-1)} \tag{9.42}$$

Satisfying the boundary condition (9.40) by (9.41), we get

$$a_0(\theta) = 2\sin^2\theta + \sin\theta \tag{9.43}$$

Now using the relation (9.42), we have the following set of expressions for the coefficients $a_2(\theta)$, $a_3(\theta)$, $a_4(\theta)$, etc.:

$$\left.\begin{aligned} a_2(\theta) &= -\frac{1}{2!}(4\cos 2\theta - \sin\theta) \\ a_3(\theta) &= -\frac{a_1^{ii}}{3!} \\ a_4(\theta) &= -\frac{1}{4!}(16\cos 2\theta - \sin\theta) \\ a_5(\theta) &= \frac{a_1^{iv}}{5!} \\ a_6(\theta) &= -\frac{1}{6!}(64\cos 2\theta - \sin\theta), \text{ etc.} \end{aligned}\right\} \tag{9.44}$$

Substituting (9.44) in the expansion of the solution (9.41), we obtain after straightforward simplifications

$$\begin{aligned} T(z,\theta) = 1 - \cos 2\theta &\left[1 + \frac{(2z)^2}{2!} + \frac{(2z)^4}{4!} + \frac{(2z)^6}{6!} + \cdots\right] \\ &+ \sin\theta \left[1 + \frac{z^2}{2!} + \frac{z^4}{4!} + \frac{z^6}{6!} + \cdots\right] \\ &+ \left[a_1 z - a_1^{ii}\frac{z^3}{3!} + a_1^{iv}\frac{z^5}{5!} - \cdots\right] \end{aligned} \tag{9.45}$$

or in polar coordinates

$$\begin{aligned} T(r,\theta) = 1 - \cos 2\theta &\left[1 + \frac{(2\log r)^2}{2!} + \frac{(2\log r)^4}{4!} + \frac{(2\log r)^6}{6!} + \cdots\right] \\ &+ \sin\theta \left[1 + \frac{(\log r)^2}{2!} + \frac{(\log r)^4}{4!} + \frac{(\log r)^6}{6!} + \cdots\right] \\ &+ \left[a_1(\log r) - a_1^{ii}\frac{(\log r)^3}{3!} + a_1^{iv}\frac{(\log r)^5}{5!} - \cdots\right] \end{aligned} \tag{9.46}$$

Now we have to choose a function $a_1(\theta)$ looking at the solution (9.46), and we set this function in the form given by

$$a_1(\theta) = k_1 \cos 2\theta + k_2 \sin\theta + k_3 \tag{9.47}$$

where k_1, k_2, and k_3 are the constants to be prescribed such that the solution (9.46) should be bounded at $r = 0$. Differentiating (9.47) with respect to θ, we have the following expressions:

$$\begin{aligned} a_1^{ii} &= -(4k_1 \cos 2\theta + k_2 \sin\theta) \\ a_1^{iv} &= \quad (16k_1 \cos 2\theta + k_2 \sin\theta) \\ a_1^{vi} &= -(64k_1 \cos 2\theta + k_2 \sin\theta), \text{ etc.} \end{aligned} \tag{9.48}$$

Substituting (9.47) and (9.48) in (9.46), we get

$$\begin{aligned} T(r,\theta) = 1 &- \cos 2\theta \left[1 + \frac{(2\log r)^2}{2!} + \frac{(2\log r)^4}{4!} + \frac{(2\log r)^6}{6!} + \cdots\right] \\ &+ \sin\theta \left[1 + \frac{(\log r)^2}{2!} + \frac{(\log r)^4}{4!} + \frac{(\log r)^6}{6!} + \cdots\right] \\ &+ \frac{k_1}{2} \cos 2\theta \left[\frac{(2\log r)}{1!} + \frac{(2\log r)^3}{3!} + \frac{(2\log r)^5}{5!} + \cdots\right] \\ &+ k_2 \sin\theta \left[\frac{(\log r)}{1!} + \frac{(\log r)^3}{3!} + \frac{(\log r)^5}{5!} + \cdots\right] \\ &+ k_3 \log r \end{aligned} \tag{9.49}$$

Now we have to prescribe the values of k_1, k_2, and k_3 so that the solution (9.49) is finite at $r = 0$. To do this, we look at the nature of the solution and prescribe the values of these constants as

$$k_1 = -2, \quad k_2 = 1, \quad k_3 = 0 \tag{9.50}$$

Then the relation (9.47) takes the form

$$a_1(\theta) = \sin\theta - 2\cos 2\theta \tag{9.51}$$

and the expression (9.49), being bounded at $r = 0$, becomes

$$T(r,\theta) = 1 - r^2 \cos 2\theta + r \sin\theta \tag{9.52}$$

which is the exact solution of the problem.

Problem 3. Use the modified decomposition method to solve the Dirichlet problem

$$\frac{\partial^2 T}{\partial r^2} + \frac{1}{r}\frac{\partial T}{\partial r} + \frac{1}{r^2}\frac{\partial^2 T}{\partial \theta^2} = 0, \quad 0 < r < 1, \quad 0 \le \theta \le 2\pi \tag{9.53}$$

$$\frac{\partial T}{\partial r}(1, \theta) = 2\cos 2\theta \tag{9.54}$$

for a circular disc.

Solution. Proceeding exactly in the same way as in Problem 1, we write the solution of the equation (9.53) as

$$T(z, \theta) = a_0(\theta) + z a_1(\theta) + \sum_{m=2}^{\infty} a_m(\theta) z^m \tag{9.55}$$

where

$$\left.\begin{aligned} a_m(\theta) &= -\frac{\partial^2}{\partial \theta^2} a_{m-2}(\theta) \frac{1}{m(m-1)} \\ z &= \log r \end{aligned}\right\} \tag{9.56}$$

Using the second relation of (9.56), the boundary condition (9.54) becomes

$$\frac{\partial T(0, \theta)}{\partial z} = 2\cos^2 \theta \tag{9.57}$$

Then satisfying this boundary condition by (9.55), we find $a_1(\theta)$ to be

$$a_1(\theta) = 2\cos 2\theta \tag{9.58}$$

From the first relation of (9.56), we get the following expressions for the coefficients:

$$\begin{aligned} a_2(\theta) &= -\frac{a_0^{ii}}{2!} \\ a_3(\theta) &= \frac{8\cos 2\theta}{3!} \\ a_4(\theta) &= \frac{a_0^{iv}}{4!} \\ a_5(\theta) &= \frac{32\cos 2\theta}{5!}, \text{ etc.} \end{aligned} \tag{9.59}$$

Combining (9.55), (9.58), and (9.59), we obtain

$$T(z, \theta) = \left[a_0 - a_0^{ii}\frac{z^2}{2!} + a_0^{iv}\frac{z^4}{4!} - \cdots\right] + \cos 2\theta \left[\frac{(2z)}{1!} + \frac{(2z)^3}{3!} + \frac{(2z)^5}{5!} + \cdots\right] \tag{9.60}$$

Now our task is to choose the function $a_0(\theta)$ so that the solution (9.60) is finite at $r = 0$. Thus, we set the function $a_0(\theta)$ as

$$a_0(\theta) = k_1 \cos 2\theta + k_2 \tag{9.61}$$

where k_1 and k_2 are the constants of which k_1 is to be prescribed.

Using the relation (9.61) in (9.60), we obtain

$$T(z,\theta) = k_2 + k_1 \cos 2\theta \left[1 + \frac{(2z)^2}{2!} + \frac{(2z)^4}{4!} + \cdots\right]$$
$$+ \cos 2\theta \left[\frac{(2z)}{1!} + \frac{(2z)^3}{3!} + \frac{(2z)^5}{5!} + \cdots\right] \tag{9.62}$$

Since $T(z, \theta)$ is finite at the center of the disc, we prescribe $k_1 = 1$, and the solution (9.62) takes the form

$$T(z,\theta) = k_2 + \cos 2\theta \left[1 + \frac{(2z)}{1!} + \frac{(2z)^2}{2!} + \cdots\right] = k_2 + e^{2z}\cos 2\theta$$

or in polar coordinates

$$T(r,\theta) = k_2 + r^2 \cos 2\theta \tag{9.63}$$

which is a closed-form solution of the problem.

9.5 Laplace's Equation for a Circular Annulus

This section deals with the discussion of Laplace's equation for an annular region bounded by the two circles. The top and bottom faces of a circular annulus are insulated. The outer and inner radii of two circles are a and b, respectively. The boundary conditions at the two circular edges of the circular annulus are specified. The temperature distribution, which reaches a steady state in the annular region, is governed by Laplace's equation in polar coordinates (r, θ) and is expressed by the mathematical model

$$\frac{\partial^2 T}{\partial r^2} + \frac{1}{r}\frac{\partial T}{\partial r} + \frac{1}{r^2}\frac{\partial^2 T}{\partial \theta^2} = 0, \quad b < r < a,\ 0 \le \theta \le 2\pi \tag{9.64}$$

$$T(a,\theta) = f_1(\theta), \quad T(b,\theta) = f_2(\theta) \tag{9.65}$$

9.5.1 Problems of Circular Annulus

Problem 4. Use the modified decomposition method to solve the problem for an annulus.

$$\frac{\partial^2 T}{\partial r^2} + \frac{1}{r}\frac{\partial T}{\partial r} + \frac{1}{r^2}\frac{\partial^2 T}{\partial \theta^2} = 0, \quad 1 < r < 2, 0 \le \theta \le 2\pi \tag{9.66}$$

$$T(1,\theta) = \frac{1}{2} + \sin\theta \tag{9.67}$$

$$T(2,\theta) = \frac{1}{2} + \frac{1}{2}\log 2 + \cos\theta \tag{9.68}$$

Solution. To solve the partial differential equation (9.66), we use the transformation $z = \log r$, and the above mathematical model takes the form

$$\frac{\partial^2 T}{\partial z^2} + \frac{\partial^2 T}{\partial \theta^2} = 0 \tag{9.69}$$

$$T(0,\theta) = \frac{1}{2} + \sin\theta \tag{9.70}$$

$$T(\log 2,\theta) = \frac{1}{2} + \frac{1}{2}\log 2 + \cos\theta \tag{9.71}$$

Using the modified decomposition method, we write the solution of the equation (9.69) in the form

$$T(z,\theta) = a_0(\theta) + z a_1(\theta) + \sum_{m=2}^{\infty} a_m(\theta) z^m \tag{9.72}$$

where

$$a_m(\theta) = -\frac{\partial^2}{\partial\theta^2} a_{m-2}(\theta) \frac{1}{m(m-1)} \tag{9.73}$$

Satisfying the boundary condition (9.70) by (9.72), we obtain

$$a_0(\theta) = \frac{1}{2} + \sin\theta \tag{9.74}$$

Using the relation (9.74), we have from (9.73)

$$\begin{aligned}
a_2(\theta) &= \frac{\sin\theta}{2!} \\
a_3(\theta) &= -\frac{a_1^{ii}}{3!} \\
a_4(\theta) &= \frac{\sin\theta}{4!} \\
a_5(\theta) &= \frac{a_1^{iv}}{5!} \\
a_6(\theta) &= \frac{\sin\theta}{6!} \text{ etc.}
\end{aligned} \tag{9.75}$$

Combining (9.72) and (9.75), we get

$$\begin{aligned}
T(z,\theta) = \frac{1}{2} &+ \sin\theta\left[1 + \frac{z^2}{2!} + \frac{z^4}{4!} + \frac{z^6}{6!} + \cdots\right] \\
&+ \left[a_1 z - a_1^{ii}\frac{z^3}{3!} + a_1^{iv}\frac{z^5}{5!} - \cdots\right]
\end{aligned} \tag{9.76}$$

Then we satisfy the boundary condition (9.71) by (9.76) and obtain

$$\frac{1}{2}\log 2+\cos\theta-\frac{5}{4}\sin\theta=a_1(\log 2)-a_1^{ii}\frac{(\log 2)^3}{3!}+a^{iv}\frac{(\log 2)^5}{5!}-\cdots \tag{9.77}$$

Now we have to find out the function $a_1(\theta)$ so that the relation (9.77) should be valid. To do this, we look at the relation and see that this relation contains sine and cosine functions. Therefore, we set $a_1(\theta)$ in the form given by

$$a_1(\theta)=k_1\sin\theta+k_2\cos\theta+k_3 \tag{9.78}$$

where the constants k_1, k_2, and k_3 are to be evaluated. Substituting (9.78) in (9.77), we get

$$\frac{1}{2}\log 2+\cos\theta-\frac{5}{4}\sin\theta=\frac{3}{4}(k_1\sin\theta+k_2\cos)+k_3\log 2$$

which, on comparison, gives

$$k_1=-\frac{5}{3},\quad k_2=\frac{4}{3},\quad k_3=\frac{1}{2} \tag{9.79}$$

and the relation (9.78) becomes

$$a_1(\theta)=\frac{4}{3}\cos\theta-\frac{5}{3}\sin\theta+\frac{1}{2} \tag{9.80}$$

The use of (9.80) in (9.76) gives the expression for $T(z,\theta)$ as

$$T(z,\theta)=\frac{1}{2}+\frac{1}{2}(e^z+e^{-z})\sin\theta+\frac{1}{6}(e^z-e^{-z})(4\cos\theta-5\sin\theta)+\frac{1}{2}z \tag{9.81}$$

or in polar form

$$T(r,\theta)=\frac{1}{2}+\frac{1}{2}\log r+\frac{1}{3}\left(\frac{4}{r}-r\right)\sin\theta+\frac{2}{3}\left(r-\frac{1}{r}\right)\cos\theta \tag{9.82}$$

which is the exact solution of the problem.

Problem 5. Use the modified decomposition method to solve the following Newmann problem for an annulus:

$$\frac{\partial^2 T}{\partial r^2}+\frac{1}{r}\frac{\partial T}{\partial r}+\frac{1}{r^2}\frac{\partial^2 T}{\partial\theta^2}=0,\quad 1<r<2,\quad 0\le\theta\le 2\pi \tag{9.83}$$

$$\frac{\partial T(1,\theta)}{\partial r}=1 \tag{9.84}$$

$$\frac{\partial T(2,\theta)}{\partial r}=\frac{1}{2}+\frac{3}{4}\cos\theta \tag{9.85}$$

Solution. Let us introduce the transformation $z = \log r$. Using this transformation, we convert the mathematical model described by (9.83), (9.84), and (9.85) into the following model:

$$\frac{\partial^2 T}{\partial z^2} + \frac{\partial^2 T}{\partial \theta^2} = 0, \quad 1 < r < 2, \quad 0 \le \theta \le 2\pi \tag{9.86}$$

$$\frac{\partial T(0,\theta)}{\partial z} = 1 \tag{9.87}$$

$$\frac{\partial T(\log 2,\theta)}{\partial z} = 1 + \frac{3}{2}\cos\theta \tag{9.88}$$

Now we apply the modified decomposition method to the equation (9.86) and obtain its solution in the form

$$T(z,\theta) = a_0(\theta) + z a_1(\theta) + \sum_{m=2}^{\infty} a_m(\theta) z^m \tag{9.89}$$

where

$$a_m(\theta) = -\frac{\partial^2}{\partial \theta^2} a_{m-2}(\theta) \frac{1}{m(m-1)} \tag{9.90}$$

Then satisfying the boundary condition (9.87) by the solution (9.89), we have

$$a_1(\theta) = 1 \tag{9.91}$$

Using the relation (9.91), we have from (9.90)

$$a_3 = a_5 = a_7 = \cdots\cdots = 0$$

$$a_2(\theta) = -\frac{a_0^{ii}}{2!},\ a_4 = \frac{a_0^{iv}}{4!}, \ \text{etc.} \tag{9.92}$$

Substituting (9.92) in (9.89), we obtain

$$T(z,\theta) = a_0 + z - a_0^{ii}\frac{z^2}{2!} + a_0^{iv}\frac{z^4}{4!} - a_0^{vi}\frac{z^6}{6!} + \cdots \tag{9.93}$$

Satisfying the boundary condition (9.88) by the solution (9.93), we have

$$1 + \frac{3}{2}\cos\theta = 1 - a_0^{ii}\frac{\log 2}{1!} + a_0^{iv}\frac{(\log 2)^3}{3!} - a_0^{vi}\frac{(\log 2)^5}{5!} + \cdots \tag{9.94}$$

Now we have to choose the function $a_0(\theta)$ to complete the solution of the problem. To do that, we assume $a_0(\theta)$ in the form

$$a_0(\theta) = k_1\cos\theta + k_2 \tag{9.95}$$

where k_1 and k_2 are constants of which k_1 is to be determined. Then putting (9.95) in (9.94), we get

$$1 + \frac{3}{2}\cos\theta = 1 + \frac{3k_1}{4}\cos\theta \tag{9.96}$$

whence we obtain $k_1 = 2$, and the function (9.95) becomes

$$a_0(\theta) = 2\cos\theta + k_2 \tag{9.97}$$

Substituting (9.96) in (9.93), we have the resulting expression for $T(z, \theta)$, and it is given by

$$T(z, \theta) = k_2 + (e^z + e^{-z})\cos\theta + z \tag{9.98}$$

or in polar coordinates

$$T(r, \theta) = k_2 + \log r + \left(r + \frac{1}{r}\right)\cos\theta \tag{9.99}$$

which is a closed-form solution of the problem.

The problems discussed above by means of modified decomposition can also be solved with the help of the ordinary or regular decomposition method, and the readers are advised to solve these problems with the help of the latter method.

The decomposition method is a very powerful method in the modern age. This method can also be applied to any type of differential equation and can provide an approximate series solution that demands to be parallel to any modern supercomputer.

References

1. Haldar, K.: Application of Modified Decomposition Method to Laplace's Equation in Two-Dimensional Polar Coordinates, *Bull. Cal. Math. Soc.*, *98*(6), 571–582 (2006).
2. Wazwaz, A. B.: *Partial Differential Equations: Methods and Applications*, A. A. Bakema (2002).

10

Flow Near a Rotating Disc in a Fluid at Rest

10.1 Introduction

The fundamental problems that exist in different branches of sciences are all nonlinear in character. These problems are described mathematically by the differential equations subject to certain initial and boundary conditions. Out of these problems, the fluid dynamical problems are described by the Navier–Stokes equations, and it is very difficult to find out the solutions of these equations because of the insurmountable mathematical troubles. The most powerful method, known as the decomposition method, can only avoid these difficulties. In this chapter, we adopt this method to solve the problems of fluid flow exactly near a rotating disc in a fluid at rest.

10.2 Equations of Motion

Consider the flow of viscous fluid around a circular disk that rotates with uniform angular velocity ω in a fluid at rest. This is an example of an exact solution of Navier–Stokes equations. The layer adjacent to the disk is carried by it through friction and is then thrown outward because of the action of centrifugal force. The vacuum space is compensated by the fluid that flows from above in the axial direction perpendicular to the disk. Then this fluid is carried by the disk in turn and is again thrown outward centrifugally. This is a case of fully developed three-dimensional flow. Considering the rotational symmetry, we write the dimensional equations of a motion continuity equation in cylindrical polar coordinates for the steady flow as

$$u\frac{\partial u}{\partial r} - \frac{v^2}{r} + w\frac{\partial u}{\partial z} = -\frac{1}{\rho}\frac{\partial p}{\partial r} + \vartheta\left(\frac{\partial^2 u}{\partial r^2} + \frac{1}{r}\frac{\partial u}{\partial r} - \frac{u}{r^2} + \frac{\partial^2 u}{\partial z^2}\right) \tag{10.1}$$

$$u\frac{\partial v}{\partial r} + \frac{uv}{r} + w\frac{\partial v}{\partial z} = \vartheta\left(\frac{\partial^2 v}{\partial r^2} + \frac{1}{r}\frac{\partial v}{\partial r} - \frac{v}{r^2} + \frac{\partial^2 v}{\partial z^2}\right) \tag{10.2}$$

$$u\frac{\partial w}{\partial r} + w\frac{\partial w}{\partial z} = -\frac{1}{\rho}\frac{\partial p}{\partial z} + \vartheta\left(\frac{\partial^2 w}{\partial r^2} + \frac{1}{r}\frac{\partial w}{\partial r} + \frac{\partial^2 w}{\partial z^2}\right) \tag{10.3}$$

$$\frac{\partial u}{\partial r} + \frac{u}{r} + \frac{\partial w}{\partial z} = 0 \tag{10.3a}$$

where u, v, and w are the dimensional velocity components in the radial direction r, circumferential direction θ^*, and axial direction z, respectively; p is the pressure; ρ is the density of the fluid; and ϑ is the kinematic viscosity of fluid.

The boundary conditions are

$$u = 0, v = \omega r, w = 0 \quad \text{at } z = 0 \tag{10.4}$$

$$u = 0, v = 0 \quad \text{at } z = \alpha \tag{10.5}$$

Before we solve the system of equations ranging from (10.1) to (10.5), it is convenient to use the following transformations:

$$\left.\begin{aligned} &u = r\omega F(\eta), v = r\omega G(\eta), w = \sqrt{\vartheta\omega}H(\eta) \\ &p = p(z) = \rho\vartheta\omega P(\eta), \eta = z\sqrt{\tfrac{\omega}{\vartheta}} \end{aligned}\right\} \tag{10.6}$$

Introducing (10.6) in the system of equations, we have the transformed equations as

$$\frac{dH}{d\eta} + 2F = 0 \tag{10.7}$$

$$\frac{d^2F}{d\eta^2} = H\frac{dF}{d\eta} + F^2 - G^2 \tag{10.8}$$

$$\frac{d^2G}{d\eta^2} = H\frac{dG}{d\eta} + 2FG \tag{10.9}$$

The boundary conditions (10.4) and (10.5) are transferred into nondimensional forms as

$$F = H = 0, G = 1 \quad \text{at } \eta = 0 \tag{10.10}$$

$$F = G = 0 \quad \text{at } \eta = \alpha \tag{10.11}$$

where F, G, and H are the velocity components in the radial, circumferential, and axial directions, respectively.

Multiplying the equation (10.9) by i and then subtracting from the equation (10.8), we have an ordinary differential equation for the complex function E as

$$\frac{d^2E}{d\eta^2} = H\frac{dE}{d\eta} + E^2 \tag{10.12}$$

where

$$E = F - iG \tag{10.13}$$

The boundary conditions imposed on E are

$$E = -i \quad \text{at } \eta = 0 \tag{10.14}$$

$$E = 0 \quad \text{at } \eta = \alpha \tag{10.15}$$

The equations (10.7) and (10.12) together with their respective boundary conditions constitute the mathematical model of the problem. Now we proceed to find out the solutions of the problem in the next sections for small and large values of argument.

10.3 Solutions for the Small Value of η

Let $L_1 = \frac{d}{d\eta}$ and $L_2 = \frac{d^2}{d\eta^2}$ be the two first- and second-order ordinary differential operators whose inverse forms are

$$L_1^{-1} = \int (.)d\eta \tag{10.16}$$

$$L_2^{-1} = \iint (.)d\eta d\eta \tag{10.17}$$

Then we make use of operations of the inverse operators on the equation (10.7) after solving it for $dH/d\eta$ and on the equation (10.12). And we get

$$H(\eta) = H_o(\eta) - 2L_1^{-1}F \tag{10.18}$$

$$E(\eta) = H_o(\eta) + L_2^{-1}\left[H\frac{dE}{d\eta} + E^2\right] \tag{10.19}$$

Here $H_o(\eta)$and$E_o(\eta)$ are solutions of the homogenous equations

$$\frac{dF}{d\eta} = 0$$

$$\frac{d^2E}{d\eta^2} = 0 \tag{10.20}$$

Solving the equations given in (10.20) with the help of the respective boundary conditions (10.10) and (10.14), we have

$$H_o(\eta) = 0 \tag{10.21}$$

$$E_o(\eta) = A_1\eta - i \tag{10.22}$$

where A_1 is a complete constant and is given by

$$A_1 = \alpha - i\beta \tag{10.23}$$

Using (10.13) and (10.23) and then equating the real and imaginary parts of the resulting equation, we have the expressions for $F_o(\eta)$ and $G_o(\eta)$ as

$$\left.\begin{aligned} F_o(\eta) &= \alpha\eta \\ G_o(\eta) &= 1 + \beta\eta \end{aligned}\right\} \tag{10.24}$$

where the constants α and β involved in (10.24) are to be evaluated later on.

We now decompose the functions $H(\eta)$, $E(\eta)$, $F(\eta)$, and $G(\eta)$ in the forms given by

$$H(\eta) = \sum_{n=0}^{\propto} \lambda_1^n H_n \tag{10.25}$$

$$E(\eta) = \sum_{n=0}^{\propto} \lambda_1^n E_n \tag{10.26}$$

$$F(\eta) = \sum_{n=0}^{\propto} \lambda_1^n F_n \tag{10.27}$$

$$G(\eta) = \sum_{n=0}^{\propto} \lambda_1^n G_n \tag{10.28}$$

Then we write the parameterized forms of (10.18) and (10.19) as

$$H(\eta) = H_o(\eta) - 2\lambda_1 L_1^{-1} F \tag{10.29}$$

$$E(\eta) = E_o(\eta) + \lambda_1 L_2^{-1} \left[H\frac{dE}{d\eta} + E^2 \right] \tag{10.30}$$

Here λ_1 is not a perturbation parameter; it is used only for grouping the terms of different orders.

If we substitute the relations ranging from (10.25) to (10.27) in (10.29) and (10.30) and then compare the like-power terms of λ_1, we get

Order λ_1:

$$\left. \begin{aligned} H_1(\eta) &= -2L_1^{-1} F_0 \\ E_1(\eta) &= L_2^{-1} \left[H_0 \tfrac{dE_0}{d\eta} + E_0^2 \right] \end{aligned} \right\} \tag{10.31}$$

Order λ_1^2:

$$H_2(\eta) = -2L_1^{-1} F_1$$

$$E_2(\eta) = L_2^{-1} \left[H_0\frac{dE_1}{d\eta} + H_1\frac{dE_0}{d\eta} + 2E_0 E_1 \right] \tag{10.32}$$

Order λ_1^3:

$$H_3(\eta) = -2L_1^{-1} F_2$$

$$E_3(\eta) = L_2^{-1} \left[H_0\frac{dE_2}{d\eta} + H_1\frac{dE_1}{d\eta} + H_2\frac{dE_0}{d\eta} + E_1^2 + 2E_0 E_2 \right] \tag{10.33}$$

... ...

... ...

Again, using (10.26), (10.27), and (10.28) in (10.13) and then comparing the terms, we have

$$
\begin{aligned}
E_0 &= F_0 - iG_0 \\
E_1 &= F_1 - iG_1 \\
E_2 &= F_2 - iG_2 \\
\cdots \quad &\cdots \quad \cdots \quad \cdots \\
\cdots \quad &\cdots \quad \cdots \quad \cdots
\end{aligned} \tag{10.34}
$$

Since F_o, E_o, andH_o are known, therefore, we have from (10.31)

$$H_1(\eta) = -\alpha\eta^2 \tag{10.35}$$

and

$$E_1(\eta) = \frac{2A_1^2}{4!}\eta^4 - \frac{2A_1 i}{3!}\eta^3 - \frac{1}{2}\eta^2 \tag{10.36}$$

Using (10.23) and the second relation of (10.34) in (10.36) and then equating the real and imaginary parts, we get the expressions of $F_1(\eta)$ and $G_1(\eta)$ as

$$F_1(\eta) = \frac{(\alpha^2 - \beta^2)}{12}\eta^4 - \frac{\beta}{3}\eta^3 - \frac{1}{2}\eta^2 \tag{10.37}$$

$$G_1(\eta) = \frac{\alpha\beta}{6}\eta^4 + \frac{\alpha}{3}\eta^3 \tag{10.38}$$

Again, we can have the expressions for $H_2(\eta)$ and $E_2(\eta)$ from (10.32) as the functions involved in the right-hand side of it are known. These expressions are given by

$$H_2(\eta) = \frac{2}{3!}\eta^3 + \frac{4\beta}{4!}\eta^4 - \frac{4\left(\alpha^2 - \beta^2\right)}{5!}\eta^5 \tag{10.39}$$

$$
\begin{aligned}
E_2(\eta) = &-\frac{\alpha^2}{12}\eta^4 + 2\left[\frac{10\left(\alpha^3 - 3\alpha\beta^2\right)}{7!}\eta^7 - \frac{20\alpha\beta}{6!}\eta^6 - \frac{5\alpha}{5!}\eta^5\right] \\
&+ i\left[\frac{\alpha\beta}{12}\eta^4 - 2\left\{\frac{10\left(3\alpha^2\beta - \beta^3\right)}{7!}\eta^7 + \frac{10\left(\alpha^2 - \beta^2\right)}{6!}\eta^6 - \frac{5\beta}{5!}\eta^5 - \frac{1}{4}\eta^4\right\}\right]
\end{aligned} \tag{10.40}
$$

Using the third relation of (10.34) in (10.40) and then equating the real and imaginary parts of resulting complex relations, we get

$$F_2(\eta) = -\frac{\alpha^2}{12}\eta^4 + 2\left[\frac{10\left(\alpha^3 - 3\alpha\beta^2\right)}{7!}\eta^7 - \frac{20\alpha\beta}{6!}\eta^6 - \frac{5\alpha}{5!}\eta^5\right] \tag{10.41}$$

$$G_2(\eta) = -\frac{\alpha^2}{12}\eta^4 + 2\left[\frac{10\left(3\alpha^2\beta - \beta^3\right)}{7!}\eta^7 + \frac{10\left(\alpha^2 - \beta^2\right)}{6!}\eta^6 - \frac{5\beta}{5!}\eta^5 - \frac{1}{4}\eta^4\right] \tag{10.42}$$

Proceeding in the same way as above, we can compute the other components of the functions $H(\eta)$, $F(\eta)$, and $G(\eta)$. If we consider the four-term approximant of the solutions, then we have

$$H(\eta) = H_0(\eta) + H_1(\eta) + H_2(\eta) + H_3(\eta) \tag{10.43}$$

$$F(\eta) = F_0(\eta) + F_1(\eta) + F_2(\eta) + F_3(\eta) \tag{10.44}$$

$$G(\eta) = G_0(\eta) + G_1(\eta) + G_2(\eta) + G_3(\eta) \tag{10.45}$$

remembering that $\lambda_1 = 1$, the functions $H_3(\eta)$, $F_3(\eta)$, and $G_3(\eta)$ are given by

$$H_3(\eta) = \frac{2\alpha^2}{5!}\eta^5 - 4\left[\frac{10\alpha\left(\alpha^2 - 3\beta^2\right)}{8!}\eta^8 - \frac{20\alpha\beta}{7!}\eta^7 - \frac{5\alpha}{6!}\eta^6\right] \tag{10.46}$$

$$\begin{aligned} F_3(\eta) = {} & \frac{600\left(\alpha^4 - 6\alpha^2\beta^2 + \beta^4\right)}{10!}\eta^{10} + \frac{600\beta\left(\beta^2 - 3\alpha^2\right)}{9!}\eta^9 - \frac{300\left(\alpha^2 - \beta^2\right)}{8!}\eta^8 \\ & + \frac{16\left[5\beta - 4\alpha\left(\alpha^2 - \beta^2\right)\right]}{7!}\eta^7 + \frac{(10 + 32\alpha\beta)}{6!}\eta^6 + \frac{8\alpha}{5!}\eta^5 \end{aligned} \tag{10.47}$$

$$\begin{aligned} G_3(\eta) = {} & \frac{2400\alpha\beta\left(\alpha^2 - \beta^2\right)}{10!}\eta^{10} + \frac{600\left(\alpha^2 - 3\beta^2\right)}{9!}\eta^9 - \frac{600\alpha\beta}{8!}\eta^8 \\ & - \frac{4\left[20\alpha + \beta\left(31\alpha^2 - \beta^2\right)\right]}{7!}\eta^7 - \frac{28\alpha^2 - 4\beta^2}{6!}\eta^6 + \frac{2\beta}{5!}\eta^5 \end{aligned} \tag{10.48}$$

If we want to improve the results more, then we can add additional terms to the expressions ranging from (10.43) to (10.45).

10.4 Solutions for the Large Value of η

For large values of η, we consider again the equations (10.7) and (10.12), that is,

$$\frac{dH}{d\eta} + 2F = 0 \tag{10.49}$$

$$\frac{d^2E}{d\eta^2} = H\frac{dE}{d\eta} + E^2 \tag{10.50}$$

together with the boundary condition at infinity, that is,

$$E = F = 0 \quad \text{at } \eta = \alpha \tag{10.51}$$

Then by means of the boundary conditions, the equation (10.50) is approximated to the equation

$$\frac{d^2E}{d\eta^2} = H_\alpha \frac{dE}{d\eta} \tag{10.52}$$

where H_α is the axial velocity at infinity. On integration we have from (10.52)

$$\frac{dE}{d\eta} = C_1 e^{H_\alpha \eta} \tag{10.53}$$

which can be integrated again to give

$$E(\eta) = \frac{C_1}{H_\alpha} e^{H_\alpha \eta} + D_1 \tag{10.54}$$

For consistency with the boundary condition (10.51), we put $H_\alpha = -k_1$ and hence, $D_1 = 0$, therefore, the relation (10.54) takes the form

$$E(\eta) = -\frac{C_1}{k_1} e^{-k_1 \eta} \tag{10.55}$$

which shows that $E(\eta)$, that is, $F(\eta)$ and $G(\eta)$, are proportional to $\exp(-k_1\eta)$. So it is reasonable to conclude that each term in the series solutions of $H(\eta)$, $F(\eta)$, and $G(\eta)$ should contain an exponential function with a negative argument.

We now decompose $H(\eta)$, $F(\eta)$, and $E(\eta)$ in the following forms:

$$H(\eta) = \sum_{n=0}^{\alpha} \lambda_1^n H_n \tag{10.56}$$

$$F(\eta) = \sum_{n=0}^{\alpha} \lambda_1^n F_n \tag{10.57}$$

$$E(\eta) = \sum_{n=0}^{\alpha} \lambda_1^n E_n \tag{10.58}$$

Substituting (10.56), (10.57), and (10.58) in the equations (10.49) and (10.50) and then equating the coefficients of different powers of λ_1 to zero, we have the following pairs of equations for different orders:

Order λ_1^0:

$$\frac{dH_0}{d\eta} = 0 \tag{10.59}$$

Order λ_1^1:

$$\left.\begin{aligned} \frac{dH_1}{d\eta} + 2F_1 = 0 \\ \frac{d^2E_1}{d\eta^2} = H_0 \frac{dE_1}{d\eta} \end{aligned}\right\} \tag{10.60}$$

Order λ_1^2:

$$\left.\begin{aligned}&\frac{dH_2}{d\eta}+2F_2=0\\&\frac{d^2E_2}{d\eta^2}=H_0\frac{dE_2}{d\eta}+H_1\frac{dE_1}{d\eta}+E_1^2\end{aligned}\right\}\tag{10.61}$$

Order λ_1^3:

$$\frac{d^2H_3}{d\eta^2}+2F_3=0$$

$$\frac{d^2E_3}{d\eta^2}=H_0\frac{dE_3}{d\eta}+H_1\frac{dE_2}{d\eta}+H_2\frac{dE_1}{d\eta}+2E_1E_2$$

$$\cdots\quad\cdots\quad\cdots\quad\cdots\quad\cdots\quad\cdots\quad\cdots\quad\cdots\quad\cdots\quad\cdots$$

$$\cdots\quad\cdots\quad\cdots\quad\cdots\quad\cdots\quad\cdots\quad\cdots\quad\cdots\quad\cdots\quad\cdots\tag{10.62}$$

From (10.59) we have, on integration,

$$H_0(\eta)=constant=H_\alpha=-k_1\tag{10.63}$$

Then we solved the second relation of (10.60) using the boundary condition (10.51), that is, $E_1=0$ at $\eta=\alpha$, and we get

$$E_1(\eta)=\bar{A}e^{-k_1\eta}\tag{10.64}$$

where $\bar{A}=\alpha_1-i\beta_1$ is a complex constant. Since $E_1=F_1-iG_1$, therefore, equating the real and imaginary parts of (10.64), we have the expressions of $F_1(\eta)$ and $G_1(\eta)$ as

$$F_1(\eta)=\alpha_1e^{-k_1\eta}\tag{10.65}$$

and

$$G_1(\eta)=\beta_1e^{-k_1\eta}\tag{10.66}$$

Substituting (10.65) in the first equation of (10.60) and using the boundary condition $H_1(\eta)=0$ at $\eta=\alpha$, we get on integration

$$H_1(\eta)=\frac{2\alpha_1}{k_1}e^{-k_1\eta}\tag{10.67}$$

Since $H_o(\eta)$, $H_1(\eta)$, and $E_1(\eta)$ are known, therefore, the second equation of (10.61) can be written as

$$\frac{d^2E_2}{d\eta^2}+k_1\frac{dE_2}{d\eta}=-\left(\alpha_1^2+\beta_1^2\right)e^{-2k_1\eta}\tag{10.68}$$

Let

$$E_2(\eta)=\bar{E}_2e^{-2k_1\eta}\tag{10.69}$$

be the solution of (10.68). Using (10.69) in (10.68), we have $\bar{E}_2$ as

$$\bar{E}_2 = \frac{\alpha_1^2 + \beta_1^2}{2k_1^2} \tag{10.70}$$

and the expression for $E_2(\eta)$ in (10.69) is found to be

$$E_2(\eta) = -\frac{\alpha_1^2 + \beta_1^2}{2k_1^2} e^{-2k_1\eta} \tag{10.71}$$

whose real and imaginary parts are given by

$$F_2(\eta) = -\frac{\alpha_1^2 + \beta_1^2}{2k_1^2} e^{-2k_1\eta} \tag{10.72}$$

$$G_2(\eta) = 0 \tag{10.73}$$

as $E_2(\eta) = F_2(\eta) - iG_2(\eta)$. Substituting (10.72) in the first equation of (10.61) and then integrating we have

$$H_2(\eta) = -\frac{\alpha_1^2 + \beta_1^2}{2k_1^3} e^{-2k_1\eta} \tag{10.74}$$

Similarly, proceeding in the same way, we have from (10.62)

$$F_3(\eta) = \frac{\alpha_1 \left(\alpha_1^2 + \beta_1^2\right)}{4k_1^4} e^{-3k_1\eta} \tag{10.75}$$

$$G_3(\eta) = -\frac{\beta_1 \left(\alpha_1^2 + \beta_1^2\right)}{12k_1^4} e^{-3k_1\eta} \tag{10.76}$$

$$H_3(\eta) = \frac{\alpha_1 \left(\alpha_1^2 + \beta_1^2\right)}{6k_1^5} e^{-3k_1\eta} \tag{10.77}$$

If four-term approximants of the solutions $H(\eta)$, $F(\eta)$, and $G(\eta)$ are sufficient for the problem, then we can write for the large value of η

$$H(\eta) = H_0(\eta) + H_1(\eta) + H_2(\eta) + H_3(\eta) \tag{10.78}$$

$$F(\eta) = F_1(\eta) + F_2(\eta) + F_3(\eta) + F_4(\eta) \tag{10.79}$$

$$G(\eta) = G_1(\eta) + G_2(\eta) + G_3(\eta) + G_4(\eta) \tag{10.80}$$

where $F_4(\eta)$ and $G_4(\eta)$ are given by

$$F_4(\eta) = \frac{\left(\alpha_1^2 + \beta_1^2\right)\left(17\alpha_1^2 + \beta_1^2\right)}{144k_1^8} e^{-4k_1\eta} \tag{10.81}$$

and

$$G_4(\eta) = \frac{\beta_1 \left(\alpha_1^2 + \beta_1^2\right)(5\alpha_1 - 1)}{72k_1^8} e^{-4k_1\eta} \tag{10.82}$$

The solutions of the system of equations ranging from (10.7) to (10.9) are obtained with the help of boundary conditions (10.10) and (10.11). Von Kármán [2] first obtained the solutions of these equations by means of an approximate method. Later on Cochran [1] improved this method by considering the series expansions for the dependent variables near $\eta = 0$ and asymptotic expansions at $\eta = \alpha$ and then matching the solutions for small and large values of η at a suitable value of η.

References

1. Cochran W. G.: The Flow Due to Rotating Disc, *Proc. Camb. Phil. Soc., 30,* 365–375 (1934).
2. Von Kármán, T.: *Ueber Laminare Turbulente Reibung,* ZAMM1.

Appendix: Equations of Motion in Different Coordinate Systems

A.1 Introduction

In solving the problems of fluid dynamics, we generally use the equations of motion in three ordinary coordinate systems, namely, the Cartesian coordinate system, cylindrical polar coordinate system, and spherical polar coordinate system. But there are many problems of fluid dynamics that cannot be solved by direct use of the equations of motion in these coordinate systems. It is advantageous to use the equations of motion in orthogonal curvilinear coordinates system for solving these problems of fluid dynamics. In this appendix we shall discuss the derivations of the equations of motion from the orthogonal curvilinear coordinate system.

A.2 Curvilinear Coordinate System

Let x, y, and z be a set of Cartesian coordinates of a point P in space that has a set of curvilinear coordinates x_1, x_2, and x_3, respectively. The relation between Cartesian coordinates (x, y, z) and curvilinear coordinates (x_1, x_2, x_3) may be written as

$$\left.\begin{aligned} x &= x(x_1, x_2, x_3) \\ y &= y(y_1, y_2, y_3) \\ z &= z(z_1, z_2, z_3) \end{aligned}\right\} \tag{A.1}$$

It is assumed that x, y, and z are completely defined and continuous in the first derivatives. Therefore, the above equations may be solved for x_1, x_2, and x_3 in terms of x, y, and z, that is,

$$\left.\begin{aligned} x_1 &= x_1(x, y, z) \\ x_2 &= x_2(x, y, z) \\ x_3 &= x_3(x, y, z) \end{aligned}\right\} \tag{A.2}$$

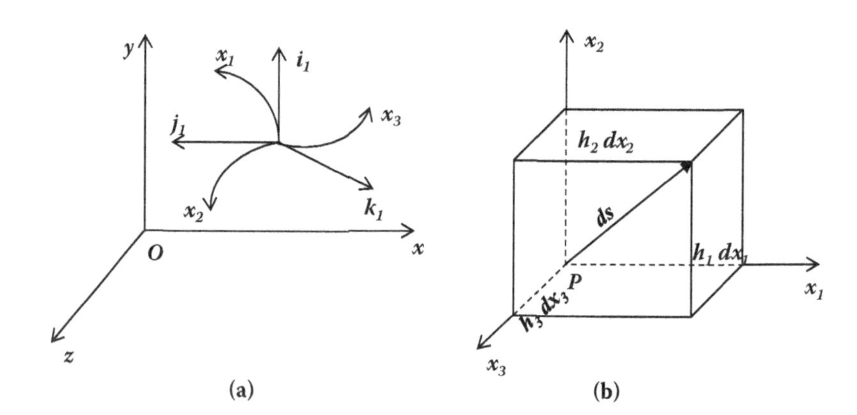

FIGURE A.1
Orthogonal curvilinear coordinate system.

The surfaces x_1 = constant, x_2 = constant, and x_3 = constant are called *levels* or *coordinate surfaces*. The intersection of any two coordinate surfaces is a curve that is called a *coordinate curve* or *coordinate line*. If the surfaces x_2 = constant and x_3 = constant intersect, we have a curve along which the coordinate x_1 varies and is known as x_1 − curve or x_1 − line. Similarly, x_2 − curve and x_3 − curve can be obtained for the pairs of surfaces (x_1, x_3) and (x_1, x_2), respectively. If the coordinate curves intersect at right angles, then the curvilinear coordinate system is called *orthogonal*. We shall restrict ourselves to orthogonal curvilinear coordinates in our discussions on solving problems of fluid dynamics.

The tangents are drawn to each of the coordinate lines at the point P (see Figure A.1(a)). These tangents are mutually perpendicular in the orthogonal coordinate system and taken as the coordinate axes at point P. These axes are positive in the directions in which x_1, x_2, and x_3 increase from point P. Let i_1, j_1, and k_1 be the unit vectors that, in general, vary from point to point indicating that these are not constant unit vectors.

The line element *ds* in Cartesian coordinate (x, y, z) in space is given by

$$(ds)^2 = (dx)^2 + (dy)^2 + (dz)^2 \tag{A.3}$$

Since each of the coordinates x, y, and z is function of coordinate lines x_1, x_2, and x_3, respectively, we have from (A.1) on differentiation

$$dx = \frac{\partial x}{\partial x_1}dx_1 + \frac{\partial x}{\partial x_2}dx_2 + \frac{\partial x}{\partial x_3}dx_3 \tag{A.4}$$

$$dy = \frac{\partial y}{\partial x_1}dx_1 + \frac{\partial y}{\partial x_2}dx_2 + \frac{\partial y}{\partial x_3}dx_3 \tag{A.5}$$

$$dz = \frac{\partial z}{\partial x_1}dx_1 + \frac{\partial z}{\partial x_2}dx_2 + \frac{\partial z}{\partial x_3}dx_3 \tag{A.6}$$

Squaring (A.4), (A.5), and (A.6) and then adding, we have

$$
\begin{aligned}
(dx)^2 + (dy)^2 + (dz)^2 = & \left[\left(\frac{\partial x}{\partial x_1}\right)^2 + \left(\frac{\partial y}{\partial x_1}\right)^2 + \left(\frac{\partial z}{\partial x_1}\right)^2\right](dx_1)^2 \\
& + \left[\left(\frac{\partial x}{\partial x_2}\right)^2 + \left(\frac{\partial y}{\partial x_2}\right)^2 + \left(\frac{\partial z}{\partial x_2}\right)^2\right](dx_2)^2 \\
& + \left[\left(\frac{\partial x}{\partial x_3}\right)^2 + \left(\frac{\partial y}{\partial x_3}\right)^2 + \left(\frac{\partial z}{\partial x_3}\right)^2\right](dx_3)^2 \\
& + 2\left[\frac{\partial x}{\partial x_1}.\frac{\partial x}{\partial x_2} + \frac{\partial y}{\partial x_1}.\frac{\partial y}{\partial x_2} + \frac{\partial z}{\partial x_1}.\frac{\partial z}{\partial x_2}\right] dx_1 dx_2 \\
& + 2\left[\frac{\partial x}{\partial x_2}.\frac{\partial x}{\partial x_3} + \frac{\partial y}{\partial x_2}.\frac{\partial y}{\partial x_3} + \frac{\partial z}{\partial x_2}.\frac{\partial z}{\partial x_3}\right] dx_2 dx_3 \\
& + 2\left[\frac{\partial x}{\partial x_3}.\frac{\partial x}{\partial x_1} + \frac{\partial y}{\partial x_3}.\frac{\partial y}{\partial x_1} + \frac{\partial z}{\partial x_3}.\frac{\partial z}{\partial x_1}\right] dx_3 dx_1
\end{aligned}
\tag{A.7}
$$

Let $\vec{r}$ be the position vector of the point P whose Cartesian coordinates are (x, y, z), and it is given by

$$
\vec{r} = xi + yj + zk \tag{A.8}
$$

where i, j, and k are constant unit vectors along x, y, and z, respectively. The partial derivatives of $\vec{r}$ with respect to x_1, x_2, and x_3, i.e., $\frac{\partial \vec{r}}{\partial x_1}$, $\frac{\partial \vec{r}}{\partial x_2}$, and $\frac{\partial \vec{r}}{\partial x_3}$ are the tangents to the coordinate lines x_1, x_2, and x_3. These are called base vectors, and their dot products taken two at a time vanish, that is,

$$
\left.
\begin{aligned}
\frac{\partial \vec{r}}{\partial x_1}.\frac{\partial \vec{r}}{\partial x_2} = 0 \\
\frac{\partial \vec{r}}{\partial x_2}.\frac{\partial \vec{r}}{\partial x_3} = 0 \\
\frac{\partial \vec{r}}{\partial x_3}.\frac{\partial \vec{r}}{\partial x_1} = 0
\end{aligned}
\right\}
\tag{A.9}
$$

which are the properties of dot products of two vectors. Using (A.8) in (A.9), we have

$$
\left.
\begin{aligned}
\frac{\partial x}{\partial x_1}.\frac{\partial x}{\partial x_2} + \frac{\partial y}{\partial x_1}.\frac{\partial y}{\partial x_2} + \frac{\partial z}{\partial x_1}.\frac{\partial z}{\partial x_2} = 0 \\
\frac{\partial x}{\partial x_2}.\frac{\partial x}{\partial x_3} + \frac{\partial y}{\partial x_2}.\frac{\partial y}{\partial x_3} + \frac{\partial z}{\partial x_2}.\frac{\partial z}{\partial x_3} = 0 \\
\frac{\partial x}{\partial x_3}.\frac{\partial x}{\partial x_1} + \frac{\partial y}{\partial x_3}.\frac{\partial y}{\partial x_1} + \frac{\partial z}{\partial x_3}.\frac{\partial z}{\partial x_1} = 0
\end{aligned}
\right\}
\tag{A.10}
$$

By virtue of (A.10), the equation (A.7) reduces to

$$(dx)^2+(dy)^2+(dz)^2=\left[\left(\frac{\partial x}{\partial x_1}\right)^2+\left(\frac{\partial y}{\partial x_1}\right)^2+\left(\frac{\partial z}{\partial x_1}\right)^2\right](dx_1)^2$$
$$+\left[\left(\frac{\partial x}{\partial x_2}\right)^2+\left(\frac{\partial y}{\partial x_2}\right)^2+\left(\frac{\partial z}{\partial x_2}\right)^2\right](dx_2)^2$$
$$+\left[\left(\frac{\partial x}{\partial x_3}\right)^2+\left(\frac{\partial y}{\partial x_3}\right)^2+\left(\frac{\partial z}{\partial x_3}\right)^2\right](dx_3)^2 \quad \text{(A.11)}$$

Then the equation (A.3) with the help of (A.11) takes the form of Figure A.1(b)

$$(ds)^2=h_1^2(dx_1)^2+h_2^2(dx_2)^2+h_3^2(dx_3)^2 \quad \text{(A.12)}$$

where

$$h_1^2=\left(\frac{\partial x}{\partial x_1}\right)^2+\left(\frac{\partial y}{\partial x_1}\right)^2+\left(\frac{\partial z}{\partial x_1}\right)^2$$
$$h_2^2=\left(\frac{\partial x}{\partial x_2}\right)^2+\left(\frac{\partial y}{\partial x_2}\right)^2+\left(\frac{\partial z}{\partial x_2}\right)^2$$
$$h_3^2=\left(\frac{\partial x}{\partial x_3}\right)^2+\left(\frac{\partial y}{\partial x_3}\right)^2+\left(\frac{\partial z}{\partial x_3}\right)^2 \quad \text{(A.13)}$$

and h_1, h_2, h_3 are known as scale factors.

Now we proceed to find out the expressions of gradient, divergence, and curl in the orthogonal curvilinear coordinate system. We know that

$$grad=\nabla=\frac{i_1}{h_1}\frac{\partial}{\partial x_1}+\frac{j_1}{h_2}\frac{\partial}{\partial x_2}+\frac{k_1}{h_3}\frac{\partial}{\partial x_3} \quad \text{(A.14)}$$

If $\vec{V}$ is the velocity vector whose components are $v_1, v_2,$ and v_3 along the coordinate axes $x_1, x_2,$ and x_3 respectively, then

$$\vec{V}=i_1v_1+j_1v_2+k_1v_3 \quad \text{(A.15)}$$

and

$$div\vec{V}=\frac{1}{h_1h_2h_3}\left[\frac{\partial}{\partial x_1}(h_2h_3v_1)+\frac{\partial}{\partial x_2}(h_3h_1v_2)+\frac{\partial}{\partial x_3}(h_1h_2v_3)\right] \quad \text{(A.16)}$$

The curl $\vec{V}$ or rot $\vec{V}$ are given by

$$Curl\vec{V} = rot\vec{V} = \nabla X\vec{V} = \begin{vmatrix} i_1 & j_1 & k_1 \\ \frac{1}{h_1}\frac{\partial}{\partial x_1} & \frac{1}{h_2}\frac{\partial}{\partial x_2} & \frac{1}{h_3}\frac{\partial}{\partial x_3} \\ v_1 & v_2 & v_3 \end{vmatrix}$$

$$= \frac{1}{h_1 h_2 h_3}\begin{vmatrix} h_1 i_1 & h_2 j_1 & h_3 k_1 \\ \frac{\partial}{\partial x_1} & \frac{\partial}{\partial x_2} & \frac{\partial}{\partial x_3} \\ h_1 v_1 & h_2 v_2 & h_3 v_3 \end{vmatrix}$$

$$= \frac{i_1}{h_2 h_3}\left[\frac{\partial}{\partial x_2}(h_3 v_3) - \frac{\partial}{\partial x_3}(h_2 v_2)\right] + \frac{j_1}{h_3 h_1}\left[\frac{\partial}{\partial x_3}(h_1 v_1) - \frac{\partial}{\partial x_1}(h_3 v_3)\right]$$

$$+ \frac{k_1}{h_1 h_2}\left[\frac{\partial}{\partial x_1}(h_2 v_2) - \frac{\partial}{\partial x_2}(h_1 v_1)\right] = n_1 i_1 + n_2 j_1 + n_3 k_1 \qquad \text{(A.17)}$$

where

$$n_1 = \frac{1}{h_2 h_3}\left[\frac{\partial}{\partial x_2}(h_3 v_3) - \frac{\partial}{\partial x_3}(h_2 v_2)\right]$$

$$n_2 = \frac{1}{h_3 h_1}\left[\frac{\partial}{\partial x_3}(h_1 v_1) - \frac{\partial}{\partial x_1}(h_3 v_3)\right]$$

$$n_3 = \frac{1}{h_1 h_2}\left[\frac{\partial}{\partial x_1}(h_2 v_2) - \frac{\partial}{\partial x_2}(h_1 v_1)\right] \qquad \text{(A.18)}$$

Let $\overline{\Phi}$ be a scalar. Then from (A.14) we have

$$\nabla\overline{\Phi} = \frac{i_1}{h_1}\frac{\partial\overline{\Phi}}{\partial x_1} + \frac{j_1}{h_2}\frac{\partial\overline{\Phi}}{\partial x_2} + \frac{k_1}{h_3}\frac{\partial\overline{\Phi}}{\partial x_3} \qquad \text{(A.19)}$$

If coefficients of i_1, j_1, and k_1 in (A.19) are denoted by l_1, l_2, and l_3, respectively that is,

$$l_1 = \frac{1}{h_1}\frac{\partial\overline{\Phi}}{\partial x_1}$$

$$l_2 = \frac{1}{h_2}\frac{\partial\overline{\Phi}}{\partial x_2}$$

$$l_3 = \frac{1}{h_3}\frac{\partial\overline{\Phi}}{\partial x_3} \qquad \text{(A.20)}$$

then equation (A.19) becomes

$$\nabla\overline{\Phi} = \vec{l} \tag{A.21}$$

where

$$\vec{l} = l_1 i_1 + l_2 j_1 + l_3 k_1 \tag{A.22}$$

l_1, l_2, and l_3 are the components of $\vec{l}$. Therefore,

$$\nabla\vec{l} = div\vec{l} = \frac{1}{h_1 h_2 h_3}\left[\frac{\partial}{\partial x_1}(h_2 h_3 l_1) + \frac{\partial}{\partial x_2}(h_3 h_1 l_2) + \frac{\partial}{\partial x_3}(h_1 h_2 l_3)\right] \tag{A.23}$$

Using (A.20) and (A.21) in (A.23), we get

$$\nabla^2\overline{\Phi} = \frac{1}{h_1 h_2 h_3}\left[\frac{\partial}{\partial x_1}\left(\frac{h_2 h_3}{h_1}\frac{\partial\overline{\Phi}}{\partial x_1}\right) + \frac{\partial}{\partial x_2}\left(\frac{h_3 h_1}{h_2}\frac{\partial\overline{\Phi}}{\partial x_2}\right) + \frac{\partial}{\partial x_3}\left(\frac{h_1 h_2}{h_3}\frac{\partial\overline{\Phi}}{\partial x_3}\right)\right] \tag{A.24}$$

The rate of strain components are given by

$$e_{x_1x_1} = \frac{1}{h_1}\frac{\partial v_1}{\partial x_1} + \frac{v_2}{h_1 h_2}\frac{\partial h_1}{\partial x_2} + \frac{v_3}{h_3 h_1}\frac{\partial h_1}{\partial x_3}$$

$$e_{x_2x_2} = \frac{1}{h_2}\frac{\partial v_2}{\partial x_2} + \frac{v_3}{h_2 h_3}\frac{\partial h_2}{\partial x_3} + \frac{v_1}{h_1 h_2}\frac{\partial h_2}{\partial x_1}$$

$$e_{x_3x_3} = \frac{1}{h_3}\frac{\partial v_3}{\partial x_3} + \frac{v_1}{h_3 h_1}\frac{\partial h_3}{\partial x_1} + \frac{v_2}{h_2 h_3}\frac{\partial h_3}{\partial x_2}$$

$$e_{x_2x_3} = \frac{h_3}{h_2}\frac{\partial}{\partial x_2}\left(\frac{v_3}{h_3}\right) + \frac{h_2}{h_3}\frac{\partial}{\partial x_3}\left(\frac{v_2}{h_2}\right)$$

$$e_{x_3x_1} = \frac{h_1}{h_3}\frac{\partial}{\partial x_3}\left(\frac{v_1}{h_1}\right) + \frac{h_3}{h_1}\frac{\partial}{\partial x_1}\left(\frac{v_3}{h_3}\right)$$

$$e_{x_1x_2} = \frac{h_2}{h_1}\frac{\partial}{\partial x_1}\left(\frac{v_2}{h_2}\right) + \frac{h_1}{h_2}\frac{\partial}{\partial x_2}\left(\frac{v_1}{h_1}\right) \tag{A.25}$$

A.3 Navier–Stokes Equations of Motion in the Orthogonal Curvilinear Coordinate System

The equations of motion of viscous incompressible fluid for solving the problems of fluid dynamics are the *continuity equation* and *momentum equations.*

It is convenient to use the orthogonal curvilinear coordinates for solving the fluid flow problems. In this section we derive the Navier–Stokes equations in terms of orthogonal curvilinear coordinates. Then these equations are to be transferred to those in the other coordinate systems, such as the Cartesian coordinate system, cylindrical polar coordinate system, spherical polar coordinate system, toroidal coordinate system, etc. To do this, we consider the vector forms of the equations of motion that are given by

$$div\vec{V} = 0 \tag{A.26}$$

$$\rho\frac{D\vec{V}}{Dt} = \rho\vec{F} - \nabla p + \mu\nabla^2\vec{V} \tag{A.27}$$

where ρ is the density of fluid, $\vec{F}$ is the body force whose components are F_1, F_2, F_3, p is the pressure, μ is the coefficient of viscosity, (v_1, v_2, v_3) are the components of velocity vector $\vec{V}$, and

$$\frac{D}{Dt} = \frac{\partial}{\partial t} + (\vec{V}.\nabla) \tag{A.28}$$

which is known as material derivative.

The equation of energy is also given by

$$\rho C_v\frac{DT}{Dt} = \frac{\partial Q}{\partial t} + k\nabla^2 T + \overline{\Phi} \tag{A.29}$$

Here C_v is the specific heat at constant volume, T is the temperature, Q is the heat added per unit mass of fluid, k is the thermal conductivity, $\nabla^2 T$ is the Laplacian of T obtained from (A.24) replacing $\overline{\Phi}$ by T, and $\overline{\Phi}$ is the dissipation function in orthogonal curvilinear coordinates for an incompressible fluid, and it is given by

$$\overline{\Phi} = \mu\left[2\left({e^2}_{x_1x_1} + {e^2}_{x_2x_2} + {e^2}_{x_3x_3}\right) + {e^2}_{x_2x_3} + {e^2}_{x_3x_1} + {e^2}_{x_1x_2}\right. \tag{A.30}$$

where the strain components involved in (A.30) are given by (A.25).

Now we consider two important vector identities and these are given by

$$\nabla(\vec{A}.\vec{B}) = (\vec{B}.\nabla)\vec{A} + (\vec{A}.\nabla)\vec{B} + \vec{B}\times(\nabla\times\vec{A}) + \vec{A}\times(\nabla\times\vec{B}) \tag{A.31}$$

and

$$\nabla(\nabla.\vec{A}) = \nabla\times(\nabla\times\vec{A}) + \nabla^2\vec{A} \tag{A.32}$$

where $\vec{A}$ and $\vec{B}$ are two arbitrary vectors. If we write $\vec{A} = \vec{B} = \vec{V}$, then the identities (A.31) and (A.32) become

$$(\vec{V}.\nabla)\vec{V} = \nabla\left(\frac{1}{2}\vec{V}.\vec{V}\right) - \vec{V}\times curl\vec{V} \tag{A.33}$$

and

$$\nabla^2 \vec{V} = grad(div\vec{V}) - curl(curl\vec{V}) \tag{A.34}$$

where

$$\left.\begin{aligned} grad &= \nabla \\ div\vec{V} &= \nabla\vec{V} \\ curl\vec{V} &= \nabla \times \vec{V} \end{aligned}\right\} \tag{A.35}$$

Using (A.28), (A.33), and (A.34) in (A.27), we have the vector form of the equation of motion as

$$\rho\left[\frac{\partial \vec{V}}{\partial t} + \nabla\left(\frac{1}{2}\vec{V}.\vec{V}\right) - \vec{V} \times curl\vec{V}\right] = \rho\vec{F} - \nabla p + \mu[grad(div\vec{V}) - curl(curl\vec{V})] \tag{A.36}$$

Next, we propose to find out the equations of motion in orthogonal curvilinear coordinates. For this purpose we write

$$\nabla\left(\frac{1}{2}\vec{V}.\vec{V}\right) = \frac{i_1}{h_1}\frac{\partial}{\partial x_1}\left[\frac{1}{2}\left(v_1{}^2 + v_2{}^2 + v_3{}^2\right)\right] + \frac{j_1}{h_2}\frac{\partial}{\partial x_2}\left[\frac{1}{2}\left(v_1{}^2 + v_2{}^2 + v_3{}^2\right)\right] + \frac{k_1}{h_3}\frac{\partial}{\partial x_3}\left[\frac{1}{2}\left(v_1{}^2 + v_2{}^2 + v_3{}^2\right)\right] \tag{A.37}$$

$$\vec{V} \times curl\vec{V} = i_1(v_2n_3 - v_3n_2) + j_1(v_3n_1 - v_1n_3) + k_1(v_1n_2 - v_2n_1) \tag{A.38}$$

where $n_1, n_2,$ and n_3 are given by (A.18). Subtracting (A.38) from (A.37) we have

$$\begin{aligned} \nabla\left(\frac{1}{2}\vec{V}.\vec{V}\right) - \vec{V} \times curl\vec{V} = {} & i_1\left[\frac{1}{h_1}\frac{\partial}{\partial x_1}\left(\frac{1}{2}v_1{}^2\right) - (v_2n_3 - v_3n_2)\right] \\ & + j_1\left[\frac{1}{h_2}\frac{\partial}{\partial x_2}\left(\frac{1}{2}v_2{}^2\right) - (v_3n_1 - v_1n_3)\right] \\ & + k_1\left[\frac{1}{h_3}\frac{\partial}{\partial x_3}\left(\frac{1}{2}v_3{}^2\right) - (v_1n_2 - v_2n_1)\right] \end{aligned} \tag{A.39}$$

Again

$$grad(div\vec{V}) = \frac{i_1}{h_1}\frac{\partial}{\partial x_1}(div\vec{V}) + \frac{j_1}{h_2}\frac{\partial}{\partial x_2}(div\vec{V}) + \frac{k_1}{h_3}\frac{\partial}{\partial x_3}(div\vec{V}) \tag{A.40}$$

where $div\vec{V}$ is given by (A.16). Then we write

$$curl(curl\vec{V}) = \nabla \times (i_1 n_1 + j_1 n_2 + k_1 n_3)$$
$$= \begin{vmatrix} i_1 & j_1 & k_1 \\ \frac{1}{h_1}\frac{\partial}{\partial x_1} & \frac{1}{h_2}\frac{\partial}{\partial x_2} & \frac{1}{h_3}\frac{\partial}{\partial x_3} \\ n_1 & n_2 & n_3 \end{vmatrix} = \frac{i_1}{h_1 h_2 h_3} \begin{vmatrix} h_1 i_1 & h_2 j_1 & h_3 k_1 \\ \frac{\partial}{\partial x_1} & \frac{\partial}{\partial x_2} & \frac{\partial}{\partial x_3} \\ h_1 n_1 & h_2 n_2 & h_3 n_3 \end{vmatrix}$$
$$= \frac{i_1}{h_2 h_3}\left[\frac{\partial}{\partial x_2}(h_3 n_3) - \frac{\partial}{\partial x_3}(h_2 n_2)\right] + \frac{j_1}{h_3 h_1}\left[\frac{\partial}{\partial x_3}(h_1 n_1) - \frac{\partial}{\partial x_1}(h_3 n_3)\right]$$
$$+ \frac{k_1}{h_1 h_2}\left[\frac{\partial}{\partial x_1}(h_2 n_2) - \frac{\partial}{\partial x_2}(h_1 n_1)\right] = g_1 i_1 + g_2 j_1 + g_3 k_1 \tag{A.41}$$

where

$$\left.\begin{aligned} g_1 &= \frac{1}{h_2 h_3}\left[\frac{\partial}{\partial x_2}(h_3 n_3) - \frac{\partial}{\partial x_3}(h_2 n_2)\right] \\ g_2 &= \frac{1}{h_3 h_1}\left[\frac{\partial}{\partial x_3}(h_1 n_1) - \frac{\partial}{\partial x_1}(h_3 n_3)\right] \\ g_3 &= \frac{1}{h_1 h_2}\left[\frac{\partial}{\partial x_1}(h_2 n_2) - \frac{\partial}{\partial x_2}(h_1 n_1)\right] \end{aligned}\right\} \tag{A.42}$$

Subtracting (A.41) from (A.40), we get

$$grad(div\vec{V}) - curl(curl\vec{V}) = i_1\left[\frac{1}{h_1}\frac{\partial}{\partial x_1}(div\vec{V}) - g_1\right]$$
$$+ j_1\left[\frac{1}{h_2}\frac{\partial}{\partial x_2}(div\vec{V}) - g_2\right] + k_1\left[\frac{1}{h_3}\frac{\partial}{\partial x_3}(div\vec{V}) - g_3\right] \tag{A.43}$$

Using (A.39) and (A.43) in (A.36) and then equating the coefficients of i_1, j_1, and k_1 on both sides of the resulting equation, we have the component expressions of the equation of motion in the orthogonal curvilinear coordinates as

$$\rho\left[\frac{\partial v_1}{\partial t} + \frac{1}{h_1}\frac{\partial}{\partial x_1}\left\{\frac{1}{2}\left(v_1^2 + v_2^2 + v_3^2\right)\right\} - (n_3 v_2 - n_2 v_3)\right]$$
$$= \rho F_1 - \frac{1}{h_1}\frac{\partial p}{\partial x_1} + \mu\left[\frac{1}{h_1}\frac{\partial}{\partial x_1}(div\vec{V}) - g_1\right] \tag{A.44}$$

$$\rho\left[\frac{\partial v_2}{\partial t} + \frac{1}{h_2}\frac{\partial}{\partial x_2}\left\{\frac{1}{2}\left(v_1^2 + v_2^2 + v_3^2\right)\right\} - (n_1 v_3 - n_3 v_1)\right]$$
$$= \rho F_2 - \frac{1}{h_2}\frac{\partial p}{\partial x_2} + \mu\left[\frac{1}{h_2}\frac{\partial}{\partial x_2}(div\vec{V}) - g_2\right] \tag{A.45}$$

$$\rho\left[\frac{\partial v_3}{\partial t}+\frac{1}{h_3}\frac{\partial}{\partial x_3}\left\{\frac{1}{2}\left({v_1}^2+{v_2}^2+{v_3}^2\right)\right\}-(n_2v_1-n_1v_2)\right]$$
$$=\rho F_3-\frac{1}{h_3}\frac{\partial p}{\partial x_3}+\mu\left[\frac{1}{h_3}\frac{\partial}{\partial x_3}(div\vec{V})-g_3\right] \tag{A.46}$$

and the continuity equation in the orthogonal curvilinear coordinates is obtained from (A.16) by putting $div\vec{V}=0$, that is,

$$\frac{\partial}{\partial x_1}(h_2h_3v_1)+\frac{\partial}{\partial x_2}(h_3h_1v_2)+\frac{\partial}{\partial x_3}(h_1h_2v_3)=0 \tag{A.47}$$

Then we write

$$(\vec{V}.\nabla)T=\frac{v_1}{h_1}\frac{\partial T}{\partial x_1}+\frac{v_2}{h_2}\frac{\partial T}{\partial x_2}+\frac{v_3}{h_3}\frac{\partial T}{\partial x_3} \tag{A.48}$$

and

$$\nabla^2T=\frac{1}{h_1h_2h_3}\left[\frac{\partial}{\partial x_1}\left(\frac{h_2h_3}{h_1}\frac{\partial T}{\partial x_1}\right)+\frac{\partial}{\partial x_2}\left(\frac{h_3h_1}{h_2}\frac{\partial T}{\partial x_2}\right)+\frac{\partial}{\partial x_3}\left(\frac{h_1h_2}{h_3}\frac{\partial T}{\partial x_3}\right)\right] \tag{A.49}$$

The equation (A.29), together with (A.30), (A.48), and (A.49), constitutes the equation of energy, which can be solved by means of the decomposition technique under certain conditions.

A.4 Navier–Stokes Equations of Motion in the Cartesian Coordinate System

In this section, we derive the Navier–Stokes equations of motion of a viscous incompressible fluid in Cartesian coordinates from those of an orthogonal curvilinear coordinate system. If (x, y, z) are the coordinates of any point in space and (x_1, x_2, x_3) are the orthogonal curvilinear coordinates of that point, then the relations between these two coordinates are

$$x=x_1,\ y=y_1,\ z=x_3 \tag{A.50}$$

and

$$v_1=v_x,\ v_2=v_y,\ v_3=v_z \tag{A.51}$$

where v_x, v_y, and v_z are the components of the velocity vector in the directions of x, y, and z increasing. Then using (A.50) in (A.13), we have

$$h_1=h_2=h_3=1 \tag{A.52}$$

and the equations (A.16) with the help of (A.51) take the forms

$$\left.\begin{aligned} n_1 &= \frac{\partial v_z}{\partial y} - \frac{\partial v_y}{\partial z} \\ n_2 &= \frac{\partial v_x}{\partial z} - \frac{\partial v_z}{\partial x} \\ n_3 &= \frac{\partial v_y}{\partial x} - \frac{\partial v_x}{\partial y} \end{aligned}\right\} \tag{A.53}$$

By virtue of (A.50), (A.51), (A.52), and (A.53), we write the following relations:

$$\left.\begin{aligned} \frac{1}{h_1}\frac{\partial}{\partial x_1}\left\{\frac{1}{2}\left(v_1^2+v_2^2+v_3^2\right)\right\} - (n_3v_2 - n_2v_3) &= -\left(v_y\frac{\partial v_x}{\partial y} + v_z\frac{\partial v_x}{\partial z}\right) \\ \frac{1}{h_2}\frac{\partial}{\partial x_2}\left\{\frac{1}{2}\left(v_1^2+v_2^2+v_3^2\right)\right\} - (n_1v_3 - n_3v_1) &= -\left(v_z\frac{\partial v_y}{\partial z} + v_x\frac{\partial v_y}{\partial x}\right) \\ \frac{1}{h_3}\frac{\partial}{\partial x_3}\left\{\frac{1}{2}\left(v_1^2+v_2^2+v_3^2\right)\right\} - (n_2v_1 - n_1v_2) &= -\left(v_x\frac{\partial v_z}{\partial x} + v_y\frac{\partial v_z}{\partial y}\right) \end{aligned}\right\} \tag{A.54}$$

Again, from (A.42) we get

$$\left.\begin{aligned} g_1 &= \frac{\partial}{\partial y}\left(\frac{\partial v_y}{\partial x} - \frac{\partial v_x}{\partial y}\right) - \frac{\partial}{\partial z}\left(\frac{\partial v_x}{\partial z} - \frac{\partial v_y}{\partial x}\right) \\ g_2 &= \frac{\partial}{\partial z}\left(\frac{\partial v_z}{\partial y} - \frac{\partial v_y}{\partial z}\right) - \frac{\partial}{\partial x}\left(\frac{\partial v_y}{\partial x} - \frac{\partial v_x}{\partial y}\right) \\ g_3 &= \frac{\partial}{\partial x}\left(\frac{\partial v_x}{\partial z} - \frac{\partial v_z}{\partial x}\right) - \frac{\partial}{\partial y}\left(\frac{\partial v_z}{\partial y} - \frac{\partial v_y}{\partial z}\right) \end{aligned}\right\} \tag{A.55}$$

Using the relations ranging from (A.50) to (A.52), we get the divergence (A.16) as

$$div\vec{V} = \frac{\partial v_x}{\partial x} + \frac{\partial v_y}{\partial y} + \frac{\partial v_z}{\partial z} \tag{A.56}$$

Differentiating (A.16) with respect to x_1, then multiplying by $1/h_1$ and finally adding to $-g_1$, we have

$$\frac{1}{h_1}\frac{\partial}{\partial x_1}(div\vec{V}) - g_1 = \frac{\partial}{\partial x}(div\vec{V} - g_1) = \frac{\partial^2 v_x}{\partial x^2} + \frac{\partial^2 v_x}{\partial y^2} + \frac{\partial^2 v_x}{\partial z^2} \tag{A.57}$$

Similarly, proceeding in the same way we get

$$\frac{1}{h_2}\frac{\partial}{\partial x_2}(div\vec{V}) - g_2 = \frac{\partial}{\partial y}(div\vec{V} - g_2) = \frac{\partial^2 v_y}{\partial x^2} + \frac{\partial^2 v_y}{\partial y^2} + \frac{\partial^2 v_y}{\partial z^2} \tag{A.58}$$

and

$$\frac{1}{h_3}\frac{\partial}{\partial x_3}(div\vec{V}) - g_3 = \frac{\partial}{\partial z}(div\vec{V} - g_3) = \frac{\partial^2 v_z}{\partial x^2} + \frac{\partial^2 v_z}{\partial y^2} + \frac{\partial^2 v_z}{\partial z^2} \tag{A.59}$$

Using (A.54), (A.57), (A.58), and (A.59), the equations ranging from (A.44) to (A.46), after somewhat straightforward calculations, become

$$\frac{\partial v_x}{\partial t} + v_x \frac{\partial v_x}{\partial x} + v_y \frac{\partial v_x}{\partial y} + v_z \frac{\partial v_x}{\partial z} = F_x - \frac{1}{\rho} \frac{\partial p}{\partial x} + \nu \left[\frac{\partial^2 v_x}{\partial x^2} + \frac{\partial^2 v_x}{\partial y^2} + \frac{\partial^2 v_x}{\partial z^2} \right] \tag{A.60}$$

$$\frac{\partial v_y}{\partial t} + v_x \frac{\partial v_y}{\partial x} + v_y \frac{\partial v_y}{\partial y} + v_z \frac{\partial v_y}{\partial z} = F_y - \frac{1}{\rho} \frac{\partial p}{\partial y} + \nu \left[\frac{\partial^2 v_y}{\partial x^2} + \frac{\partial^2 v_y}{\partial y^2} + \frac{\partial^2 v_y}{\partial z^2} \right] \tag{A.61}$$

$$\frac{\partial v_z}{\partial t} + v_x \frac{\partial v_z}{\partial x} + v_y \frac{\partial v_z}{\partial y} + v_z \frac{\partial v_z}{\partial z} = F_z - \frac{1}{\rho} \frac{\partial p}{\partial z} + \nu \left[\frac{\partial^2 v_z}{\partial x^2} + \frac{\partial^2 v_z}{\partial y^2} + \frac{\partial^2 v_z}{\partial z^2} \right] \tag{A.62}$$

where $\nu = \mu/\rho$ is the kinematic viscosity and $\vec{F} = F_x i_1 + F_y j_1 + F_z k_1$, F_x, F_y, F_z are the components of the body force $\vec{F}$ in the directions of x, y, and z, respectively. The continuity equation (A.47) takes the form

$$\frac{\partial v_x}{\partial x} + \frac{\partial v_y}{\partial y} + \frac{\partial v_z}{\partial z} = 0 \tag{A.63}$$

The equations ranging from (A.60) to (A.63) are the Cartesian equations of motion and continuity equation, which are derived from those of the orthogonal curvilinear coordinates system.

Then we proceed to find out the equation of energy in the Cartesian coordinates and for this purpose we write below the expressions of $(\vec{V}.\nabla)T$, the Laplacian of T obtained by changing $\vec{\Phi} = T$ in (A.24), and the dissipation function $\underline{\Phi}$ defined in (A.30). Using (A.50), (A.51), and (A.52), we have

$$(\vec{V}.\nabla)T = v_x \frac{\partial T}{\partial x} + v_y \frac{\partial T}{\partial y} + v_z \frac{\partial T}{\partial z} \tag{A.64}$$

$$\nabla^2 T = \frac{\partial^2 T}{\partial x^2} + \frac{\partial^2 T}{\partial y^2} + \frac{\partial^2 T}{\partial z^2} \tag{A.65}$$

$$\underline{\Phi} = \mu \left[2\left(e_{xx}^2 + e_{yy}^2 + e_{zz}^2 \right) + e_{yz}^2 + e_{zx}^2 + e_{xy}^2 \right] \tag{A.66}$$

where the rate of strain components involved in the expression of $\vec{\underline{\Phi}}$ may be obtained using (A.50), (A.51), and (A.52) in (A.25). These are given by

$$\left.\begin{aligned} e_{xx} &= \frac{\partial v_x}{\partial x}, \quad & e_{yz} &= \frac{\partial v_z}{\partial y} + \frac{\partial v_y}{\partial z} \\ e_{yy} &= \frac{\partial v_y}{\partial y}, \quad & e_{zx} &= \frac{\partial v_x}{\partial z} + \frac{\partial v_z}{\partial x} \\ e_{zz} &= \frac{\partial v_z}{\partial z}, \quad & e_{xy} &= \frac{\partial v_y}{\partial x} + \frac{\partial v_x}{\partial y} \end{aligned}\right\} \tag{A.67}$$

Substituting (A.64), (A.65), and (A.66) in (A.29), we can get the energy equation whose solution may be obtained by solving it under certain conditions.

For nondimensional forms of the equation, we consider the following transformation:

$$\overline{x} = \frac{x}{L}, \quad \overline{y} = \frac{y}{L}, \quad z = \frac{\overline{z}}{L}$$

$$\overline{v_x} = \frac{v_x}{\overline{v}}, \quad \overline{v_y} = \frac{v_y}{\overline{v}}, \quad \overline{v_z} = \frac{v_z}{\overline{v}}$$

$$\overline{t} = \frac{t}{\overline{\overline{T}}}, \quad \overline{p} = \frac{p}{\overline{\overline{P}}}, \quad \overline{F_x} = \frac{F_x}{F}, \quad \overline{F_y} = \frac{F_y}{F}, \quad \overline{F_z} = \frac{F_z}{F} \tag{A.68}$$

Here $\overline{v}$, L, $\overline{T}$, $\overline{P}$, and F are, respectively the characteristics velocity, length, time, pressure, and force. To nondimensionalize the equations ranging from (A.60) to (A.62), we substitute the transformations (A.68) and remove the bars over the variables and get the nondimensionalized equations of motion as

$$\frac{L}{\overline{v}\overline{\overline{T}}}\frac{\partial v_x}{\partial t} + v_x\frac{\partial v_x}{\partial x} + v_y\frac{\partial v_x}{\partial y} + v_z\frac{\partial v_x}{\partial z} = C_F F_x - C_p\frac{\partial p}{\partial x} + \frac{1}{R_e}\nabla^2 v_x \tag{A.69}$$

$$\frac{L}{\overline{v}\overline{\overline{T}}}\frac{\partial v_y}{\partial t} + v_x\frac{\partial v_y}{\partial x} + v_y\frac{\partial v_y}{\partial y} + v_z\frac{\partial v_y}{\partial z} = C_F F_y - C_p\frac{\partial p}{\partial y} + \frac{1}{R_e}\nabla^2 v_y \tag{A.70}$$

$$\frac{L}{\overline{v}\overline{\overline{T}}}\frac{\partial v_z}{\partial t} + v_x\frac{\partial v_z}{\partial x} + v_y\frac{\partial v_z}{\partial y} + v_z\frac{\partial v_z}{\partial z} = C_F F_z - C_p\frac{\partial p}{\partial z} + \frac{1}{R_e}\nabla^2 v_z \tag{A.71}$$

where ∇^2 is the Laplacian operator defined by

$$\nabla^2 = \frac{\partial^2}{\partial x^2} + \frac{\partial^2}{\partial y^2} + \frac{\partial^2}{\partial z^2} \tag{A.72}$$

Here $L/\vec{v}\vec{T}$ is a parameter and C_F, C_P, R_e are the force coefficient, pressure coefficient, and Reynolds number, respectively, defined by

$$\left.\begin{aligned} C_F &= \frac{FL}{\overline{v}^2} \\ C_p &= \frac{\overline{P}}{\rho\overline{v}^2} \\ R_e &= \frac{\overline{v}L}{\vartheta} \end{aligned}\right\} \tag{A.73}$$

A.5 Navier–Stokes Equations of Motion in the Cylindrical Polar Coordinate System

The equations of motion in cylindrical polar coordinates may be obtained from the equations of motion in the orthogonal curvilinear coordinates. Let P be any position of a point in space whose Cartesian and polar coordinates are (x, y, z) and (r, θ, z), respectively, with respect to the origin O. Draw a perpendicular PN from P upon the xy-plane, which passes through the origin O and perpendicular to the z-axis. Let N be the foot of PN in this plane, and from N draw the perpendicular NM on the x-axis such that $OM = x$ and $NM = y$. Then from Figure A.2, we have the relations between the coordinates (x, y, z) and (r, θ, z) as

$$\left.\begin{aligned} x &= r\cos\theta \\ y &= r\sin\theta \\ z &= z \end{aligned}\right\} \tag{A.74}$$

where

$$\left.\begin{aligned} ON &= r \\ OM &= x \\ NM &= y \end{aligned}\right\} \tag{A.75}$$

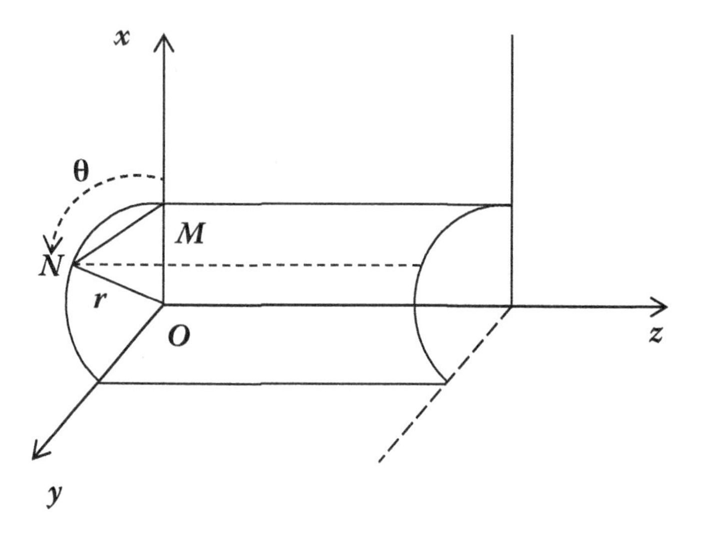

FIGURE A.2
Cylindrical polar coordinate system.

Let v_r, v_θ, and v_z be the components of velocity vector $\vec{V}$ where v_r is in the direction of r, v_θ is in the direction of θ and perpendicular to v_r, and v_z is in the direction of z and perpendicular to the plane, which contains v_r and v_θ.

It is assumed that

$$\left.\begin{aligned} x_1 &= r \\ x_2 &= \theta \\ x_3 &= z \end{aligned}\right\} \tag{A.76}$$

and

$$\left.\begin{aligned} v_1 &= v_r \\ v_2 &= v_\theta \\ v_3 &= v_z \end{aligned}\right\} \tag{A.77}$$

Using (A.74) and (A.76) in (A.13), we find the scale factors as

$$\left.\begin{aligned} h_1 &= 1 \\ h_2 &= r \\ h_3 &= 1 \end{aligned}\right\} \tag{A.78}$$

Substituting the relations ranging from (A.76) to (A.78) in (A.18), we have

$$\left.\begin{aligned} n_1 &= \frac{1}{r}\left[\frac{\partial v_z}{\partial \theta} - \frac{\partial}{\partial z}(r v_\theta)\right] \\ n_2 &= \frac{\partial v_r}{\partial z} - \frac{\partial v_z}{\partial r} \\ n_3 &= \frac{1}{r}\left[\frac{\partial}{\partial r}(r v_\theta) - \frac{\partial v_r}{\partial \theta}\right] \end{aligned}\right\} \tag{A.79}$$

Therefore, multiplying n_3 and n_2 of equation (A.79) by v_2 and v_3, respectively, and then subtracting the latter from the former, we have

$$n_3 v_2 - n_2 v_3 = -\frac{v_\theta}{r}\frac{\partial v_r}{\partial \theta} - v_z\frac{\partial v_r}{\partial z} + \frac{v_\theta^2}{r} + v_\theta\frac{\partial v_\theta}{\partial r} + v_z\frac{\partial v_z}{\partial r} \tag{A.80}$$

Likewise, proceeding in the same way, we get

$$n_1 v_3 - n_3 v_1 = -v_r\frac{\partial v_\theta}{\partial r} - v_z\frac{\partial v_\theta}{\partial z} - \frac{v_r v_\theta}{r} + \frac{v_z}{r}\frac{\partial v_z}{\partial \theta} + \frac{v_r}{r}\frac{\partial v_r}{\partial \theta} \tag{A.81}$$

and

$$n_2 v_1 - n_1 v_2 = -v_r\frac{\partial v_z}{\partial r} - \frac{v_\theta}{r}\frac{\partial v_z}{\partial \theta} + v_r\frac{\partial v_r}{\partial z} + v_\theta\frac{\partial v_\theta}{\partial z} \tag{A.82}$$

Then we write

$$\frac{1}{h_1}\frac{\partial}{\partial x_1}\left\{\frac{1}{2}\left(v_1^2+v_2^2+v_3^2\right)\right\}=v_r\frac{\partial v_r}{\partial r}+v_\theta\frac{\partial v_\theta}{\partial r}+v_z\frac{\partial v_z}{\partial r} \tag{A.83}$$

$$\frac{1}{h_2}\frac{\partial}{\partial x_2}\left\{\frac{1}{2}\left(v_1^2+v_2^2+v_3^2\right)\right\}=\frac{1}{r}\left(v_r\frac{\partial v_r}{\partial \theta}+v_\theta\frac{\partial v_\theta}{\partial \theta}+v_z\frac{\partial v_z}{\partial \theta}\right) \tag{A.84}$$

$$\frac{1}{h_3}\frac{\partial}{\partial x_3}\left\{\frac{1}{2}\left(v_1^2+v_2^2+v_3^2\right)\right\}=v_r\frac{\partial v_r}{\partial z}+v_\theta\frac{\partial v_\theta}{\partial z}+v_z\frac{\partial v_z}{\partial z} \tag{A.85}$$

Subtracting (A.80) from (A.83), (A.81) from (A.84), and (A.82) from (A.85), we get

$$\frac{1}{h_1}\frac{\partial}{\partial x_1}\left\{\frac{1}{2}\left(v_1^2+v_2^2+v_3^2\right)\right\}-(n_3v_2-n_2v_3)=v_r\frac{\partial v_r}{\partial r}+\frac{v_\theta}{r}\frac{\partial v_r}{\partial \theta}+v_z\frac{\partial v_r}{\partial z}-\frac{v_\theta^2}{r} \tag{A.86}$$

$$\frac{1}{h_2}\frac{\partial}{\partial x_2}\left\{\frac{1}{2}\left(v_1^2+v_2^2+v_3^2\right)\right\}-(n_1v_3-n_3v_1)=v_r\frac{\partial v_\theta}{\partial r}+\frac{v_\theta}{r}\frac{\partial v_\theta}{\partial \theta}+v_z\frac{\partial v_\theta}{\partial z}+\frac{v_rv_\theta}{r} \tag{A.87}$$

$$\frac{1}{h_3}\frac{\partial}{\partial x_3}\left\{\frac{1}{2}\left(v_1^2+v_2^2+v_3^2\right)\right\}-(n_2v_1-n_1v_2)=v_r\frac{\partial v_z}{\partial r}+\frac{v_\theta}{r}.\frac{\partial v_z}{\partial \theta}+v_z\frac{\partial v_z}{\partial z} \tag{A.88}$$

Again, from (A.42)

$$g_1=\frac{1}{r}\left[\frac{\partial n_3}{\partial \theta}-\frac{\partial}{\partial z}(rn_2)\right] \tag{A.89}$$

$$g_2=\frac{\partial n_1}{\partial z}-\frac{\partial n_3}{\partial r} \tag{A.89a}$$

$$g_3=\frac{1}{r}\left[\frac{\partial}{\partial r}(rn_2)-\frac{\partial n_1}{\partial \theta}\right] \tag{A.89b}$$

and from (A.16), the divergence of the vector $\vec{V}$ is given by

$$div\vec{V}=\frac{1}{r}\left[\frac{\partial}{\partial r}(rv_1)+\frac{\partial v_\theta}{\partial \theta}+\frac{\partial}{\partial z}(rv_2)\right] \tag{A.89c}$$

Differentiating (A.89c) with respect to x_1, x_2, and x_3, respectively, and then dividing each by h_1, h_2, and h_3, we get

$$\frac{1}{h_1}\frac{\partial}{\partial x_1}(div\vec{V}) = \frac{\partial}{\partial r}\left[\frac{1}{r}\left\{\frac{\partial}{\partial r}(rv_r) + \frac{\partial v_\theta}{\partial \theta} + \frac{\partial}{\partial z}(rv_z)\right\}\right] \tag{A.89d}$$

$$\frac{1}{h_2}\frac{\partial}{\partial x_2}(div\vec{V}) = \frac{1}{r}\frac{\partial}{\partial \theta}\left[\frac{1}{r}\left\{\frac{\partial}{\partial r}(rv_r) + \frac{\partial v_\theta}{\partial \theta} + \frac{\partial}{\partial z}(rv_z)\right\}\right] \tag{A.89e}$$

$$\frac{1}{h_3}\frac{\partial}{\partial x_3}(div\vec{V}) = \frac{\partial}{\partial z}\left[\frac{1}{r}\left\{\frac{\partial}{\partial r}(rv_r) + \frac{\partial v_\theta}{\partial \theta} + \frac{\partial}{\partial z}(rv_z)\right\}\right] \tag{A.89f}$$

Subtracting (A.89) from (A.89d), we get after cumbersome calculations

$$\frac{1}{h_1}\frac{\partial}{\partial x_1}(div\vec{V}) - g_1 = \nabla^2 v_r - \frac{v_r}{r^2} - \frac{2}{r^2}\frac{\partial v_\theta}{\partial \theta} \tag{A.89g}$$

where ∇^2 is defined by

$$\nabla^2 = \frac{\partial^2}{\partial r^2} + \frac{1}{r}\frac{\partial}{\partial r} + \frac{1}{r^2}\frac{\partial^2}{\partial \theta^2} + \frac{\partial^2}{\partial z^2} \tag{A.89h}$$

Similarly, proceeding in the same way we have on subtractions of (A.89a) from (A.89e) and (A.89b) from (A.89f)

$$\frac{1}{h_2}\frac{\partial}{\partial x_2}(div\vec{V}) - g_2 = \nabla^2 v_\theta - \frac{v_\theta}{r^2} + \frac{2}{r^2}\frac{\partial v_r}{\partial \theta} \tag{A.89i}$$

and

$$\frac{1}{h_3}\frac{\partial}{\partial x_3}(div\vec{V}) - g_3 = \nabla^2 v_z \tag{A.89j}$$

Then with the help of the relations ranging from (A.86) to (A.88) and (A.89g) to (A.89j) except (A.89h), the equations of motion (A.44), (A.45), and (A.46) become

$$\frac{Dv_r}{Dt} - \frac{v_\theta^2}{r} = F_r - \frac{1}{\rho}\frac{\partial p}{\partial r} + \nu\left[\nabla^2 v_r - \frac{v_r}{r^2} - \frac{2}{r^2}\frac{\partial v_\theta}{\partial \theta}\right] \tag{A.90}$$

$$\frac{Dv_\theta}{Dt} + \frac{v_r v_\theta}{r} = F_\theta - \frac{1}{\rho r}\frac{\partial p}{\partial \theta} + \nu\left[\nabla^2 v_\theta - \frac{v_\theta}{r^2} + \frac{2}{r^2}\frac{\partial v_r}{\partial \theta}\right] \tag{A.91}$$

$$\frac{Dv_z}{Dt} = F_z - \frac{1}{\rho}\frac{\partial p}{\partial z} + \nu\nabla^2 v_z \tag{A.92}$$

where

$$\frac{D}{Dt} = \frac{\partial}{\partial t} + v_r \frac{\partial}{\partial r} + \frac{v_\theta}{r}\frac{\partial}{\partial \theta} + v_z \frac{\partial}{\partial z} \tag{A.93}$$

Here F_r, F_θ, and F_z are the components of the body force $\vec{F}$ given by

$$\vec{F} = F_r i_1 + F_\theta j_1 + F_z k_1 \tag{A.94}$$

in the directions of $r, \theta,$ and z increasing. The equations (A.90), (A.91), and (A.92) are the equations of motion in the cylindrical polar coordinates derived from the equations of motion in the orthogonal curvilinear coordinates. The equation of continuity in polar coordinates may be obtained from (A.92) by putting $div\vec{V} = 0$, that is,

$$\frac{\partial}{\partial r}(r v_r) + \frac{\partial v_\theta}{\partial \theta} + \frac{\partial}{\partial z}(r v_z) = 0 \tag{A.95}$$

The rate of strain components in the orthogonal curvilinear coordinates given by (A.25) are converted into cylindrical polar forms by means of relations (A.76), (A.77), and (A.78), and these are given below:

$$\begin{aligned}
e_{rr} &= \frac{\partial v_r}{\partial r} \\
e_{\theta\theta} &= \frac{1}{r}\frac{\partial v_r}{\partial r} + \frac{v_r}{r} \\
e_{zz} &= \frac{\partial v_z}{\partial z} \\
e_{\theta z} &= \frac{1}{r}\frac{\partial v_z}{\partial r} + r\frac{\partial}{\partial z}\left(\frac{v_\theta}{r}\right) \\
e_{zr} &= \frac{\partial v_r}{\partial z} + \frac{\partial v_z}{\partial r} \\
e_{r\theta} &= r\frac{\partial}{\partial r}\left(\frac{v_\theta}{r}\right) + \frac{1}{r}\frac{\partial v_r}{\partial \theta}
\end{aligned} \tag{A.96}$$

The dissipation function $\underline{\Phi}$ in the orthogonal curvilinear coordinates given by (A.30) is

$$\underline{\Phi} = \mu[2\,({e^2}_{rr} + {e^2}_{\theta\theta} + {e^2}_{zz} + {e^2}_{\theta z} + {e^2}_{zr} + {e^2}_{r\theta}] \tag{A.97}$$

where the rate of strain components involved in (A.97) are given by (A.96).

Then we write from (A.48) and (A.49) the expressions of $(\vec{V}\nabla)T$ and the Laplacian of T in terms of cylindrical polar coordinates as

$$(\vec{V}\nabla)T = v_r\frac{\partial T}{\partial r} + \frac{v_\theta}{r}\frac{\partial T}{\partial \theta} + v_z\frac{\partial T}{\partial z} \tag{A.98}$$

and

$$\nabla^2 T = \frac{1}{r}\left[\frac{\partial}{\partial r}\left(r\frac{\partial T}{\partial r}\right) + \frac{\partial}{\partial \theta}\left(\frac{1}{r}\frac{\partial T}{\partial \theta}\right) + \frac{\partial}{\partial z}\left(r\frac{\partial T}{\partial z}\right)\right] \tag{A.99}$$

The equation (A.29), together with the equations ranging from (A.97) to (A.99), constitutes the energy equation that will be solved under certain boundary conditions.

If we consider the steady flow that is free from body force and rotation, then the velocity component in the θ direction should vanish. Also the t-derivative and θ-derivative terms must be discarded from the continuity equations of motion. As a result the system of equations reduces to the forms

$$v_r\frac{\partial v_r}{\partial r} + v_z\frac{\partial v_r}{\partial z} = -\frac{1}{\rho}\frac{\partial p}{\partial r} + \nu\left[\nabla_1^2 v_r - \frac{v_r}{r^2}\right] \tag{A.100}$$

$$v_r\frac{\partial v_z}{\partial r} + v_z\frac{\partial v_z}{\partial z} = -\frac{1}{\rho}\frac{\partial p}{\partial z} + \nu\nabla_1^2 v_z \tag{A.101}$$

$$\frac{\partial}{\partial r}(rv_r) + \frac{\partial}{\partial z}(rv_z) = 0 \tag{A.102}$$

where ∇_1^2 is defined by

$$\nabla_1^2 = \frac{\partial^2}{\partial r^2} + \frac{1}{r}\frac{\partial}{\partial r} + \frac{\partial^2}{\partial z^2} \tag{A.103}$$

For nondimensionalizing the equations (A.100), (A.101), and (A.102), we consider the following transformation:

$$\overline{v_r} = \frac{v_r}{u_o}, \quad \overline{v_z} = \frac{v_z}{u_o}, \quad \overline{r} = \frac{r}{r_o}, \quad \overline{z} = \frac{z}{r_o}, \quad \overline{p} = \frac{p}{\rho u_o^2} \tag{A.104}$$

Here $\overline{v_r}, \overline{v_z}, \overline{r}, \overline{z}, \overline{p}$ are dimensionless quantities and r_o, u_o are the characteristic length and velocity, respectively. Substituting (A.103) in (A.100), (A.101), and (A.102) and then removing the bars we get the nondimensional equations of motion as

$$v_r\frac{\partial v_r}{\partial r} + v_z\frac{\partial v_r}{\partial z} = -\frac{\partial p}{\partial r} + \frac{1}{R_e}\left[\nabla^2 v_r - \frac{v_r}{r^2}\right] \tag{A.105}$$

$$v_r \frac{\partial v_z}{\partial r} + v_z \frac{\partial v_z}{\partial z} = -\frac{\partial p}{\partial z} + \frac{1}{R_e} \nabla^2 v_z \tag{A.106}$$

$$\frac{\partial}{\partial r}(r v_r) + \frac{\partial}{\partial z}(r v_z) = 0 \tag{A.107}$$

where R_e is the Reynolds number defined by

$$R_e = \frac{u_o}{\vartheta} \tag{A.108}$$

A.6 Navier–Stokes Equations of Motion in the Spherical Polar Coordinates System

In this section, we derive the Navier–Stokes equations of motion, continuity equation, and energy equation in spherical polar coordinates from the orthogonal curvilinear coordinate system.

Let P be any position of the point in space with (x, y, z) and (r, θ, φ) as the Cartesian and polar coordinates, respectively. From P draw a perpendicular PN on a plane called the *central plane* that passes through the origin O and is perpendicular to the z-axis. Let N be the foot of PN that lies on the central plane, and then join ON. Again, draw a line from N parallel to the y-axis, meeting it at M, and a perpendicular line from P on the z-axis meeting at D. $OP = r$ is the radius vector, which makes an azimuthal angle θ with the z-axis, and φ is the angle made by the projection of the radius vector on the central plane. Then, we have from Figure A.3

$$\left.\begin{aligned} PD &= r\cos\theta \\ ON &= r\sin\theta \\ OM &= r\sin\theta\cos\varphi \\ NM &= r\sin\theta\sin\varphi \end{aligned}\right\} \tag{A.109}$$

Therefore, the relations between the Cartesian and spherical polar coordinates are

$$\left.\begin{aligned} x &= OM = r\sin\theta\cos\varphi \\ y &= NM = r\sin\theta\sin\varphi \\ z &= PD = r\cos\theta \end{aligned}\right\} \tag{A.110}$$

Let v_r, v_θ, and v_φ be the components of velocity vector $\vec{V}$ where v_r is in the direction of r, v_θ is in the direction of θ perpendicular to v_r, and v_φ is in the direction of φ, respectively.

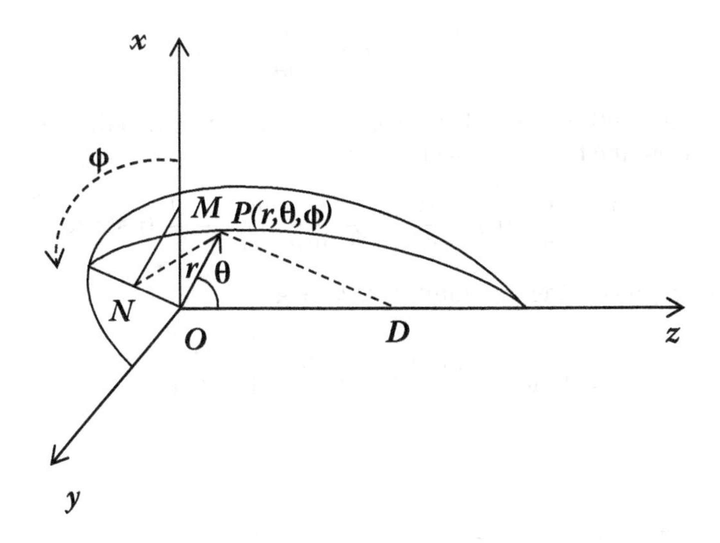

FIGURE A.3
Spherical polar coordinate system.

It is assumed that

$$\left.\begin{aligned} x_1 &= r \\ x_2 &= \theta \\ x_3 &= \varphi \end{aligned}\right\} \tag{A.111}$$

and

$$\left.\begin{aligned} v_1 &= v_r \\ v_2 &= v_\theta \\ v_3 &= v_\varphi \end{aligned}\right\} \tag{A.112}$$

Using (A.110) and (A.111) in (A.13), we find the sale factors $h_1, h_2,$ and h_3 are

$$\left.\begin{aligned} h_1 &= 1 \\ h_2 &= r \\ h_3 &= r \sin\theta \end{aligned}\right\} \tag{A.113}$$

Substituting the relations ranging from (A.111) to (A.113) in (A.18), we have

$$n_1 = \frac{1}{r^2 \sin\theta}\left[\frac{\partial}{\partial\theta}\left(r \sin\theta . v_\varphi\right) - \frac{\partial}{\partial\varphi}\left(r v_\theta\right)\right] \tag{A.114}$$

$$n_2 = \frac{1}{r \sin\theta}\left[\frac{\partial v_r}{\partial\varphi} - \frac{\partial}{\partial r}\left(r \sin\theta . v_\varphi\right)\right] \tag{A.115}$$

$$n_3 = \frac{1}{r}\left[\frac{\partial}{\partial r}(r v_\theta) - \frac{\partial v_r}{\partial \theta}\right] \tag{A.116}$$

Therefore, multiplying (A.116) by v_2 and (A.115) by v_3, and then subtracting the latter from the former, we get

$$n_3 v_2 - n_2 v_3 = \frac{v_\theta}{r}\left[\frac{\partial}{\partial \theta}(r v_\theta) - \frac{\partial v_r}{\partial \theta}\right] - \frac{v_\varphi}{r\sin\theta}\left[\frac{\partial v_r}{\partial \varphi} - \frac{\partial}{\partial r}\left(r\sin\theta v_\varphi\right)\right] \tag{A.117}$$

Likewise, proceeding in a similar way, we get

$$n_1 v_3 - n_3 v_1 = \frac{v_\varphi}{r\sin\theta}\left[\frac{\partial}{\partial \theta}\left(r\sin\theta v_\varphi\right) - \frac{\partial}{\partial \varphi}(r v_\theta)\right] - \frac{v_r}{r}\left[\frac{\partial}{\partial r}(r v_\theta) - \frac{\partial v_r}{\partial \theta}\right] \tag{A.118}$$

and

$$\begin{aligned} n_2 v_1 - n_1 v_2 = {} & \frac{v_r}{r\sin\theta}\left[\frac{\partial v_r}{\partial \varphi} - \frac{\partial}{\partial r}\left(r\sin\theta v_\varphi\right)\right] \\ & - \frac{v_\theta}{r^2\sin\theta}\left[\frac{\partial}{\partial \theta}\left(r\sin\theta v_\varphi\right) - \frac{\partial}{\partial \varphi}(r v_\theta)\right] \end{aligned} \tag{A.119}$$

Then we write

$$\begin{aligned} \frac{1}{h_1}\frac{\partial}{\partial x_1}\left\{\frac{1}{2}\left(v_1^2 + v_2^2 + v_3^2\right)\right\} &= \frac{\partial}{\partial r}\left\{\frac{1}{2}\left(v_r^2 + v_\theta^2 + v_\varphi^2\right)\right\} \\ &= v_r\frac{\partial v_r}{\partial r} + v_\theta\frac{\partial v_\theta}{\partial r} + v_\varphi\frac{\partial v_\varphi}{\partial r} \end{aligned} \tag{A.120}$$

$$\begin{aligned} \frac{1}{h_2}\frac{\partial}{\partial x_2}\left\{\frac{1}{2}\left(v_1^2 + v_2^2 + v_3^2\right)\right\} &= \frac{1}{r}\frac{\partial}{\partial \theta}\left\{\frac{1}{2}\left(v_r^2 + v_\theta^2 + v_\varphi^2\right)\right\} \\ &= \frac{1}{r}\left[v_r\frac{\partial v_r}{\partial \theta} + v_\theta\frac{\partial v_\theta}{\partial \theta} + v_\varphi\frac{\partial v_\varphi}{\partial \theta}\right] \end{aligned} \tag{A.121}$$

$$\begin{aligned} \frac{1}{h_3}\frac{\partial}{\partial x_3}\left\{\frac{1}{2}\left(v_1^2 + v_2^2 + v_3^2\right)\right\} &= \frac{1}{r\sin\theta}\frac{\partial}{\partial \varphi}\left\{\frac{1}{2}\left(v_r^2 + v_\theta^2 + v_\varphi^2\right)\right\} \\ &= \frac{1}{r\sin\theta}\left[v_r\frac{\partial v_r}{\partial \varphi} + v_\theta\frac{\partial v_\theta}{\partial \varphi} + v_\varphi\frac{\partial v_\varphi}{\partial \varphi}\right] \end{aligned} \tag{A.122}$$

Subtracting (A.117) from (A.120), (A.118) from (A.121), and (A.119) from (A.122), we get

$$\begin{aligned} & \frac{1}{h_1}\frac{\partial}{\partial x_1}\left\{\frac{1}{2}\left(v_1^2 + v_2^2 + v_3^2\right)\right\} - (n_3 v_2 - n_2 v_3) \\ &= v_r\frac{\partial v_r}{\partial r} + \frac{v_\theta}{r}\frac{\partial v_r}{\partial \theta} + \frac{v_\varphi}{r\sin\theta}\frac{\partial v_r}{\partial \varphi} - \frac{v_\theta^2 + v_\varphi^2}{r} \end{aligned} \tag{A.123}$$

$$\frac{1}{h_2}\frac{\partial}{\partial x_2}\left\{\frac{1}{2}\left(v_1^2+v_2^2+v_3^2\right)\right\}-(n_1v_3-n_3v_1)$$
$$=v_r\frac{\partial v_\theta}{\partial r}+\frac{v_\theta}{r}\frac{\partial v_\theta}{\partial \theta}+\frac{v_\varphi}{r\sin\theta}\frac{\partial v_\theta}{\partial \varphi}+\frac{v_r v_\theta}{r}-\frac{v_\varphi^2}{r}\cot\theta \tag{A.124}$$

$$\frac{1}{h_3}\frac{\partial}{\partial x_3}\left\{\frac{1}{2}\left(v_1^2+v_2^2+v_3^2\right)\right\}-(n_2v_1-n_1v_2)$$
$$=v_r\frac{\partial v_\varphi}{\partial r}+\frac{v_\theta}{r}\frac{\partial v_\varphi}{\partial \theta}+\frac{v_\varphi}{r\sin\theta}\frac{\partial v_\varphi}{\partial \varphi}+\frac{v_r v_\varphi}{r}-\frac{v_\theta v_\varphi}{r}\cot\theta \tag{A.125}$$

Therefore,

$$\frac{\partial v_1}{\partial t}+\frac{1}{h_1}\frac{\partial}{\partial x_1}\left\{\frac{1}{2}\left(v_1^2+v_2^2+v_3^2\right)\right\}-(n_3v_2-n_2v_3)=\frac{Dv_r}{Dt}-\frac{v_\theta^2+v_\varphi^2}{r} \tag{A.126}$$

$$\frac{\partial v_2}{\partial t}+\frac{1}{h_2}\frac{\partial}{\partial x_2}\left\{\frac{1}{2}\left(v_1^2+v_2^2+v_3^2\right)\right\}-(n_1v_3-n_3v_1)=\frac{Dv_\theta}{Dt}+\frac{v_r v_\theta}{r}-\frac{v_\varphi^2}{r}\cot\theta \tag{A.127}$$

$$\frac{\partial v_3}{\partial t}+\frac{1}{h_3}\frac{\partial}{\partial x_3}\left\{\frac{1}{2}\left(v_1^2+v_2^2+v_3^2\right)\right\}-(n_2v_1-n_1v_2)=\frac{Dv_\varphi}{Dt}+\frac{v_r v_\varphi}{r}-\frac{v_\theta v_\varphi}{r}\cot\theta \tag{A.128}$$

where

$$\frac{D}{Dt}=\frac{\partial}{\partial t}+v_r\frac{\partial}{\partial r}+\frac{v_\theta}{r}\frac{\partial}{\partial \theta}+\frac{v_\varphi}{r\sin\theta}\frac{\partial}{\partial \varphi} \tag{A.129}$$

Again, from (A.42)

$$g_1=\frac{1}{r^2\sin\theta}\frac{\partial}{\partial\theta}\left[\sin\theta\left\{\frac{\partial}{\partial r}(rv_\theta)-\frac{\partial v_r}{\partial\theta}\right\}\right]$$
$$-\frac{1}{r^2\sin\theta}\frac{\partial}{\partial\varphi}\left[\frac{1}{\sin\theta}\left\{\frac{\partial v_r}{\partial\varphi}-\frac{\partial}{\partial r}\left(r\sin v_\varphi\right)\right\}\right] \tag{A.130}$$

$$g_2=\frac{1}{r\sin\theta}\frac{\partial}{\partial\varphi}\left[\frac{1}{r^2\sin\theta}\left\{\frac{\partial}{\partial\theta}\left(r\sin\theta v_\varphi\right)-\frac{\partial}{\partial\varphi}(rv_\theta)\right\}\right]$$
$$-\frac{1}{r\sin\theta}\frac{\partial}{\partial r}\left[\sin\theta\left\{\frac{\partial}{\partial r}(rv_\theta)-\frac{\partial v_r}{\partial\theta}\right\}\right] \tag{A.131}$$

$$g_3 = \frac{1}{r}\frac{\partial}{\partial r}\left[\frac{1}{\sin\theta}\left\{\frac{\partial v_r}{\partial \varphi} - \frac{\partial}{\partial r}\left(r\sin\theta v_\varphi\right)\right\}\right]$$
$$- \frac{1}{r}\frac{\partial}{\partial \theta}\left[\frac{1}{r^2\sin\theta}\left\{\frac{\partial}{\partial \theta}\left(r\sin\theta v_\varphi\right) - \frac{\partial}{\partial \varphi}\left(r v_\theta\right)\right\}\right] \quad (A.132)$$

The divergence (A.16) of the vector $\vec{V}$ in terms of spherical polar coordinates is given by

$$div\vec{V} = \frac{1}{r^2\sin\theta}\left[\frac{\partial}{\partial r}\left(r^2\sin\theta v_r\right) + \frac{\partial}{\partial \theta}\left(r\sin\theta v_\theta\right) + \frac{\partial}{\partial \varphi}\left(r v_\varphi\right)\right] \quad (A.133)$$

Differentiating (A.133) with respect to x_1, then dividing both sides of the resulting equation by h_1 and finally using the relations (A.111) and (A.113), we have

$$\frac{1}{h_1}\frac{\partial}{\partial x_1}\left(div\vec{V}\right) = \frac{\partial}{\partial r}\left[\frac{1}{r^2\sin\theta}\left\{\frac{\partial}{\partial r}\left(r^2\sin\theta v_r\right)\right.\right.$$
$$\left.\left. + \frac{\partial}{\partial \theta}\left(r\sin\theta v_\theta\right) + \frac{\partial}{\partial \varphi}\left(r v_\varphi\right)\right\}\right] \quad (A.134)$$

Subtracting (A.130) from (A.134), we have after somewhat straightforward calculations

$$\frac{1}{h_1}\frac{\partial}{\partial x_1}(div\vec{V}) - g_1 = \nabla^2 v_r - \frac{2v_r}{r^2} - \frac{2}{r^2}\frac{\partial v_\theta}{\partial \theta}$$
$$- \frac{2\cot\theta}{r^2}v_\theta - \frac{2}{r^2\sin\theta}\frac{\partial v_\varphi}{\partial \varphi} \quad (A.135)$$

where

$$\nabla^2 = \frac{1}{r^2}\left[\frac{\partial}{\partial r}\left(r^2\frac{\partial}{\partial r}\right) + \frac{1}{\sin\theta}\frac{\partial}{\partial \theta}\left(\sin\theta\frac{\partial}{\partial \theta}\right) + \frac{1}{\sin^2\theta}\frac{\partial^2}{\partial \varphi^2}\right] \quad (A.136)$$

Similarly, proceeding exactly in the same way as above, we can write

$$\frac{1}{h_2}\frac{\partial}{\partial x_2}(div\vec{V}) - g_2 = \nabla^2 v_\theta + \frac{2}{r^2}\frac{\partial v_r}{\partial \theta} - \frac{v_\theta}{r^2\sin^2\theta} - \frac{2\cos\theta}{r^2\sin^2\theta}\frac{\partial v_\varphi}{\partial \varphi} \quad (A.137)$$

and

$$\frac{1}{h_3}\frac{\partial}{\partial x_3}(div\vec{V}) - g_3 = \nabla^2 v_\varphi - \frac{v_\varphi}{r^2\sin^2\theta} + \frac{2}{r^2\sin^2\theta}\frac{\partial v_r}{\partial \varphi} \quad (A.138)$$

Using the relations ranging from (A.126) to (A.128) and the relations ranging from (A.135) to (A.138) except (A.136), we have from (A.44), (A.45), and (A.46) the equations of motion in spherical polar coordinates after cumbersome calculations as

$$\frac{Dv_r}{Dt} + \frac{v_\theta^2 + v_\varphi^2}{r} = F_r - \frac{1}{\rho}\frac{\partial p}{\partial r}$$
$$+ \nu\left[\nabla^2 v_r - \frac{2v_r}{r^2} - \frac{2}{r^2}\frac{\partial v_\theta}{\partial \theta} - \frac{2v_\theta}{r^2}\cot\theta - \frac{2}{r^2\sin\theta}\frac{\partial v_\varphi}{\partial \varphi}\right]$$
$$(A.139)$$

$$\frac{Dv_\theta}{Dt}+\frac{v_rv_\theta}{r}-\frac{v_\varphi^2\cot\theta}{r}=F_\theta-\frac{1}{\rho r}\frac{\partial p}{\partial\theta}+\nu\left[\nabla^2v_\theta-\frac{2}{r^2}\frac{\partial v_r}{\partial\theta}-\frac{v_\theta}{r^2\sin^2\theta}-\frac{2\cos\theta}{r^2\sin^2\theta}\frac{\partial v_\varphi}{\partial\varphi}\right] \tag{A.140}$$

$$\frac{Dv_\varphi}{Dt}+\frac{v_rv_\varphi}{r}+\frac{v_\theta v_\varphi}{r}\cot\theta=F_r-\frac{1}{\rho r\sin\theta}\frac{\partial p}{\partial\varphi}+\nu\left[\nabla^2v_\varphi-\frac{v_\varphi}{r^2\sin^2\varphi}+\frac{2}{r^2\sin\theta}\frac{\partial v_r}{\partial\varphi}\right] \tag{A.141}$$

where

$$\vec{F}=F_ri_1+F_\theta j_1+F_\varphi k_1 \tag{A.142}$$

F_r, F_θ, and F_φ being the components of body force $\vec{F}$ in the directions of $r, \theta,$ and φ increasing. The equation of continuity in terms of spherical polar coordinates is given by (A.47) as

$$\frac{1}{r}\frac{\partial}{\partial r}\left(r^2v_r\right)+\frac{1}{\sin\theta}\frac{\partial}{\partial\theta}(\sin\theta v_\theta)+\frac{1}{\sin\theta}\frac{\partial v_\varphi}{\partial\varphi}=0 \tag{A.143}$$

Now we proceed to find out the energy equation from (A.29) in the spherical polar coordinates. To do this we write the rate of strain components (A.25) using the relatives ranging from (A.111) to (A.113), and these are given by

$$\begin{aligned}
e_{rr}&=\frac{\partial v_r}{\partial r}\\
e_{\theta\theta}&=\frac{1}{r}\frac{\partial v_\theta}{\partial\theta}+\frac{v_r}{r}\\
e_{\varphi\varphi}&=\frac{1}{r\sin\theta}\frac{\partial v_\varphi}{\partial\varphi}+\frac{v_r}{r\sin\theta}+\frac{v_\theta}{r}\cot\theta\\
e_{\theta\varphi}&=\sin\theta\frac{\partial}{\partial\theta}\left(\frac{\partial v_\varphi}{r\sin\theta}\right)+\frac{1}{\sin\theta}\frac{\partial}{\partial\varphi}\left(\frac{v_\theta}{r}\right)\\
e_{\varphi r}&=\frac{1}{r\sin\theta}\frac{\partial v_r}{\partial\varphi}+r\sin\theta\frac{\partial}{\partial r}\left(\frac{\partial v_\varphi}{r\sin\theta}\right)\\
e_{r\theta}&=r\frac{\partial}{\partial r}\left(\frac{v_\theta}{r}\right)+\frac{1}{r}\frac{\partial v_r}{\partial\theta}
\end{aligned} \tag{A.144}$$

The dissipation function $\underline{\overline{\emptyset}}$ is given by (A.30) as

$$\underline{\overline{\Phi}}=\mu\left[2\left({e_{rr}}^2+{e_{\theta\theta}}^2+{e_{\varphi\varphi}}^2\right)+{e_{\theta\varphi}}^2+{e_{\varphi r}}^2+{e_{r\theta}}^2\right] \tag{A.145}$$

where the rate of strain components involved in (A.145) are given by (A.144).

Then we write the Laplacian of T in spherical polar coordinates and it is given by (A.24) changing $\overline{\Phi}$ by T as

$$\nabla^2 T = \frac{1}{r^2 \sin\theta}\left[\frac{\partial}{\partial r}\left(r^2 \sin\theta \frac{\partial T}{\partial r}\right) + \frac{\partial}{\partial \theta}\left(\sin\theta \frac{\partial T}{\partial \theta}\right) + \frac{\partial}{\partial \varphi}\left(\frac{1}{\sin\theta}\frac{\partial T}{\partial \varphi}\right)\right] \quad \text{(A.146)}$$

and from (A.48)

$$(\vec{V}.\nabla)T = v_r \frac{\partial T}{\partial r} + \frac{v_\theta}{r}\frac{\partial T}{\partial \theta} + \frac{v_\varphi}{r \sin\theta}\frac{\partial T}{\partial \varphi} \quad \text{(A.147)}$$

The equation (A.29) together with the equations (A.145), (A.146), and (A.147) constitutes the energy equation in spherical polar coordinates.

A.7 Navier–Stokes Equations of Motion in a Curved Pipe with Constant Curvature

In this section, we derive the Navier–Stokes equations of motion from the orthogonal curvilinear coordinates system when fluid flows through a curved pipe of a circular cross section coiled in the form of a circle of radius R.

Let C be the center of the cross section of the pipe made by a plane through C and OZ. This plane makes an angle φ with the fixed-axial plane. Now CO is drawn perpendicular to OZ and lies in the plane through C, which is perpendicular to OZ. The length of CO is R, which is the radius of a circle in which the pipe is coiled. The circumference of the circle of radius R traced out by C is called the *central line*, and the plane through O and perpendicular to OZ is called the *central plane*.

Let the position of any point in the cross section of the pipe be P whose orthogonal coordinates are (r, θ, φ) with respect to C as the origin. A perpendicular from P to OC is drawn meeting the latter at N such that

$$\left.\begin{aligned} CN &= r\cos\theta \\ PN &= r\sin\theta \\ CP &= r \end{aligned}\right\} \quad \text{(A.148)}$$

If x, y, s_1 are the orthogonal coordinates of the point with respect to the origin O, then the relations between the coordinates (r, θ, φ) and (x, y, s_1) are given by Figure A.4

$$\begin{aligned} x &= OC + CN = R + r\cos\theta \\ y &= PN = r\sin\theta \\ s_1 &= (OC + CN)\varphi = (R + r\cos\theta)\varphi \end{aligned} \quad \text{(A.149)}$$

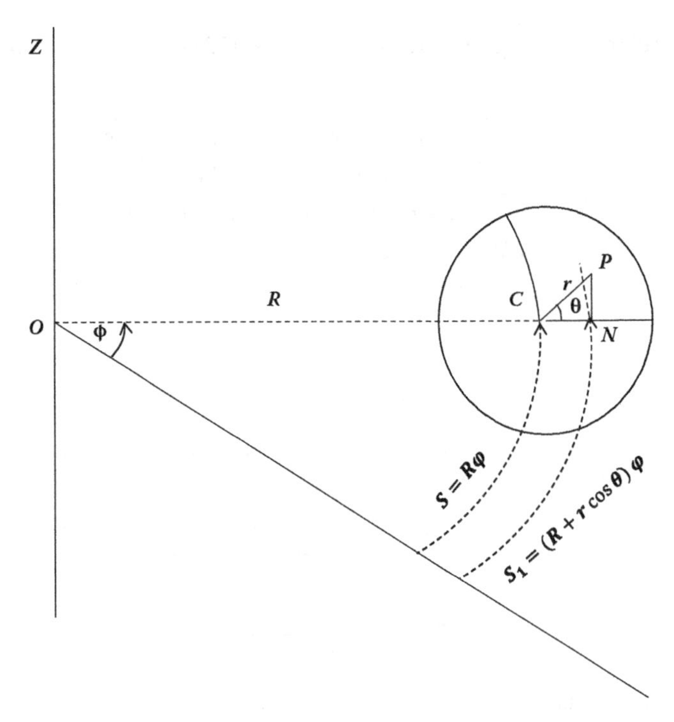

FIGURE A.4
Toroidal coordinate system.

Let U, V, and W be the components of the velocity vector with respect to C where U is in the direction of r, V is in the direction of θ perpendicular to U, and W is in the direction of φ, respectively. Then we assume that

$$\left.\begin{aligned} x_1 &= r \\ x_2 &= \theta \\ x_3 &= \varphi \end{aligned}\right\} \tag{A.150}$$

and

$$\left.\begin{aligned} v_1 &= U \\ v_2 &= V \\ v_3 &= W \end{aligned}\right\} \tag{A.151}$$

Substituting (A.149) and (A.150) in (A.13), we get the scale factors h_1, h_2, and h_3 as

$$\begin{aligned} h_1 &= 1 \\ h_2 &= r \\ h_3 &= R + r\cos\theta \end{aligned} \tag{A.152}$$

Using the relations ranging from (A.150) to (A.152) in (A.18), we have

$$n_1 = \frac{1}{r}\frac{\partial W}{\partial \theta} - \frac{W\sin\theta}{R+r\cos\theta} - \frac{R}{R+r\cos\theta}.\frac{\partial V}{\partial s} \tag{A.153}$$

$$n_2 = \frac{R}{R+r\cos\theta}.\frac{\partial U}{\partial s} - \frac{\partial W}{\partial r} - \frac{W\cos\theta}{R+r\cos\theta} \tag{A.154}$$

$$n_3 = \frac{\partial V}{\partial r} + \frac{V}{r} - \frac{1}{r}.\frac{\partial U}{\partial \theta} \tag{A.155}$$

where

$$s = R\varphi \tag{A.156}$$

Therefore, multiplying (A.155) by v_2 and (A.154) by v_3 and then subtracting the latter from the former, we have

$$\begin{aligned} n_3v_2 - n_2v_3 = {} & V\frac{\partial V}{\partial r} + \frac{V^2}{r} - \frac{V}{r}\frac{\partial U}{\partial \theta} - \frac{WR}{R+r\cos\theta}\frac{\partial U}{\partial s} \\ & + W\frac{\partial W}{\partial r} + \frac{W^2\cos\theta}{R+r\cos\theta} \end{aligned} \tag{A.157}$$

Likewise, proceeding in the same way, we get

$$\begin{aligned} n_1v_3 - n_3v_1 = {} & \frac{W}{r}\frac{\partial W}{\partial \theta} - \frac{W^2\sin\theta}{R+r\cos\theta} - \frac{RW}{R+r\cos\theta}\frac{\partial V}{\partial s} \\ & - U\frac{\partial V}{\partial r} - \frac{UV}{r} + \frac{U}{r}\frac{\partial U}{\partial \theta} \end{aligned} \tag{A.158}$$

and

$$\begin{aligned} n_2v_1 - n_1v_2 = {} & \frac{RU}{R+r\cos\theta}\frac{\partial U}{\partial s} - U\frac{\partial W}{\partial r} - \frac{UV\cos\theta}{R+r\cos\theta} - \frac{V}{r}\frac{\partial W}{\partial \theta} \\ & + \frac{VW\sin\theta}{R+r\cos\theta} + \frac{RV}{R+r\cos\theta}\frac{\partial V}{\partial s} \end{aligned} \tag{A.159}$$

Then we write

$$\begin{aligned} \frac{1}{h_1}\frac{\partial}{\partial x_1}\left\{\frac{1}{2}\left(v_1^2+v_2^2+v_3^2\right)\right\} &= \frac{\partial}{\partial r}\left\{\frac{1}{2}\left(U^2+V^2+W^2\right)\right\} \\ &= U\frac{\partial U}{\partial r} + V\frac{\partial V}{\partial r} + W\frac{\partial W}{\partial r} \end{aligned} \tag{A.160}$$

$$\begin{aligned} \frac{1}{h_2}\frac{\partial}{\partial x_2}\left\{\frac{1}{2}\left(v_1^2+v_2^2+v_3^2\right)\right\} &= \frac{1}{r}\frac{\partial}{\partial \theta}\left\{\frac{1}{2}\left(U^2+V^2+W^2\right)\right\} \\ &= \frac{1}{r}\left[U\frac{\partial U}{\partial \theta} + V\frac{\partial V}{\partial \theta} + W\frac{\partial W}{\partial \theta}\right] \end{aligned} \tag{A.161}$$

$$\frac{1}{h_3}\frac{\partial}{\partial x_3}\left\{\frac{1}{2}\left(U^2+V^2+W^2\right)\right\} = \frac{1}{R+r\cos\theta}\frac{\partial}{\partial\phi}\left\{\frac{1}{2}\left(U^2+V^2+W^2\right)\right\}$$
$$= \frac{1}{R+r\cos\theta}\left[U\frac{\partial U}{\partial\phi}+V\frac{\partial V}{\partial\phi}+W\frac{\partial W}{\partial\phi}\right] \tag{A.162}$$

Subtracting (A.157) from (A.160), we have

$$\frac{1}{h_1}\frac{\partial}{\partial x_1}\left\{\frac{1}{2}\left({v_1}^2+{v_2}^2+{v_3}^2\right)\right\}-(n_3v_2-n_2v_3)$$
$$=U\frac{\partial U}{\partial r}+\frac{V}{r}\frac{\partial U}{\partial\theta}-\frac{V^2}{r}+\frac{WR}{R+r\cos\theta}.\frac{\partial U}{\partial s}-\frac{W^2\cos\theta}{R+r\cos\theta} \tag{A.163}$$

Similarly, subtracting (A.158) from (A.161) and (A.159) from (A.162), we get

$$\frac{1}{h_2}\frac{\partial}{\partial x_2}\left\{\frac{1}{2}\left({v_1}^2+{v_2}^2+{v_3}^2\right)\right\}-(n_1v_3-n_3v_1)$$
$$=U\frac{\partial V}{\partial r}+\frac{V}{r}\frac{\partial V}{\partial\theta}+\frac{RW}{R+r\cos\theta}.\frac{\partial V}{\partial s}+\frac{UV}{r}+\frac{W^2\sin\theta}{R+r\cos\theta} \tag{A.164}$$

and

$$\frac{1}{h_3}\frac{\partial}{\partial x_3}\left\{\frac{1}{2}\left({v_1}^2+{v_2}^2+{v_3}^2\right)\right\}-(n_2v_1-n_1v_2)$$
$$=U\frac{\partial W}{\partial r}+\frac{UW\cos\theta}{R+r\cos\theta}+\frac{V}{r}\frac{\partial W}{\partial\theta}-\frac{VW\sin\theta}{R+r\cos\theta}+\frac{RW}{R+r\cos\theta}.\frac{\partial W}{\partial s} \tag{A.165}$$

Therefore,

$$\frac{\partial v_1}{\partial t}+\frac{1}{h_1}\frac{\partial}{\partial x_1}\left\{\frac{1}{2}(v_1^2+v_2^2+v_3^2)\right\}-(n_3v_2-n_2v_3)$$
$$=\frac{DU}{Dt}-\frac{V^2}{r}-\frac{W^2\cos\theta}{R+r\cos\theta} \tag{A.166}$$

$$\frac{\partial v_2}{\partial t}+\frac{1}{h_2}\frac{\partial}{\partial x_2}\left\{\frac{1}{2}\left(v_1^2+v_2^2+v_3^2\right)\right\}-(n_1v_3-n_3v_1)$$
$$=\frac{DV}{Dt}+\frac{UV}{r}+\frac{W^2\sin\theta}{R+r\cos\theta} \tag{A.167}$$

$$\frac{\partial v_3}{\partial t}+\frac{1}{h_3}\frac{\partial}{\partial x_3}\left\{\frac{1}{2}(v_1^2+v_2^2+v_3^2)\right\}-(n_2v_1-n_1v_2)$$
$$=\frac{DW}{Dt}+\frac{U\cos\theta-V\sin\theta}{R+r\cos\theta}W \tag{A.168}$$

where

$$\frac{D}{Dt} = \frac{\partial}{\partial t} + U\frac{\partial}{\partial r} + \frac{V}{r}\frac{\partial}{\partial \theta} + \frac{RW}{R + r\cos\theta}.\frac{\partial}{\partial s} \tag{A.169}$$

The divergence (A.16) of the velocity vector $\vec{V}$ in terms of the toroidal coordinates $[r, \theta, \varphi]$ is given by

$$div\vec{V} = \frac{\partial U}{\partial r} + \frac{U}{r} + \frac{U\cos\theta - V\sin\theta}{R + r\cos\theta} + \frac{1}{r}\frac{\partial V}{\partial \theta} + \frac{R}{R + r\cos\theta}.\frac{\partial W}{\partial s} \tag{A.170}$$

Differentiating (A.170) with respect to x_1, then dividing both sides of the resulting equation by h_1 and finally using the relations (A.150) and (A.152), we get

$$\frac{1}{h_1}.\frac{\partial}{\partial x_1}(div\vec{V}) = \frac{\partial}{\partial r}\left[\frac{\partial U}{\partial r} + \frac{U}{r} + \frac{U\cos\theta - V\sin\theta}{R + r\cos\theta} + \frac{1}{r}\frac{\partial V}{\partial \theta} + \frac{R}{R + r\cos\theta}.\frac{\partial W}{\partial s}\right] \tag{A.171}$$

Proceeding exactly in the same way as above, we have

$$\frac{1}{h_2}.\frac{\partial}{\partial x_2}(div\vec{V}) = \frac{1}{r}\frac{\partial}{\partial \theta}\left[\frac{\partial U}{\partial r} + \frac{U}{r} + \frac{U\cos\theta - V\sin\theta}{R + r\cos\theta} + \frac{1}{r}\frac{\partial V}{\partial \theta} + \frac{R}{R + r\cos\theta}.\frac{\partial W}{\partial s}\right] \tag{A.172}$$

and

$$\frac{1}{h_3}.\frac{\partial}{\partial x_3}(div\vec{V}) = \frac{1}{R + r\cos\theta}\frac{\partial}{\partial \varphi}\left[\frac{\partial U}{\partial r} + \frac{U}{r} + \frac{U\cos\theta - V\sin\theta}{R + r\cos\theta} + \frac{1}{r}\frac{\partial V}{\partial \theta} + \frac{R}{R + r\cos\theta}.\frac{\partial W}{\partial s}\right] \tag{A.173}$$

Again from (A.42)

$$g_1 = \frac{1}{r(R + \cos\theta)}\left[\frac{\partial}{\partial \theta}\left\{(R + r\cos\theta)\left(\frac{\partial V}{\partial r} + \frac{V}{r} - \frac{1}{r}\frac{\partial U}{\partial \theta}\right)\right\} - \frac{\partial}{\partial \phi}\left(\frac{Rr}{R + r\cos\theta}.\frac{\partial U}{\partial s} - r\frac{\partial W}{\partial r} - \frac{Wr\cos\theta}{R + r\cos\theta}\right)\right] \tag{A.174}$$

$$g_2 = \frac{1}{r(R + \cos\theta)}\left[\frac{\partial}{\partial \phi}\left(\frac{1}{r}\frac{\partial W}{\partial \theta} - \frac{W\sin\theta}{R + r\cos\theta} - \frac{R}{R + r\cos\theta}\frac{\partial V}{\partial s}\right) - \frac{\partial}{\partial r}\left\{(R + r\cos\theta)\left(\frac{\partial V}{\partial r} + \frac{V}{r} - \frac{1}{r}\frac{\partial U}{\partial \theta}\right)\right\}\right] \tag{A.175}$$

$$g_3 = \frac{1}{r}\left[\frac{\partial}{\partial r}\left\{r\left(\frac{R}{R+r\cos\theta}\frac{\partial U}{\partial s} - \frac{\partial W}{\partial r} - \frac{W\cos\theta}{R+r\cos\theta}\right)\right\} - \frac{\partial}{\partial \theta}\left(\frac{1}{r}\frac{\partial W}{\partial \theta} - \frac{W\sin\theta}{R+r\cos\theta} - \frac{R}{R+r\cos\theta}\frac{\partial V}{\partial s}\right)\right] \quad \text{(A.176)}$$

Subtracting (A.174) from (A.171), we have after cumbersome calculations

$$\begin{aligned}\frac{1}{h_1}\frac{\partial}{\partial x_1}(div\vec{V}) - g_1 &= \nabla^2 U + \frac{\cos\theta}{R+r\cos\theta}\frac{\partial U}{\partial r} - \frac{\sin\theta}{r\,(R+r\cos\theta)}\frac{\partial U}{\partial \theta} - \left\{\frac{1}{r^2} + \frac{\cos^2\theta}{(R+r\cos\theta)^2}\right\}U \\ &\quad - \frac{2}{r^2}\frac{\partial V}{\partial \theta} + \frac{V\sin\theta}{R+r\cos\theta} \times \left(\frac{1}{r} + \frac{\cos\theta}{R+r\cos\theta}\right) \\ &\quad + \left[\frac{R^2}{(R+r\cos\theta)^2}\frac{\partial^2 U}{\partial s^2} - \frac{2R\cos\theta}{(R+r\cos\theta)^2}\frac{\partial W}{\partial s}\right]\end{aligned} \quad \text{(A.177)}$$

Similarly, subtracting (A.175) from (A.172) and (A.176) from (A.173), we get

$$\begin{aligned}\frac{1}{h_2}\frac{\partial}{\partial x_2}(div\vec{V}) - g_2 &= \nabla^2 V + \frac{\cos\theta}{R+r\cos\theta}\frac{\partial V}{\partial r} - \frac{\sin\theta}{r(R+r\cos\theta)}\frac{\partial V}{\partial \theta} - \left\{\frac{1}{r^2} + \frac{\sin^2\theta}{(R+r\cos\theta)^2}\right\}V \\ &\quad + \frac{2}{r^2}\frac{\partial U}{\partial \theta} + \frac{R}{(R+r\cos\theta)^2}\frac{\partial^2 V}{\partial s^2} + \left(\cos\theta - \frac{1}{r}\right)\frac{U\sin\theta}{(R+r\cos\theta)^2}\end{aligned} \quad \text{(A.178)}$$

and

$$\begin{aligned}\frac{1}{h_3}\frac{\partial}{\partial x_3}(div\vec{V}) - g_3 &= \nabla^2 W + \frac{\cos\theta}{R+r\cos\theta}\frac{\partial W}{\partial r} - \frac{\sin\theta}{r(R+r\cos\theta)}\frac{\partial W}{\partial \theta} - \frac{W}{(R+r\cos\theta)^2} \\ &\quad + \frac{2R\cos\theta}{(R+r\cos\theta)^2}\frac{\partial U}{\partial s} - \frac{2R\sin\theta}{(R+r\cos\theta)^2}\frac{\partial V}{\partial s} + \frac{R^2}{(R+r\cos\theta)^2}\frac{\partial^2 W}{\partial s^2}\end{aligned} \quad \text{(A.179)}$$

where the Laplacian operator ∇^2 is defined by

$$\nabla^2 = \frac{\partial^2}{\partial r^2} + \frac{1}{r}\frac{\partial}{\partial r} + \frac{1}{r^2}\frac{\partial^2}{\partial \theta^2} \quad \text{(A.180)}$$

Using the equations ranging from (A.166) to (A.168) and the equations ranging from (A.177) to (A.179), we have from (A.44), (A.45), and (A.46) the equations of motion in toroidal coordinates after somewhat straightforward calculations

$$
\begin{aligned}
&\frac{DU}{Dt} - \frac{V^2}{r} - \frac{W^2\cos\theta}{R + r\cos\theta} \\
&\quad = f_r - \frac{1}{\rho}\frac{\partial p}{\partial r} + \nu\left[\nabla^2 U + \frac{\cos\theta}{R + r\cos\theta}\frac{\partial U}{\partial r} - \frac{\sin\theta}{r\,(R + r\cos\theta)}\frac{\partial U}{\partial \theta}\right. \\
&\qquad - \left\{\frac{1}{r^2} + \frac{\cos^2\theta}{(R + r\cos\theta)^2}\right\}U - \frac{2}{r^2}\frac{\partial V}{\partial \theta} + \frac{V\sin\theta}{R + r\cos\theta}\left\{\frac{1}{r} + \frac{\cos\theta}{R + r\cos\theta}\right\} \\
&\qquad \left. + \left\{\frac{R^2}{(R + r\cos\theta)^2}\frac{\partial^2 U}{\partial s^2} - \frac{2R\cos\theta}{(R + r\cos\theta)^2}\frac{\partial W}{\partial s}\right\}\right]
\end{aligned}
\tag{A.181}
$$

$$
\begin{aligned}
&\frac{DV}{Dt} + \frac{UV}{r} + \frac{W^2\sin\theta}{R + r\cos\theta} \\
&\quad = f_\theta - \frac{1}{\rho r}\frac{\partial p}{\partial \theta} + \nu\left[\nabla^2 V + \frac{\cos\theta}{R + r\cos\theta}\frac{\partial V}{\partial r} - \frac{\sin\theta}{r\,(R + r\cos\theta)}\frac{\partial V}{\partial \theta}\right. \\
&\qquad - \left\{\frac{1}{r^2} + \frac{\sin^2\theta}{(R + r\cos\theta)^2}\right\}V + \left(\cos\theta - \frac{1}{r}\right)\frac{U\sin\theta}{(R + r\cos\theta)^2} \\
&\qquad \left. + \frac{R}{(R + r\cos\theta)^2}\frac{\partial^2 V}{\partial s^2}\right]
\end{aligned}
\tag{A.182}
$$

and

$$
\begin{aligned}
&\frac{DW}{Dt} + \frac{U\cos\theta - V\sin\theta}{R + r\cos\theta} \\
&\quad = f_\varphi - \frac{1}{\rho\,(R + r\cos\theta)}\frac{\partial p}{\partial \varphi} \\
&\qquad + \nu\left[\nabla^2 W + \frac{\cos\theta}{R + r\cos\theta}\frac{\partial W}{\partial r} - \frac{\sin\theta}{r\,(R + r\cos\theta)}\frac{\partial W}{\partial \theta} - \frac{W}{(R + r\cos\theta)^2}\right. \\
&\qquad \left. + \frac{2R\cos\theta}{(R + r\cos\theta)^2}\frac{\partial U}{\partial s} - \frac{2R\sin\theta}{(R + r\cos\theta)^2}\frac{\partial V}{\partial s} + \frac{R^2}{(R + r\cos\theta)^2}\frac{\partial^2 W}{\partial s^2}\right]
\end{aligned}
\tag{A.183}
$$

where

$$
\vec{F} = f_r i_1 + f_\theta j_1 + f_\varphi k_1 \tag{A.184}
$$

Here f_r, f_θ, and f_φ are the components of the body force $\vec{F}$ in the direction of $r, \theta,$ and φ increasing. The equation of continuity can be obtained from

(A.47) or (A.170) by putting $div\vec{V} = 0$, that is,

$$\frac{1}{r}\frac{\partial}{\partial r}(rU) + \frac{U\cos\theta - V\sin\theta}{R + r\cos\theta} + \frac{1}{r}\frac{\partial V}{\partial \theta} + \frac{R}{R + r\cos\theta}\frac{\partial W}{\partial s} = 0 \quad \text{(A.185)}$$

If we consider the incompressible steady laminar and fully developed flow of fluid in the curved pipe coiled in the form of a circle, then the time derivative terms in the momentums equations should be discarded. Besides this, the velocity components U, V, and W except the pressure are all independent of φ. If the body force $\vec{F}$ is neglected, then the system of equations reduces to the following forms:

$$\begin{aligned} &U\frac{\partial U}{\partial r} + \frac{V}{r}\frac{\partial U}{\partial \theta} - \frac{V^2}{r} - \frac{W^2\cos\theta}{R + r\cos\theta} \\ &= -\frac{1}{\rho}\frac{\partial p}{\partial r} + \nu\left[\nabla^2 U + \frac{\cos\theta}{R + r\cos\theta}\frac{\partial U}{\partial r} - \frac{\sin\theta}{r(R + r\cos\theta)}\frac{\partial U}{\partial \theta}\right. \\ &\left. - \left\{\frac{1}{r^2} + \frac{\cos^2\theta}{(R + r\cos\theta)^2}\right\}U - \frac{2}{r^2}\frac{\partial V}{\partial \theta} + \frac{V\sin\theta}{R + r\cos\theta}\left(\frac{1}{r} + \frac{\cos\theta}{R + r\cos\theta}\right)\right] \end{aligned} \quad \text{(A.186)}$$

$$\begin{aligned} &U\frac{\partial V}{\partial r} + \frac{V}{r}\frac{\partial V}{\partial \theta} + \frac{UV}{r} + \frac{W^2\cos\theta}{R + r\cos\theta} \\ &= -\frac{1}{\rho r}\frac{\partial p}{\partial \theta} + \nu\left[\nabla^2 V + \frac{\cos\theta}{R + r\cos\theta}\frac{\partial V}{\partial r} - \frac{\sin\theta}{r(R + r\cos\theta)}\frac{\partial V}{\partial \theta}\right. \\ &\left. - \left\{\frac{1}{r^2} + \frac{\sin^2\theta}{(R + r\cos\theta)^2}\right\}V + \frac{2}{r^2}\frac{\partial U}{\partial \theta} + \left(\cos\theta - \frac{1}{r}\right)\frac{U\sin\theta}{(R + r\cos\theta)^2}\right] \end{aligned} \quad \text{(A.187)}$$

$$\begin{aligned} &U\frac{\partial W}{\partial r} + \frac{V}{r}\frac{\partial W}{\partial \theta} + \frac{U\cos\theta - V\sin\theta}{R + r\cos\theta}W = -\frac{1}{\rho(R + r\cos\theta)}\frac{\partial p}{\partial \varphi} \\ &+ \nu\left[\nabla^2 W + \frac{\cos\theta}{R + r\cos\theta}\frac{\partial W}{\partial r} - \frac{\sin\theta}{r(R + r\cos\theta)}\frac{\partial W}{\partial \theta} - \frac{W}{(R + r\cos\theta)^2}\right] \end{aligned} \quad \text{(A.188)}$$

$$\frac{1}{r}\frac{\partial}{\partial r}(rU) + \frac{1}{r}\frac{\partial V}{\partial \theta} + \frac{U\cos\theta - V\sin\theta}{R + r\cos\theta} = 0 \quad \text{(A.189)}$$

Now we proceed to find out the energy equation from (A.29) in terms of the toroidal coordinates for the flow in a curved pipe coiled in the form of a circle. To do this we write the rate of strain components (A.25) using the relations ranging from (A.150) to (A.152), and these are given by

$$
\begin{aligned}
e_{rr} &= \frac{\partial U}{\partial r} \\
e_{\theta\theta} &= \frac{1}{r}\frac{\partial V}{\partial \theta} + \frac{U}{r} \\
e_{\varphi\varphi} &= \frac{1}{R + r\cos\theta}\frac{\partial W}{\partial \varphi} + \frac{U}{R + r\cos\theta} - \frac{V\sin\theta}{R + r\cos\theta} \\
e_{\theta\varphi} &= \frac{R + r\cos\theta}{r}\frac{\partial}{\partial\theta}\left(\frac{W}{R + r\cos\theta}\right) + \frac{r}{R + r\cos\theta}\frac{\partial}{\partial\varphi}\left(\frac{V}{r}\right) \\
e_{\varphi r} &= \frac{1}{R + r\cos\theta}\frac{\partial U}{\partial \varphi} + (R + r\cos\theta)\frac{\partial}{\partial r}\left(\frac{W}{R + r\cos\theta}\right) \\
e_{r\theta} &= r\frac{\partial}{\partial r}\left(\frac{V}{r}\right) + \frac{1}{r}\frac{\partial U}{\partial \theta}
\end{aligned}
\tag{A.190}
$$

The dissipation function $\underline{\Phi}$ is given by (A.30) and its terms of toroidal coordinates as

$$
\underline{\Phi} = \mu[2\left(e^2{}_{rr} + e^2{}_{\theta\theta} + e^2{}_{\varphi\varphi} + e^2{}_{\theta\varphi} + e^2{}_{\varphi r} + e^2{}_{r\theta}\right]
\tag{A.191}
$$

where the rate of strain components involved in (A.191) are given by (A.190).

Then we write the Laplacian of T in terms of toroidal coordinates changing $\underline{\Phi}$ by T in (A.24) as

$$
\begin{aligned}
\nabla^2 T = {} & \frac{1}{r(R + r\cos\theta)}\left[\frac{\partial}{\partial r}\left\{r(R + r\cos\theta)\frac{\partial T}{\partial r}\right\} + \frac{\partial}{\partial \theta}\left\{\left(\frac{R + r\cos\theta}{r}\right)\frac{\partial T}{\partial \theta}\right\}\right. \\
& \left. + \frac{\partial}{\partial \varphi}\left\{\left(\frac{r}{R + r\cos\theta}\right)\frac{\partial T}{\partial \varphi}\right\}\right]
\end{aligned}
\tag{A.192}
$$

and from (A.48)

$$
(\vec{V}.\nabla)T = U\frac{\partial T}{\partial r} + \frac{V}{r}\frac{\partial T}{\partial \theta} + \frac{W}{R + r\cos\theta}\frac{\partial T}{\partial \varphi}
\tag{A.193}
$$

The equation (A.29) with the help of the relations ranging from (A.191) to (A.193) constitutes the energy equation in terms of coordinates (r, θ, φ).

A.8 Navier–Stokes Equations of Motion in a Curved Pipe with Varying Curvature

We begin with laminar flow of viscous incompressible flow in a sinusoidally curved pipe of a uniform circular cross section whose center line is a

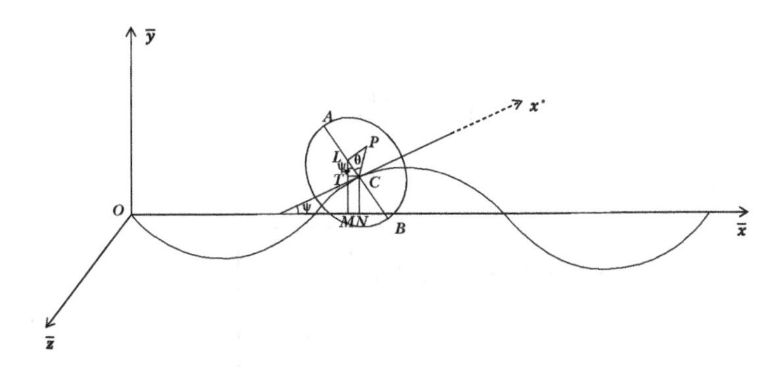

FIGURE A.5
Coordinate system.

twodimensional sinusoidal curve in the $\overline{x}\,\overline{y}$ plane of the $\overline{x}\,\overline{y}\,\overline{z}$ coordinate system. The pipe center line is described by

$$\bar{y} = \bar{A}\sin\bar{K}\bar{x} \tag{A.194}$$

where $\overline{A}$ is the dimensional amplitude of the waviness of the pipe axis and K is the dimensional wave number of waviness of the pipe center-line curve. In Figure A.5, let AB with lies in the $\overline{x}\,\overline{y}$ plane of the $\overline{x}\,\overline{y}\,\overline{z}$ coordinate system be the diameter of a normal section of the pipe at the point C on the pipe center line and in this section let P be any point whose circular coordinates are r^*, θ where r^* makes an angle θ with the diameter AB. The third coordinate axis x^* is taken in the mean flow direction along the tangent at C to the center-line profile of the pipe and the coordinate r^*, θ, x^* system is always orthogonal.

Let LMN and T be the feet of the perpendicular drawn from P on AB, from L and C on the x-axis, and from C on LM, respectively. And all these points lie in the $\overline{x}\,\overline{y}$ plane. If $(\overline{x}\,\overline{y})$ and (x^*, y^*) are the coordinates of L and C with respect to the origin O in the $\overline{x}\,\overline{y}$ plane, then we have

$$\begin{aligned} CL &= r^*\cos\theta \\ OM &= \bar{x},\ LM = \bar{y} \\ ON &= x^*, CN = y^* \\ y^* &= A^*\sin K^* x^* \end{aligned} \tag{A.195}$$

These relations in (A.195) are valid only for the centers of all normal sections of the pipe. Now from the Figure A.5 we get

$$x^* - \bar{x} = ON - OM = MN = CT = CL\sin\psi = r^*\cos\theta\sin\psi \tag{A.196}$$

$$\bar{y} - y^* = LM - CN = LM - TM = LT = CL\cos\psi = r^*\cos\theta\cos\psi \tag{A.197}$$

where ψ is the angle made by the tangent at C with the positive direction of $\bar{x}$-axis.

Now

$$\tan\psi = \frac{dy^*}{dx^*}$$

Therefore,

$$\left.\begin{aligned} \sin\psi &= \frac{dy^*}{dx^*}\bigg/\sqrt{1+\left(\frac{dy^*}{dx^*}\right)^2} \\ \cos\psi &= 1\bigg/\sqrt{1+\left(\frac{dy^*}{dx^*}\right)^2} \end{aligned}\right\} \tag{A.198}$$

Using the last relation of (A.195) and (A.198), we write (A.196) and (A.197) as

$$\bar{x} = x^* - \frac{K^*A^*r^*\cos\theta\cos K^*x^*}{S^*} \tag{A.199}$$

$$\bar{y} = A^*\sin K^*x^* + \frac{r^*\cos\theta}{S^*} \tag{A.200}$$

where

$$S^* = \left[1+\left(\frac{dy^*}{dx^*}\right)^2\right]^{1/2} \tag{A.201}$$

Therefore, the attached orthogonal coordinates related to the rectilinear coordinates are given by (A.199) and (A.200) together with the projection $r^*\sin\theta$, which lies in the normal cross section and is parallel to $\bar{z}$. This is denoted by

$$\bar{z} = r^*\sin\theta \tag{A.202}$$

If R_c^* is the radius of curvature at the point C for the center-line curve

$$y^* = A^*\sin K^*x^* \tag{A.203}$$

then using differential calculus we have R_c^* as

$$R_c^* = \left[1+\left(\frac{dy}{dx}\right)^2\right]^{3/2} \bigg/ \frac{d^2y^*}{dx^*} = \frac{S^{*3}}{K^{*2}A^*\sin K^*x^*} \tag{A.204}$$

considering only the magnitude of the radius of curvature. Since

$$R_c^*K_c^* = 1 \tag{A.205}$$

where K_c^* is the curvature of the pipe center-line profile at C, then

$$K_c^* = \frac{1}{R_c^*} = \frac{K^{*2} A^* \sin K^* x^*}{S^{*3}} \tag{A.206}$$

Again, since r^*, θ, x^* are the toroidal coordinates, therefore we consider here

$$\left.\begin{aligned} x_1 &= r^* \\ x_2 &= \theta \\ x_1 &= x^* \end{aligned}\right\} \tag{A.207}$$

Now we proceed to find out the continuity equation and the equations of motion. Before doing this, we consider the expressions of scale factors $s, h_1, h_2,$ and h_3 from (A.13). Using (A.199), (A.200), (A.202), and (A.207) in these expressions and then simplifying the scale factors are found to be

$$\left.\begin{aligned} h_1 &= 1 \\ h_2 &= r^* \\ h_3 &= h^* = S^* \left(1 + K_c^* r^* \cos\theta\right) \end{aligned}\right\} \tag{A.208}$$

Let u^*, v^*, w^* be the velocity components of velocity vector $\vec{V}$ in the directions of r^*, θ, x^* increasing, and we write

$$\left.\begin{aligned} v_1 &= u^* \\ v_2 &= v^* \\ v_3 &= w^* \end{aligned}\right\} \tag{A.209}$$

Using the relations ranging from (A.207) to (A.209), we get from (A.18)

$$n_1 = \frac{1}{r^*}\frac{\partial w^*}{\partial \theta} - \frac{S^* K_c^* \sin\theta}{1 + K_c^* r^* \cos\theta} w^* - \frac{1}{h^*}\frac{\partial v^*}{\partial x^*} \tag{A.210}$$

$$n_2 = \frac{1}{h^*}\frac{\partial u^*}{\partial x^*} - \frac{\partial w^*}{\partial r^*} - \frac{S^* K_c^* \cos\theta}{h^*} w^* \tag{A.211}$$

$$n_3 = \frac{\partial v^*}{\partial r^*} + \frac{v^*}{r^*} - \frac{1}{r^*}\frac{\partial u^*}{\partial \theta} \tag{A.212}$$

Multiplying (A.212) by v_2 and (A.210) by v_3 and then subtracting one from the other, we get with the help of (A.209)

$$\begin{aligned} n_3 v_2 - n_2 v_3 = v^* \frac{\partial v^*}{\partial r^*} + \frac{v^{*2}}{r^*} - \frac{v^*}{r^*}\frac{\partial u^*}{\partial \theta} + w^* \frac{\partial w^*}{\partial r^*} \\ - \frac{w^*}{h^*}\frac{\partial u^*}{\partial x^*} + \frac{S^* K_c^* \cos\theta}{h^*} w^{*2} \end{aligned} \tag{A.213}$$

Similarly, proceeding in the same way as above, we have

$$n_1v_3 - n_3v_1 = \frac{w^*}{r^*}\frac{\partial w^*}{\partial \theta} - \frac{S^*K_c^*\sin\theta}{h^*}w^{*2} - \frac{w^*}{h^*}\frac{\partial v^*}{\partial x^*} - u^*\frac{\partial v^*}{\partial r^*} - \frac{u^*v^*}{r^*} + \frac{u^*}{r^*}\frac{\partial u^*}{\partial \theta} \tag{A.214}$$

and

$$n_2v_1 - n_1v_2 = \frac{u^*}{h^*}\frac{\partial u^*}{\partial x^*} - u^*\frac{\partial w^*}{\partial r^*} - \frac{S^*K_c^*\cos\theta}{h^*}u^*w^* - \frac{v^*}{r^*}\frac{\partial w^*}{\partial \theta} + \frac{S^*K_c^*\sin\theta}{h^*}v^*w^* + \frac{v^*}{h^*}\frac{\partial v^*}{\partial x^*} \tag{A.215}$$

Again, we write

$$\frac{1}{h_1}\frac{\partial}{\partial x_1}\left\{\frac{1}{2}\left(v_1^2+v_2^2+v_3^2\right)\right\} = u^*\frac{\partial u^*}{\partial r^*} + v^*\frac{\partial v^*}{\partial r^*} + w^*\frac{\partial w^*}{\partial r^*} \tag{A.216}$$

$$\frac{1}{h_2}\frac{\partial}{\partial x_2}\left\{\frac{1}{2}\left(v_1^2+v_2^2+v_3^2\right)\right\} = \frac{1}{r^*}\left[u^*\frac{\partial u^*}{\partial \theta} + v^*\frac{\partial v^*}{\partial \theta} + w^*\frac{\partial w^*}{\partial \theta}\right] \tag{A.217}$$

$$\frac{1}{h_3}\frac{\partial}{\partial x_3}\left\{\frac{1}{2}\left(v_1^2+v_2^2+v_3^2\right)\right\} = \frac{1}{h^*}\left[u^*\frac{\partial u^*}{\partial x^*} + v^*\frac{\partial v^*}{\partial x^*} + w^*\frac{\partial w^*}{\partial x^*}\right] \tag{A.218}$$

The subtractions of (A.213) from (A.216), (A.214) from (A.217), and (A.215) from (A.218) give

$$\frac{1}{h_1}\frac{\partial}{\partial x_1}\left\{\frac{1}{2}\left(v_1^2+v_2^2+v_3^2\right)\right\} - (n_3v_2 - n_2v_3) = u^*\frac{\partial u^*}{\partial r^*} + \frac{v^*}{r^*}\frac{\partial u^*}{\partial \theta} - \frac{v^{*2}}{r^*} + \frac{w^*}{h^*}\frac{\partial u^*}{\partial x^*} - \frac{S^*K_c^*\cos\theta}{h^*}w^{*2} \tag{A.219}$$

$$\frac{1}{h_2}\frac{\partial}{\partial x_2}\left\{\frac{1}{2}\left(v_1^2+v_2^2+v_3^2\right)\right\} - (n_1v_3 - n_3v_1) = u^*\frac{\partial v^*}{\partial r^*} + \frac{v^*}{r^*}\frac{\partial v^*}{\partial \theta} + \frac{w^*}{h^*}\frac{\partial v^*}{\partial x^*} + \frac{u^*v^*}{h^*} + \frac{S^*K_c^*\sin\theta}{h^*}w^{*2} \tag{A.220}$$

$$\frac{1}{h_3}\frac{\partial}{\partial x_3}\left\{\frac{1}{2}\left(v_1^2+v_2^2+v_3^2\right)\right\} - (n_2v_1 - n_1v_2) = u^*\frac{\partial w^*}{\partial r^*} + \frac{v^*}{r^*}\frac{\partial w^*}{\partial \theta} + \frac{w^*}{h^*}\frac{\partial w^*}{\partial x^*} + \frac{S^*K_c^*w^*}{h^*}\left(u^*\cos\theta - v^*\sin\theta\right) \tag{A.221}$$

$$= u^*h^* \frac{\partial}{\partial r^*}\left(\frac{w^*}{h^*}\right) + \frac{v^*h^*}{r^*}\frac{\partial}{\partial \theta}\left(\frac{w^*}{h^*}\right) + w^* \frac{\partial}{\partial x^*}\left(\frac{w^*}{h^*}\right)$$
$$+ \frac{w^{*2}}{h^{*2}}\frac{\partial h^*}{\partial x^*} + \frac{2S^*K_c^*w^*}{h^*}\left(u^*\cos\theta - v^*\sin\theta\right) \tag{A.222}$$

The divergence of velocity vector $\vec{V}$ can be obtained by (A.16), and it is given by

$$div\vec{V} = \frac{1}{r^*}\frac{\partial}{\partial r^*}(r^*u^*) + \frac{1}{r^*}\frac{\partial v^*}{\partial \theta} + \frac{1}{h^*}\frac{\partial w^*}{\partial x^*} + \frac{S^*K_c^*}{h^*}(u^*\cos\theta - v^*\sin\theta) \tag{A.223}$$

which can be written as

$$div\vec{V} = \frac{1}{r^*}\frac{\partial}{\partial r^*}\left(r^*u^*\right) + \frac{1}{r^*}\frac{\partial v^*}{\partial \theta} + \frac{1}{\partial x^*}\left(\frac{\partial w^*}{h^*}\right)$$
$$+ \frac{w^*}{h^{*2}}\frac{\partial h^*}{\partial x^*} + \frac{S^*K_c^*}{h^*}\left(u^*\cos\theta - v^*\sin\theta\right) \tag{A.224}$$

Differentiating (A.223) with respect to x_1, then dividing both sides of the resulting equation by h_1 and finally using relations ranging from (A.207) to (A.209), we get

$$\frac{1}{h_1}\frac{\partial}{\partial x_1}(div\vec{V}) = \frac{\partial}{\partial r^*}(div\vec{V}) \tag{A.225}$$

Similarly, proceeding in the same way as above, we have

$$\frac{1}{h_2}\frac{\partial}{\partial x_2}(div\vec{V})$$
$$= \frac{1}{r^*}\frac{\partial^2 u^*}{\partial r^*\partial \theta} + \frac{1}{r^{*2}}\frac{\partial u^*}{\partial \theta} + \frac{1}{r^{*2}}\frac{\partial^2 v^*}{\partial \theta^2} + \frac{1}{h^*r^*}\frac{\partial^2 w^*}{\partial \theta \partial x^*} + \frac{S^*K_c^*\sin\theta}{h^{*2}}\frac{\partial w^*}{\partial x^*}$$
$$+ \frac{S^*K_c^*}{h^*r^*}\left(\frac{\partial u^*}{\partial \theta}\cos\theta - \frac{\partial v^*}{\partial \theta}\sin\theta\right) + \frac{S^{*2}K_c^{*2}\sin\theta}{h^{*2}}\left(u^*\cos\theta - v^*\sin\theta\right)$$
$$- \frac{S^*K_c^*}{h^*r^*}\left(u^*\cos\theta + v^*\sin\theta\right) \tag{A.226}$$

and

$$\frac{1}{h_3}\frac{\partial}{\partial x_3}(div\vec{V}) = \frac{1}{h^*}\left[\frac{\partial^2 u^*}{\partial r^* x^*} + \frac{1}{r^*}\frac{\partial u^*}{\partial x^*} + \frac{1}{r^*}\frac{\partial^2 w^*}{\partial \theta \partial x^*} + \frac{\partial^2}{\partial x^{*2}}\left(\frac{w^*}{h^*}\right)\right.$$
$$+ \frac{\partial}{\partial x^*}\left(\frac{S^*K_c^*}{h^*}\right)\left(u^*\cos\theta - v^*\sin\theta\right)$$
$$\left. + \frac{S^*K_c^*}{h^*}\left(\frac{\partial u^*}{\partial x^*}\cos\theta - \frac{\partial v^*}{\partial x^*}\sin\theta\right)\frac{\partial}{\partial x^*}\left(\frac{w^*}{h^*}\frac{\partial h^*}{\partial x^*}\right)\right] \tag{A.227}$$

Again, from (A.42) we have using (A.207), (A.208), and the relations ranging from (A.210) to (A.212)

$$g_1 = \frac{1}{h^{*3}}\frac{\partial h^*}{\partial x^*}\frac{\partial u^*}{\partial x^*} - \frac{1}{h^{*2}}\frac{\partial^2 u^*}{\partial x^{*2}} - \frac{1}{r^{*2}}\frac{\partial^2 u^*}{\partial \theta^2} + \frac{S^* K_c^* \sin\theta}{h^* r^*}\frac{\partial u^*}{\partial \theta} + \frac{1}{r^*}\frac{\partial^2 v^*}{\partial x^* \partial \theta}$$
$$+ \frac{1}{r^{*2}}\frac{\partial v^*}{\partial \theta} - \frac{S^* K_c^* \sin\theta}{h^*}\left(\frac{\partial v^*}{\partial r^*} + \frac{v^*}{r^*}\right) + \frac{1}{h^*}\frac{\partial^2 w^*}{\partial r^* \partial x^*} + \frac{S^* K_c^* \cos\theta}{h^{*2}}\frac{\partial w^*}{\partial x^*}$$
$$- \frac{1}{h^{*3}}\left[\frac{\partial}{\partial x^*}\left(S^* K_c^* h^*\right) - 2S^* K_c^* \frac{\partial h^*}{\partial x^*}\right]\cos\theta w^* \tag{A.228}$$

$$g_2 = \frac{1}{h^* r^*}\frac{\partial^2 w^*}{\partial \theta \partial x^*} - \frac{1}{h^*}\left[S^* K_c^* \frac{\partial}{\partial x^*}\left(\frac{w^*}{h^*}\right) + \frac{w^*}{h^{*2}}\frac{\partial}{\partial x^*}\left(S^* K_c^* h^*\right)\right.$$
$$\left. - \frac{S^* K_c^* w^*}{h^{*2}}\frac{\partial h^*}{\partial x^*}\right]\sin\theta - \frac{1}{h^{*2}}\frac{\partial^2 v^*}{\partial x^{*2}} + \frac{1}{h^{*3}}\frac{\partial h^*}{\partial x^*}\frac{\partial v^*}{\partial x^*}$$
$$- \frac{S^* K_c^*}{h^*}\left[\frac{\partial v^*}{\partial r^*} + \frac{v^*}{r^*} - \frac{1}{r^*}\frac{\partial u^*}{\partial \theta}\right]\cos\theta$$
$$- \left[\frac{\partial^2 v^*}{\partial x^{*2}} + \frac{1}{r^*}\frac{\partial v^*}{\partial r^*} - \frac{v^*}{r^{*2}} - \frac{1}{r^*}\frac{\partial^2 uu^*}{\partial r^* \partial \theta} + \frac{1}{r^{*2}}\frac{\partial u^*}{\partial \theta}\right] \tag{A.229}$$

$$g_3 = \frac{1}{h^* r^*}\frac{\partial u^*}{\partial x^*} + \frac{1}{h^*}\frac{\partial^2 u^*}{\partial r^* \partial x^*} - \frac{S^* K_c^* \cos\theta}{h^{*2}}\frac{\partial u^*}{\partial x^*} - h^* \frac{\partial^2}{\partial r^{*2}}\left(\frac{w^*}{h^*}\right)$$
$$- \frac{h^*}{r^*}\frac{\partial}{\partial r^*}\left(\frac{w^*}{h^*}\right) - \frac{h^*}{r^{*2}}\frac{\partial^2}{\partial \theta^2}\left(\frac{w^*}{h^*}\right) - 3S^* K_c^* \cos\theta \frac{\partial}{\partial r^*}\left(\frac{w^*}{h^*}\right)$$
$$+ \frac{3S^* K_c^* \sin\theta}{r^*}\frac{\partial}{\partial \theta}\left(\frac{w^*}{h^*}\right) + \frac{1}{h^* r^*}\frac{\partial^2 v^*}{\partial \theta \partial x^*} + \frac{S^* K_c^* \sin\theta}{h^{*2}}\frac{\partial v^*}{\partial x^*} \tag{A.230}$$

Subtracting (A.228) from (A.225), we have after cumbersome calculations

$$\frac{1}{h_1}\frac{\partial}{\partial x_1}(div\vec{V}) - g_1$$
$$= \nabla^{*2} u^* - \frac{u^*}{r^{*2}} - \frac{2}{r^{*2}}\frac{\partial v^*}{\partial \theta} - \frac{S^{*2} K_c^{*2}}{h^{*2}}\left(u^* \cos\theta - v^* \sin\theta\right)\cos\theta$$
$$- \frac{1}{h^{*3}}\frac{\partial u^*}{\partial x^*}\frac{\partial h^*}{\partial x^*} + \frac{S^{*2} K_c^{*2}}{h^{*2}}\left\{\frac{\partial u^*}{\partial r^*}\cos\theta + \left(\frac{v^*}{r^*} - \frac{1}{r^*}\frac{\partial u^*}{\partial \theta}\right)\sin\theta\right\}$$
$$- \frac{1}{h^*}\left\{\frac{w^*}{h^{*2}}\frac{\partial}{\partial x^*}\left(S^* K_c^* h^*\right) + 2S^* K_c^* \frac{\partial}{\partial x^*}\left(\frac{w^*}{h^*}\right)\right\}\cos\theta \tag{A.231}$$

where ∇^{*2} is defined by

$$\nabla^{*2} = \frac{\partial^2}{\partial r^{*2}} + \frac{1}{r^*}\frac{\partial}{\partial r^*} + \frac{1}{r^{*2}}\frac{\partial^2}{\partial \theta^2} + \frac{1}{h^{*2}}\frac{\partial^2}{\partial x^{*2}} \tag{A.232}$$

Similarly, we can write subtracting (A.229) from (A.226) and (A.230) from (A.227)

$$\begin{aligned}
&\frac{1}{h_2}\frac{\partial}{\partial x_2}(div\vec{V}) - g_2 \\
&= \nabla^{*2}v^* - \frac{v^*}{r^{*2}} + \frac{2}{r^{*2}}\frac{\partial u^*}{\partial \theta} + \frac{S^{*2}K_C^2}{h^{*2}}(u^*\cos\theta - v^*\sin\theta)\sin\theta - \frac{1}{h^{*2}}\frac{\partial w^*}{\partial x^*}\frac{\partial h^*}{\partial x^*} \\
&+ \frac{S^*K_c^*}{h^*}\left\{\frac{\partial v^*}{\partial r^*}\cos\theta - \left(\frac{u^*}{r^*} + \frac{1}{r^*}\frac{\partial v^*}{\partial \theta}\right)\sin\theta\right\} \\
&+ \frac{1}{r^*}\left\{\frac{w^*}{h^{*2}}\frac{\partial}{\partial x^*}(S^*K_c^*h^*) + 2S^*K_c^*\frac{\partial}{\partial x^*}\left(\frac{w^*}{h^*}\right)\right\}\sin\theta
\end{aligned} \tag{A.233}$$

and

$$\begin{aligned}
&\frac{1}{h_3}\frac{\partial}{\partial x_3}(div\vec{V}) - g_3 \\
&= h^*\left[\nabla^{*2}\left(\frac{w^*}{h^*}\right) + \frac{1}{h^{*2}}\frac{\partial}{\partial x^*}\left(\frac{S^*K_c^*}{h^*}\right)(u^*\cos\theta - v^*\sin\theta)\right. \\
&+ \frac{2S^*K_c^*}{h^{*3}}\left(\frac{\partial u^*}{\partial x^*}\cos\theta - \frac{\partial v^*}{\partial x^*}\sin\theta\right) \\
&\left.+ \frac{3S^*K_c^*}{h^*}\left\{\frac{\partial}{\partial r^*}\left(\frac{w^*}{h^*}\right)\cos\theta - \frac{1}{r^*}\frac{\partial}{\partial \theta}\left(\frac{w^*}{h^*}\right)\sin\theta\right\} + \frac{1}{h^{*2}}\frac{\partial}{\partial x^*}\left(\frac{w^*}{h^{*2}}\frac{\partial h^*}{\partial x^*}\right)\right]
\end{aligned} \tag{A.234}$$

Using the relations ranging from (A.219) to (A.221) or (A.222) and the relations ranging from (A.231) to (A.234) except (A.232) and then considering the $t = t^*$, $p = p^*$ in (r^*, θ, x^*) coordinate system, we have the momentum equations ranging from (A.44) to (A.46) as

$$\begin{aligned}
&\frac{D^*u^*}{D^*t^*} - \frac{v^{*2}}{r^*} - \frac{S^*K_c^*\cos\theta}{h^*}w^{*2} \\
&= F_{r^*} - \frac{1}{\rho}\frac{\partial p^*}{\partial r^*} + \nu^*\left[\nabla^{*2}u^* - \frac{u^*}{r^{*2}} - \frac{2}{r^{*2}}\frac{\partial v^*}{\partial \theta}\right. \\
&- \frac{S^{*2}K_c^{*2}}{h^{*2}}(u^*\cos\theta - v^*\sin\theta)\cos\theta - \frac{1}{h^{*3}}\frac{\partial u^*}{\partial x^*}\frac{\partial h^*}{\partial x^*} \\
&+ \frac{S^*K_c^*}{h^{*2}}\left\{\frac{\partial u^*}{\partial r^*}\cos\theta + \left(\frac{v^*}{r^*} - \frac{1}{r^*}\frac{\partial u^*}{\partial \theta}\right)\sin\theta\right\} \\
&\left.- \frac{1}{h^*}\left\{\frac{w^*}{h^{*2}}\frac{\partial}{\partial x^*}(S^*K_c^*h^*) + 2S^*K_c^*\frac{\partial}{\partial x^*}\left(\frac{w^*}{h^*}\right)\right\}\cos\theta\right]
\end{aligned} \tag{A.235}$$

$$\begin{aligned}
&\frac{D^*v^*}{D^*t^*} + \frac{u^*v^*}{h^*} + \frac{S^*K_c^*\sin\theta}{h^*}w^{*2} \\
&\quad = F_\theta - \frac{1}{\rho r^*}\frac{\partial p^*}{\partial\theta} + \nu^*\left[\nabla^{*2}v^* - \frac{v}{r^{*2}} - \frac{2}{r^{*2}}\frac{\partial u^*}{\partial\theta}\right. \\
&\qquad + \frac{S^{*2}K_c^{*2}}{h^{*2}}\left(u^*\cos\theta - v^*\sin\theta\right)\sin\theta - \frac{1}{h^{*2}}\frac{\partial v^*}{\partial x^*}\frac{\partial h^*}{\partial x^*} \\
&\qquad + \frac{S^*K_c^*}{h^{*2}}\left\{\frac{\partial v^*}{\partial r^*}\cos\theta - \left(\frac{u^*}{r^*} + \frac{1}{r^*}\frac{\partial v^*}{\partial\theta}\right)\sin\theta\right\} \\
&\qquad \left. - \frac{1}{h^*}\left\{\frac{w^*}{h^{*2}}\frac{\partial}{\partial\theta}\left(S^*K_c^*h^*\right) + 2S^*K_c^*\frac{\partial}{\partial x^*}\left(\frac{w^*}{h^*}\right)\right\}\sin\theta\right]
\end{aligned} \tag{A.236}$$

$$\begin{aligned}
&\frac{D^*}{D^*t^*}\left(\frac{w^*}{h^*}\right) + \frac{w^{*2}}{h^{*3}}\frac{\partial h^*}{\partial x^*} + \frac{S^*K_c^*w^*}{h^{*2}}\left(u^*\cos\theta - v^*\sin\theta\right) \\
&= F_{x^*} - \frac{1}{\rho S^*\left(1 + K_c^*r^*\cos\theta\right)}\frac{\partial p^*}{\partial x^*} + \nu^*\left[\nabla^{*2}\left(\frac{w^*}{h^*}\right)\right. \\
&\quad + \frac{1}{h^{*2}}\frac{\partial}{\partial x^*}\left(\frac{S^*K_c^*}{h^*}\right)\left(u^*\cos\theta - v^*\sin\theta\right) + \frac{2S^*K_c^*}{h^{*3}}\left(\frac{\partial u^*}{\partial x^*}\cos\theta - \frac{\partial v^*}{\partial x^*}\sin\theta\right) \\
&\quad \left. + \frac{3S^*K_c^*}{h^*}\left\{\frac{\partial}{\partial r^*}\left(\frac{w^*}{h^*}\right)\cos\theta - \frac{1}{r^*}\frac{\partial}{\partial\theta}\left(\frac{w^*}{h^*}\right)\sin\theta\right\} + \frac{1}{h^{*2}}\frac{\partial}{\partial x^*}\left(\frac{w^*}{h^*}\frac{\partial h^*}{\partial x^*}\right)\right]
\end{aligned} \tag{A.237}$$

Here, F_{r^*}, F_θ, and F_{x^*} are the components of the body force $\vec{F}$ defined by

$$\vec{F} = F_{r^*}i_1 + F_\theta j_1 + F_{x^*}k_1 \tag{A.238}$$

in the directions of r^*, θ, and x^* increasing. The operator (D^*/D^*t^*) is defined by

$$\frac{D^*}{D^*t^*} = \frac{\partial}{\partial t^*} + u^*\frac{\partial}{\partial r^*} + \frac{v^*}{r^*}\frac{\partial}{\partial\theta} + \frac{w^*}{h^*}\frac{\partial}{\partial x^*} \tag{A.239}$$

The continuity equation may be obtained from (A.47) or (A.224) by putting $div\vec{V} = 0$, that is,

$$\frac{1}{r^*}\frac{\partial}{\partial r^*}\left(r^*u^*\right) + \frac{1}{r^*}\frac{\partial v^*}{\partial\theta} + \frac{1}{h^*}\frac{\partial w^*}{\partial h^*} + \frac{S^*K_c^*}{h^*}\left(u^*\cos\theta - v^*\sin\theta\right) = 0 \tag{A.240}$$

Now the rate of strain components that are given in (A.25) can be converted into the r^*, θ, x^* coordinate system and these are given by

$$e_{r^*r^*} = \frac{\partial u^*}{\partial r^*}$$

$$
\begin{aligned}
e_{\theta\theta} &= \frac{1}{r^*}\frac{\partial v^*}{\partial \theta} + \frac{u^*}{r^*} \\
e_{x^*x^*} &= \frac{1}{h^*}\left[\frac{\partial w^*}{\partial x^*} + S^* K_c^* \left(u^* \cos\theta - v^* \sin\theta\right)\right] \\
e_{\theta x^*} &= \frac{1}{h^*}\left[\frac{\partial v^*}{\partial x^*} + \frac{h^{*2}}{r^*}\frac{\partial}{\partial \theta}\left(\frac{w^*}{h^*}\right)\right] \\
e_{x^*r^*} &= \frac{1}{h^*}\left[\frac{\partial u^*}{\partial x^*} + h^{*2}\frac{\partial}{\partial r^*}\left(\frac{w^*}{h^*}\right)\right] \\
e_{r^*\theta} &= \frac{\partial v^*}{\partial r^*} - \frac{v^*}{r^*} + \frac{1}{r^*}\frac{\partial u^*}{\partial \theta}
\end{aligned}
\tag{A.241}
$$

Therefore, the dissipation function $\underline{\Phi}$ in (A.30) is converted into

$$
\underline{\Phi} = \mu\left[2\left(e_{r^*r^*}^2 + e_{\theta\theta}^2 + e_{x^*x^*}^2\right) + e_{\theta r^*}^2 + e_{x^*r^*}^2 + e_{r^*\theta}^2\right] \tag{A.242}
$$

where the rate of strain components involved in (A.240) are given by (A.241). Then we write from (A.48) and (A.49) the expressions of $(\vec{V}\nabla)T$ and the Laplacian of T in the r^*, θ, x^* coordinate system as

$$
(\vec{V}\nabla)T = u^*\frac{\partial T}{\partial r^*} + \frac{v^*}{r^*}\frac{\partial T}{\partial \theta} + \frac{w^*}{h^*}\frac{\partial T}{\partial x^*} \tag{A.243}
$$

and

$$
\nabla^2 T = \nabla^{*2} T + \frac{S^* K_c^*}{h^*}\left(\frac{\partial T}{\partial r^*}\cos\theta - \frac{1}{r^*}\frac{\partial T}{\partial \theta}\sin\theta\right) \tag{A.244}
$$

The equation (A.29), together with (A.242), (A.243), and (A.244), constitutes the energy equation that can be solved under certain initial and boundary conditions.

Now proceed to find out the nondimensional forms of the momentum equations and continuity equation. To do this we consider the following transformations:

$$
\left.
\begin{aligned}
& t = t^*\omega^*,\ u = \frac{u^*}{\vartheta/r_0^*},\ v = \frac{v^*}{\vartheta/r_0^*} \\
& w = \frac{w^*}{\vartheta/r_0^*},\ p = \frac{p^*}{\rho\vartheta^{*2}/r_0^{*2}} \\
& r = \frac{r^*}{r_0^*},\ x = \frac{x^*}{r_0^*},\ \frac{A^*}{r_0^*} = a
\end{aligned}
\right\}
\tag{A.245}
$$

Here, t is the nondimensional time, (u, v, w) are the nondimensional velocity components in the nondimensional directions of r, θ, and x, respectively, p is the nondimensional pressure, a is the nondimensional amplitude of the pipe axis, and ω^* is the angular frequency.

If λ^* is the dimensional wavelength, then the nondimensional form of it is

$$\lambda = \frac{\lambda^*}{r_0^*} \tag{A.246}$$

The dimensional wave number K^* of the waviness of the pipe center-line profile is

$$K^* = \frac{2\pi}{\lambda^*} \tag{A.247}$$

Using (A.246) in (A.247), we have

$$K^* = \frac{2\pi}{r_0^* \lambda} = \frac{K}{r_0^*} \tag{A.248}$$

where K is the nondimensional wave number of the waviness of the pipe center-line curve.

From (A.201) and (A.203) we have, with the help of (A.245) and (A.248),

$$S^* = \left(1 + K^{*2} A^{*2} \cos K^* x^*\right)^{1/2} = \left(1 + K^2 a^2 \cos Kx\right)^{1/2} = S \tag{A.249}$$

which shows that S is the nondimensional form of S^*. Then using (A.245), (A.248), and (A.249), we get from (A.206)

$$K_c^* = K^{*2} A \sin K^* x^* / S^{*3} = \frac{K^2 a}{r_0^*} \sin Kx / S^3 = K_0 / r_0^* \tag{A.250}$$

and from the third relation of (A.208), we have

$$h^* = S^* \left(1 + K_c^* r^* \cos\theta\right) = S\left(1 + K_c r \cos\theta\right) = h \tag{A.251}$$

The relations (A.250) and (A.251) show that K_c / r_o^*, and h are the nondimensional forms of K_c^* and h^*. Then the nondimensional forms of the terms in the left-hand side of (A.235) are given below:

$$\begin{aligned}
\frac{\partial u^*}{\partial t^*} &= \frac{\omega^* \vartheta^*}{r_0^*} \frac{\partial u}{\partial t} \\
u^* \frac{\partial u^*}{\partial r^*} &= \frac{\vartheta^{*2}}{r_0^{*3}} u \frac{\partial u}{\partial r} \\
\frac{v^{*2}}{r^*} &= \frac{\vartheta^{*2}}{r_0^{*3}} \frac{v^2}{r} \\
\frac{w^*}{h^*} \frac{\partial u^*}{\partial x^*} &= \frac{\vartheta^{*2}}{r_0^{*3}} \frac{w}{h} \frac{\partial u}{\partial x} \\
\frac{S^* K_c^*}{h^*} w^{*2} \cos\theta &= \frac{\vartheta^{*2}}{r_0^{*3}} \frac{S K_c}{h} \cos\theta
\end{aligned} \tag{A.251a}$$

Therefore, using the above transformations ranging from (A.245) to (A.251a) except (A.246) and (A.247), we see that the nondimensional form of the left-hand side of (A.235) is found to be

$$L.H.S. = \frac{\vartheta^{*2}}{r_0^{*3}}\left[\frac{Du}{Dt} - \frac{v^2}{r^2} - \frac{SK_c}{h}\cos\theta\right] \tag{A.252}$$

where

$$\frac{D}{Dt} = k^2\frac{\partial}{\partial t} + u\frac{\partial}{\partial r} + \frac{v}{r}\frac{\partial}{\partial\theta} + \frac{w}{h}\frac{\partial}{\partial x} \tag{A.253}$$

and

$$k^2 = \frac{\omega^{*2}r_0^{*2}}{\vartheta^*} \tag{A.254}$$

Again, we can also obtain the dimensionless forms of the terms in the right-hand side of (A.235), and these are given below:

$$\frac{1}{\rho}\frac{\partial p^*}{\partial r^*} = \frac{\vartheta^{*2}}{r_0^{*3}}$$

$$\frac{\partial^2 u^*}{\partial r^{*2}} = \frac{\vartheta^*}{r_0^{*3}}\frac{\partial^2 u}{\partial r^2}$$

$$\frac{1}{r^*}\frac{\partial u^*}{\partial r^*} = \frac{\vartheta^*}{r_0^{*3}}\frac{1}{r}\frac{\partial u}{\partial r}$$

$$\frac{1}{r^{*2}}\frac{\partial^2 u^*}{\partial\theta^2} = \frac{\vartheta^*}{r_0^{*3}}\frac{1}{r^2}\frac{\partial^2 u}{\partial\theta^2}$$

$$\frac{1}{h^{*2}}\frac{\partial^2 u^*}{\partial x^{*2}} = \frac{\vartheta^*}{r_0^{*3}}\frac{1}{h^2}\frac{\partial^2 u}{\partial x^2}$$

$$\frac{u^*}{r^{*2}} = \frac{\vartheta^*}{r_0^{*3}}\frac{u}{r^3}$$

$$\frac{2}{r^{*2}}\frac{\partial v^*}{\partial\theta} = \frac{\vartheta^*}{r_0^{*3}}\frac{2}{r^2}\frac{\partial v}{\partial\theta}$$

$$\frac{S^{*2}K_c^{*2}}{h^{*2}}(u^*\cos\theta - v^*\sin\theta)\cos\theta = \frac{\vartheta^*}{r_0^{*3}}\frac{S^2K_c^2}{h^2}(u\cos\theta - v\sin\theta)\cos\theta$$

$$\frac{1}{h^{*3}}\frac{\partial u^*}{\partial x^*}\frac{\partial h^*}{\partial x^*} = \frac{\vartheta^*}{r_0^{*3}}\frac{1}{h^3}\frac{\partial u}{\partial x}\frac{\partial h}{\partial x}$$

$$\frac{S^*K_c^*}{h^*}\left\{\frac{\partial u^*}{\partial r^*}\cos\theta+\left(\frac{v^*}{r^*}-\frac{1}{r^*}\frac{\partial u^*}{\partial\theta}\right)\sin\theta\right\}$$

$$=\frac{\vartheta^*}{r_0^{*3}}\left[\frac{SK_c}{h}\left\{\frac{\partial u}{\partial r}\cos\theta+\left(\frac{v}{r}-\frac{1}{r}\frac{\partial u}{\partial\theta}\right)\sin\theta\right\}\right]$$

$$\frac{1}{h^*}\left\{\frac{w^*}{h^{*2}}\frac{\partial}{\partial x^*}\left(S^*K_c^*h^*\right)+2S^*K_c^*\frac{\partial}{\partial x^*}\left(\frac{w^*}{h^*}\right)\right\}\cos\theta$$

$$=\frac{\vartheta^*}{r_0^{*3}}\left[\frac{1}{h}\left\{\frac{w}{h^2}\frac{\partial}{\partial x}\left(SK_ch\right)+2SK_c\frac{\partial}{\partial x}\left(\frac{w}{h}\right)\right\}\cos\theta\right]$$

With the help of the above transformations, the dimensionless form of the right-hand side of (A.235) in the absence of body force is

$$\begin{aligned}R.H.S. = \frac{\vartheta^*}{r_0^{*3}}\Bigg[&-\frac{\partial p}{\partial r}+\bar{\nabla}^2u-\frac{u}{r^2}-\frac{2}{r^2}\frac{\partial v}{\partial\theta}-\frac{S^2K_c^2}{h^2}(u\cos\theta-v\sin\theta)\cos\theta\\&-\frac{1}{h^3}\frac{\partial u}{\partial x}\frac{\partial h}{\partial x}+\frac{SK_c}{h}\left\{\frac{\partial u}{\partial r}\cos\theta+\left(\frac{v}{r}-\frac{1}{r}\frac{\partial u}{\partial\theta}\right)\sin\theta\right\}\\&-\frac{1}{h}\left\{\frac{w}{h^2}\frac{\partial}{\partial x}(SK_ch)+2SK_c\frac{\partial}{\partial x}\left(\frac{w}{h}\right)\right\}\cos\theta\Bigg]\end{aligned}\tag{A.255}$$

where

$$\bar{\nabla}^2=\frac{\partial^2}{\partial r^2}+\frac{1}{r}\frac{\partial}{\partial r}+\frac{1}{r^2}\frac{\partial^2}{\partial\theta^2}+\frac{1}{h^2}\frac{\partial^2}{\partial x^2}\tag{A.256}$$

Equating (A.252) and (A.255), we have the nondimensional r-momentum equation as

$$\begin{aligned}&\frac{\bar{D}u}{\bar{D}t}-\frac{v^2}{r^2}-\frac{SK_c}{h}\cos\theta\\&\quad=-\frac{\partial p}{\partial r}+\bar{\nabla}^2u-\frac{u}{r^2}-\frac{2}{r^2}\frac{\partial v}{\partial\theta}-\frac{S^2K_c^2}{h^2}(u\cos\theta-v\sin\theta)\cos\theta-\frac{1}{h^3}\frac{\partial u}{\partial x}\frac{\partial h}{\partial x}\\&\qquad+\frac{SK_c}{h}\left[\frac{\partial u}{\partial r}\cos\theta+\left(\frac{v}{r}-\frac{1}{r}\frac{\partial u}{\partial\theta}\right)\sin\theta\right]\\&\qquad-\frac{1}{h}\left[\frac{w}{h^2}\frac{\partial}{\partial x}(SK_ch)+2SK_c\frac{\partial}{\partial x}\left(\frac{w}{h}\right)\right]\cos\theta\end{aligned}\tag{A.257}$$

Proceeding in the same way as above, we may get the nondimensional forms of θ-momentum and x-momentum equations in the absence of body force. These equations are given below:

$$
\begin{aligned}
&\frac{\overline{D}v}{\overline{D}t} + \frac{uv}{r^2} + \frac{SK_c}{h} w^2 \sin\theta \\
&\quad = -\frac{1}{r}\frac{\partial p}{\partial \theta} + \bar{\nabla}^2 v - \frac{v}{r^2} - \frac{2}{r^2}\frac{\partial u}{\partial \theta} - \frac{1}{h^3}\frac{\partial v}{\partial x}\frac{\partial h}{\partial x} + \frac{S^2K_c^2}{h^2}(u\cos\theta - v\sin\theta)\sin\theta \\
&\qquad + \frac{SK_c}{h}\left[\frac{\partial v}{\partial r}\cos\theta - \left(\frac{u}{r} + \frac{1}{r}\frac{\partial v}{\partial \theta}\right)\sin\theta\right] \\
&\qquad + \frac{1}{h}\left[\frac{w}{h^2}\frac{\partial}{\partial x}(SK_c h) + 2SK_c\frac{\partial}{\partial x}\left(\frac{w}{h}\right)\sin\theta\right]
\end{aligned}
\tag{A.258}
$$

and

$$
\begin{aligned}
&\frac{\overline{D}}{\overline{D}t}\left(\frac{w}{h}\right) + \frac{w^2}{h^3}\frac{\partial h}{\partial x} + \frac{SK_c w}{h}(u\cos\theta - v\sin\theta) \\
&\quad = -\frac{1}{h^2}\frac{\partial p}{\partial x} + \bar{\nabla}^2\left(\frac{w}{h}\right) + \frac{1}{h^2}\frac{\partial}{\partial x}\left(\frac{SK_c}{h}\right)(u\cos\theta - v\sin\theta) \\
&\qquad + \frac{2}{h^3}SK_c\left(\frac{\partial u}{\partial r}\cos\theta - \frac{\partial v}{\partial x}\sin\theta\right) \\
&\qquad + \frac{3}{h}SK_c\left[\frac{\partial}{\partial r}\left(\frac{w}{h}\right)\cos\theta - \frac{1}{r}\frac{\partial}{\partial \theta}\left(\frac{w}{h}\right)\sin\theta\right] + \frac{1}{h^2}\frac{\partial}{\partial x}\left(\frac{w}{h^2}\frac{\partial h}{\partial x}\right)
\end{aligned}
\tag{A.259}
$$

The dimensional analysis of the continuity equation gives the nondimensional form of it as

$$
\frac{1}{r}\frac{\partial}{\partial r}(ru) + \frac{1}{r}\frac{\partial v}{\partial \theta} + \frac{1}{h}\frac{\partial w}{\partial x} + \frac{SK_c}{h}(u\cos\theta - v\sin\theta) = 0 \tag{A.260}
$$

Index